PROTEIN CONFORMATION: NEW RESEARCH

PROTEIN CONFORMATION: NEW RESEARCH

LINDA B. ROSWELL
Editor

Nova Science Publishers, Inc.
New York

For permission to use material from this book please contact us:
Telephone 631-231-7269; Fax 631-231-8175
Web Site: http://www.novapublishers.com

Library of Congress Cataloging-in-Publication Data

Protein conformation : new research / Linda B. Roswell (editor).
p. ; cm.
Includes bibliographical references and index.
ISBN 978-1-60456-219-4 (hardcover)
1. Proteins--Conformation. I. Roswell, Linda B.
[DNLM: 1. Protein Conformation. QU 55.9 P9661 2007]
QP551.P69592 2007
572'.6--dc22

2007047461

Published by Nova Science Publishers, Inc. ✦ New York

CONTENTS

PREFACE

Protein structure is the characteristic 3-dimensional shape of a protein, imposed upon it by the secondary and tertiary structure of the peptide chain. This stage in the structure of a protein describes the highest level of organization in overall structure assumed by multimeric proteins (aggregates of more than one polypeptide chain). This is the fourth folding level of protein building. This new book presents the latest research in the field.

Chapter 1 - A protein conformational alphabet refers to the discretized states of the three-dimensional segmental structure of protein backbones. Under a specific representation (e.g. coordinates, dihedral angles, or $C\alpha$ pseudobond angles), the structure of a segment corresponds to a point in a phase space, and hence a structure database is mapped to a point set in the phase space. Generally, the procedure to deduce finite discrete conformational states from a point set in a continuous conformational phase space is a clustering analysis. By representing the conformation of a quadrapeptide segment with the three $C\alpha$ pseudobond angles, and clustering based on the probability distribution of the angles, an alphabet of 17 letters is derived. The relationship of these conformational states with the secondary structure and their associated amino acids, and an entropic reduction of conformational and amino acidic alphabets are discussed.

Without a conformational substitution matrix the use of a conformational alphabet is very limited. A substitution matrix called CLESUM is constructed from a representative pairwise aligned structure dataset of the families of structurally similar proteins (FSSP). By restricting the local conformations to a handful of states, a protein 3D backbone structure is converted to a 1D sequence of letters akin to the amino acid sequence; many tools of sequence analysis can then be applied with certain minor modification. As an example of application, fast structure alignment based on the conformational alphabet is described. Structurally similar fragment pairs are found by merely string comparison. Interference of frequently occurring matches of helices or sheets can be largely suppressed, and a greedy strategy guided by CLESUM similarity scores becomes possible. The same framework works also for multiple structure alignment.

Chapter 2 - The Proteomic Code is a set of rules by which information in genetic material is transferred into the physico-chemical properties of amino acids and determines how individual amino acids interact with each other during folding and in specific protein–protein interactions. The Proteomic Code is part of the redundant Genetic Code. The 25-year-old history of this concept is reviewed from the first independent suggestions by Biro and Mekler, through the works of Blalock, Root-Bernstein, Siemion, Miller and others, followed by the discovery of a Common Periodic Table of Codons and Nucleic Acids in 2003 and culminating in the recent conceptualization of partial complementary coding of interacting amino acids as well as the theory of the nucleic acid-assisted protein folding. A novel cloning method for the design and production of specific and with high affinity reacting proteins (SHARP) is presented. This method is based on the concept of proteomic codes and is suitable for large-scale, industrial production of specifically interacting peptides.

Chapter 3 - In the review (including authors' experimental data) the questions of natural serum albumin conformation and its changes due to pathology are considered. Particular attention is given to a method of albumin natural conformational changes detection using a new fluorescent probe, CAPIDAN.

Human serum albumin (HSA) is a protein of human body liquids which possesses many important functions. Its primary and secondary structure are known. At the same time its tertiary structure is uncertain because of flexibility of this molecule. Crystallographic data show a heart-shaped structure with long axis of 8-9 nm but that is a special crystal-induced shape. Hydrodynamic data in water solutions show that HSA molecule is not a compact globular sphere and are in accordance with two alternative (or co-existing?) conformations: prolate ellipsoid of 11-14 nm or/and oblate ellipsoid with long axis of approximately 9 nm. The authors neutron scattering data let them suggest conformation of albumin molecule in solution to be between more short prolate ellipsoud of 11 nm or less and an oblate ellipsoid with diameter of 8-9 nm

One of the most important HSA functions is adsorption of many toxic metabolites in tissues and their transport to organs of detoxification. There are special sites in the HSA molecule for those metabolites. These sites are also able to bind many dyes, and that approach is very effective in studies of the site properties. Especially convenient are fluorescent probes proposed at 1952 and developed in further. After appearance of) several fluorescent probes for albumin study, a new probe, N-carboxyphenylimine of dimethylaminonaphthalic acid (CAPIDAN or K-35), has been proposed in 1990-s. CAPIDAN fluorescence in diluted human serum can be attributed (>96%) to HSA-bound probe molecules. A technique has been proposed to measure competition of the probe with specific markers of drug binding sites as well as with many other HSA ligands. The technique is based on fluorescence energy transfer from a covalent fluorescent label (donor) bound to HSA amino groups to CAPIDAN molecule (an acceptor).

A fluorescent test has been proposed to detect the state of HSA main binding sites in natural serum. The advantage of the proposed test is that it can be used in intact serum, at natural HSA conformation. In the majority of serum samples of healthy volunteers, CAPIDAN fluorescence intensity is proportional of HSA concentration in serum. However, significant deviations from this rule have been detected in sera of patients. The correlation was revealed between those fluorescence deviations, i.e. changes of HSA natural conformation, on one hand, and the severity of diseases, on the other hand. Examples and perspectives of detection of albumin conformational changes in clinical practice are discussed.

Chapter 4 - In this work, a difference matrix based method for analyzing and comparing conformational changes associated with substrate or inhibitor binding was implemented. The method was applied on a series of protein structures: aquaporin (AQP0, AQP1, AQPGlpF, AQPZ), calmodulin (CaM), chloride channel (ClC), cyclin-dependent kinase 2 (CDK2), cyclooxygenases (COX-1 and COX-2), factor Xa and the human immunodeficiency virus type 1 (HIV-1) protease. Various receptor conformations were analyzed, where applicable: bound vs. unbound, open vs. closed, active vs. inactive forms. This approach allowed for the comparison of conformational changes upon binding of substrate versus inhibitor or activators. Clear trends of domain motions associated with substrate binding were detected in the various conformations of each protein analyzed. The application of this tool between the structural pairs is more informative compared to the single number reported by root-mean-square-deviation (RMSD) as it provides detailed information to the mechanisms associated with ligand binding. Identification of "hot spots" in conformations can provide a better understanding of molecular recognition phenomenon and its implications on resistance, specificity and drug design.

Chapter 5 - In eukaryotes, the initiation of DNA replication is a fundamental process that occurs prior to cell division, assuring the transfer of genomic information. For replication to occur, dozens of proteins must gather in a hierarchical and timely manner to establish the pre-replication complex (pre-RC) at DNA sites where DNA replication occurs. These proteins eventually form large replisome progression complexes. The six subunits of the origin recognition complex (ORC) serve as landmarks for the initiation of pre-RC assembly on the DNA. Together with CDC6 and CDT1, the ORC recruits six mini-chromosome maintenance proteins. These factors are essential for proper DNA replication initiation and licensing by the central unit to form replisomes. Regulators of pre-RC formation, such as kinases and a unique protein called geminin, are important topics of research. For the initiation of DNA replication to progress to elongation, several other factors including CDC45, GINS, and replication protein A are required. Once the protein assembly process of the DNA replication apparatus is understood in greater detail, its regulation could be tweaked to enable the molecular dissection of other cellular functional apparatuses, such as DNA-protein interaction-based transcriptional regulation and protein-protein interaction-based signaling pathways, eventually leading to a greater understanding of complicated cellular systems at a molecular level. In this review article, the authors summarize recent knowledge of the protein assembly process that occurs during the metazoan initiation of DNA replication and discuss future perspectives.

Chapter 6 - Comparative modeling builds a three-dimensional structure of a target protein based on one or several related proteins of known structure. The underlying assumption is that the sequence homology between two proteins implies structural similarity, which in turn implies a similarity in their function. Most algorithms for aligning and scoring pairs of proteins are based on the dynamic programming techniques in conjunction with the residue-residue mutation scores. However, the pair-wise alignment techniques that use amino-acid substitution matrices and residue-type information are often unable to recognize the relationships between proteins that are known to be homologous. Profile-profile methods utilize sequence information contained in multiple sequence alignments corresponding to both target and the template family of proteins. By taking advantage of the distribution of amino-acid counts in the columns of the alignment profiles, these methods can often recognize the similarity between remotely related sequences, as evidenced by large scale benchmarking experiments such as LiveBench, CAFASP, and CASP. The authors give an overview of the existing profile-profile techniques, concentrating on the key aspects of the alignment algorithms. A novel approach to scoring pairs of profiles is presented that focuses on the alignment scoring function and the estimation of the statistical parameters for the alignment score statistics.

Chapter 7 - Pentamidine is a bisbenzamidine endowed with antiparasitic and antitumor properties, which induces cell death by apoptotic mechanisms. Its cellular targets are both DNA and proteins. In this communication, the authors analyze the interaction of pentamidine with Ubiquitin (Ub), which is a protein involved through the Ub-proteasome pathway in the turn-over of cellular proteins. Hence, the authors report herein that the interaction of pentamidine with Ub produces a 6% increase in the β-sheet content of the protein as detected by Far-UV circular dichroism spectroscopy. Moreover, fluorescence emission studies show that pentamidine induces in Ub a significant decrease in fluorescence of both Phe(4) and Phe(45) residues located on the β-cluster of the molecule but not of Tyr(59) residue located on the α-cluster. Because the β-cluster of Ub is composed by a large number of positively charged or basic aminoacids the authors data indicate that pentamidine specifically modifies the β-cluster or "basic face" of Ub. Thus, since pentamidine has two positive charges at physiological pH it is not surprising that electric repulsion may be involved in the modification of the β-cluster of Ub by pentamidine. It has been reported that the "basic" face of Ub is involved in recognition and poly-ubiquitinylation of different protein substrates. So, due to the fact that pentamidine specifically modifies the β-cluster or "basic" face of Ub, it is likely that pentamidine may inhibit the biochemical functions of Ub.

Chapter 8 - A new approach to definition of relatedness of two protein sequences has been proposed recently. The main point of this approach is finding a chain of natural "intermediate" sequence fragments between two selected fragments, in which all closest neighbors are similar at (high enough) selected threshold. The authors have found, that by taking relatively short fragments (15-25 amino acids) one can build long (more than 100 steps) chains, "walks" in the sequence space of all fragments from 112 complete bacterial proteomes (about 300,000 sequences). Such chains can be combined into the trees and networks. Belonging of two given sequences to the same "walk" or network – is a new criterion of sequence relatedness. Large networks are observed which contain fragments from different protein families with very different sequences but with the similar 3D structure, as well as with different sequences but from the proteins with the same functions. One perspective future development of this criterion of the sequence relatedness is also a new definition of relatedness of 3D-structures via the networks. At present, a single commonly used method for quantitative estimation of difference between two protein 3D-structures remains RMSD (Root Mean Square Deviation) in spite of its well known substantial limitations. The authors new approach can reveal that, sometimes, two RMSD-different but clearly similar structures correspond to sequences from the same network or sequence walk, separated by only several intermediate sequences. The authors believe, that this approach will provide a new point of view on protein structure evolution and classification as well as new potentials for protein structure prediction.

In: Protein Conformation: New Research
Editor: Linda B. Roswell, pp. 1-49
ISBN 978-1-60456-219-4

Chapter 1

PROTEIN CONFORMATIONAL ALPHABETS

Wei-Mou Zheng
Institute of Theoretical Physics, Academia Sinica
Beijing 100080, China
The Beijing Genomics Institute, Academia Sinica
Beijing 101318, China

Abstract

A protein conformational alphabet refers to the discretized states of the three-dimensional segmental structure of protein backbones. Under a specific representation (e.g. coordinates, dihedral angles, or C_α pseudobond angles), the structure of a segment corresponds to a point in a phase space, and hence a structure database is mapped to a point set in the phase space. Generally, the procedure to deduce finite discrete conformational states from a point set in a continuous conformational phase space is a clustering analysis. By representing the conformation of a quadrapeptide segment with the three C_α pseudobond angles, and clustering based on the probability distribution of the angles, an alphabet of 17 letters is derived. The relationship of these conformational states with the secondary structure and their associated amino acids, and an entropic reduction of conformational and amino acidic alphabets are discussed.

Without a conformational substitution matrix the use of a conformational alphabet is very limited. A substitution matrix called CLESUM is constructed from a representative pairwise aligned structure dataset of the families of structurally similar proteins (FSSP). By restricting the local conformations to a handful of states, a protein 3D backbone structure is converted to a 1D sequence of letters akin to the amino acid sequence; many tools of sequence analysis can then be applied with certain minor modification. As an example of application, fast structure alignment based on the conformational alphabet is described. Structurally similar fragment pairs are found by merely string comparison. Interference of frequently occurring matches of helices or sheets can be largely suppressed, and a greedy strategy guided by CLESUM similarity scores becomes possible. The same framework works also for multiple structure alignment.

Key words: structure alignment; structural codes; structural substitution matrix.

1. Introduction

Proteins fold into specific spatial conformations to perform their biological functions. Geometric reasoning plays a major role in our attempt to understand the activities of proteins.

Biologically important modules have been repeatedly employed in protein evolution by gene duplication and rearrangement mechanisms. They form components of fundamental units of structure and function. The presence of modules provides a guide to classify proteins into module-based families, and helps the structure prediction. The existence of such conservative recurrent segments presents a solid foundation for the local analysis.

The usual approaches for modelling proteins often simplify protein structures as chains of one or two interacting centers representing individual amino acids, e.g. the polypeptide backbones. Compared with a random chain, the conformational space accessible by a protein backbone fragment is rather limited. The patterns of hydrogen bonds between the main-chain peptide groups define various types of secondary structure. The first elements of regular secondary structure suggested in 1951 by Pauling and coworkers (Pauling and Corey, 1951) are the alpha helix and beta sheet, which are located in specific regions of the Ramachandran plot. Helices and beta strands form a basis for protein classification. However, segments of a single secondary structure may vary significantly in their 3D structures. In particular, the coil is not a true secondary structure but is a class of conformations that indicate the absence of regular secondary structure. A refined objective classification of segments may provide an improved 'coarse-grained' description of protein structure with only a small number of discrete conformational states. Such a description directly reflects the conformational heterogeneity in the phase space of protein structure and drastically reduces the number of degrees of freedom so that it is easier for us to understand the modular protein architecture and discover the correlation between sequences and structures. Our ability to predict the structure can thus be greatly enhanced.

The history of using discretized conformational states or structural alphabets for protein structures is very long and may be traced to at least as far back as Liquori's concept of a stereochemical alphabet (Liquori, 1969). We mention here a few typical cases; for a more complete review on structural alphabets see Offmann *et al.*, 2007.

Generally, the procedure to deduce finite discrete conformational states from continuous states in a conformational phase space is a clustering analysis. There has been a variety of different ways of clustering using various descriptors for structure. The original descriptor in the structure database Protein Data Bank (PDB) is the Cartesian coordinates of atoms. Many studies to investigate the classification of protein fragments use the backbone (ϕ, ψ) dihedral angles, or angles of C_α psuedobonds or distances derived from the positions of C_α atoms. Due to the anticorrelation between ϕ and ψ (McCammon *et al.*, 1977; Flocco and Mowbray, 1995), there may be instances where a big change in both ϕ and ψ does not represent an obvious change in the C_α pseudobond angles, but a reorientation of the peptide in question.

In an early work by Robson and Pain (1974) the phase space of (ϕ, ψ) was partitioned into a three-letter HPL alphabet by a survey on the contour plot of some information measure. Rooman, Kocher and Wodak (1991) intuitively divide the ϕ-ψ space into 6 regions, which corresponds to a partitioning based on the Ramachandran plot. Matsuda *et al.* (1997) introduced a backbone conformation encoding which is a uniform partition of the 2D phase space of C_α bending-torsion angles by mapping the phase space onto the surfaces of a icosahedron. Park and Levitt (1995) represented the polypeptide chain by a sequence of rigid fragments that were chosen from a library of representative fragments, and concatenated without any degrees of freedom. The average deviation of the global-fit approximations over the training set was taken as the objective function for optimizing the finite representative fragments. The cluster states there are representative points of the phase space.

Most authors used standard methods for clustering analysis to generate discrete structure states. For example, Micheletti and coworkers (2000) used a similarity threshold in terms of root mean square deviation (RMSD) of optimal superposition for two fragments to locate the center of the most populated cluster as the first representative oligomer or oligon. By successively removing fragments of clusters found from the training set and locating new oligons, the clustering gives simultaneously both the subdivision in clusters and the optimal representatives, which were singled out in order of importance.

Hidden Markov models (HMMs; Rabiner, 1989), possessing a rigorous but flexible mathematical structure, have been used in a variety of computational biology problems such as sequence motif recognition (Fujiwara et al., 1994), gene finding (Burge and Karlin, 1997), protein secondary structure prediction (Asai *et al.*, 1993; Zheng, 2004), and multiple sequence alignments (Krogh *et al.*, 1994). The HMMs have also been used for identifying the modular framework for the protein backbone (Edgoose *et al.*, 1998; Camproux *et al.*, 1999). In these HMMs, conformation states are represented by probability distributions, which are much finer than a simple partition of the phase space or a set of representative points. The HMMs also take into account the sequential connections between conformational states, which hence involves a large number of parameters and makes the model training a tough task. Furthermore, it is not so convenient to assign structure codes to a short segment with HMMs.

To construct a structural alphabet we have to balance accuracy with simplicity. Prediction of protein structure depends on the accuracy and complexity of the models used. A model must be as simple as possible to reduce the conformational space to be searched for a 'correct' conformation, while a model of low complexity tends to have a lower accuracy. A model is required to represent the actual geometry of protein conformations accurately enough, but a complex model is prone to over-fitting of the observed data.

Here, using a description of protein backbone tertiary structure with the psuedobond angles of successive C_α atoms, we construct finite structural states as our conformational letters (CLs) according to the density peaks of the probability distribution in the phase space spanned by the pseudobond angles, and examine their feasibility in characterizing the short segment polypeptide backbone conformation. In order to use the structural codes in the structure analysis we derive a substitution matrix for these conformational states from a representative set of pairwise aligned structures of the families of structurally similar proteins (FSSP) database by Holm and Sander (1994; 1997; 1998a).

By restricting the local conformations to a handful of states, protein 3D backbone structures are converted to a 1D sequence of letters akin to the amino acid sequence; many tools of sequence analysis can then be applied with certain minor modification. Equipped with the substitution matrix, tools for sequence alignments may be adopted to compare structures after converting them into sequences of the conformational letters. The concept of the structural alphabet is local, while that of the structure alignment is global. Since the local similarity is a necessary condition for global alignment, similar fragment pairs (SFPs) of two structures form the basis of the global alignment. Such SFPs can now be extracted from structures by just simple string comparison, and the similarity score can be used to sort the importance of SFPs for further development of a greedy algorithm. The same idea is also valid for the multiple structure alignment, where an SFP is replaced by a similar fragment block (SFB). We shall propose fast, efficient and reliable algorithms for pairwise and multiple alignments of protein structures, and briefly discuss other applications of structural alphabets.

2. Conformational Letters

Among the variety of abstract representating forms for protein 3D structure, one that is frequently encountered is the protein virtual backbone. The C_α atoms of residues are chosen as the representative points of the backbone chain. In this representation, two adjacent residues in a protein sequence are virtually bonded, forming a pseudobond. Our conformational letters are discretized states of fragments of such pseudobond chains (Zheng and Liu, 2005).

2.1. Pseudobond Angles

Since atom coordinates are not invariant under transformation, intrinsic variables like angles and distances are more convenient as descriptors of conformations. The relation between C_α coordinates and pseudobond angles is rather straightforward, and pseudobond angles have a more direct geometric meaning than distances. The virtual bond bending angle θ defined for three contiguous points (a, b, c) is the angle between the vectors $\mathbf{r}_{ab} = \mathbf{r}_b - \mathbf{r}_a$ and $\mathbf{r}_{bc}$, i.e. $\theta = \mathbf{r}_{ab} \cdot \mathbf{r}_{bc}/(|r_{ab} r_{bc}|)$. The range of θ is $[0, \pi]$. The virtual bond torsion angle τ defined for four noncoplanar contiguous points (a, b, c, d) is the dihedral angle between the planes abc and bcd. The range of τ is $(-\pi, \pi]$, and its sign is the same as the mixed product $(\mathbf{r}_{ab} \times \mathbf{r}_{bc}) \cdot \mathbf{r}_{cd}$. In fact, we may adopt a wider range of τ under the equivalence relation that τ_1 and τ_2 are equivalent if $\tau_1 = \tau_2 \pmod{2\pi}$. For the four-residue segment $abcd$, by taking a as the origin, b on the x-axis, and c on the xy-plane, the number of independent relative coordinates is 6. The assumption of the fixed pseudobond length, which is 3.8 Å for the dominating *trans* peptide, further reduces the number of degrees of freedom to 3. These new independent 'coordinates' correspond to the angles $(\theta_{abc}, \tau_{abcd}, \theta_{bcd})$. We shall assign the angle triplet $(\theta_{abc}, \tau_{abcd}, \theta_{bcd}) \equiv (\theta_b, \tau_c, \theta_c)$ to residue c, the third of the four-residue segment. (An argument for this assignment will be presented later.) Elongating the segment by one residue e will add two more angles τ_{bcde} and θ_{cde}. Generally, for a structure of n residues, we have $n - 2$ bending angles and $n - 3$ torsion angles, $2n - 5$ in total.

Bending and torsion angles of a chain correspond to curvature and torsion of a curve. The relative coordinates of the chain $\{\mathbf{r}_0, \mathbf{r}_1, \cdots \mathbf{r}_n\}$ can be recovered from their $2n - 3$ angles $\{\theta_1; \tau_2, \theta_2; \cdots; \tau_{n-1}, \theta_{n-1}\}$. By convention, we set the origin at $\mathbf{r}_0$, put $\mathbf{r}_1$ along the x-axis, and add $\tau_1 = 0$. Introducing the rotation matrices R_θ and R_τ (with respect to the z-

and x-axis, respectively):

$$R_\theta = \begin{pmatrix} \cos\theta & -\sin\theta & 0 \\ \sin\theta & \cos\theta & 0 \\ 0 & 0 & 1 \end{pmatrix}, \; R_\tau = \begin{pmatrix} 1 & 0 & 0 \\ 0 & \cos\tau & -\sin\tau \\ 0 & \sin\tau & \cos\tau \end{pmatrix}, \text{ and } \mathbf{d} = \begin{pmatrix} 1 \\ 0 \\ 0 \end{pmatrix}, \quad (1)$$

the position $\mathbf{r}_k$ for $k \geq 1$ is determined by

$$T_0 = I, \; \mathbf{r}_0 = 0 \cdot \mathbf{d}, \; T_k = T_{k-1} R_{\tau_k} R_{\theta_k}, \; \mathbf{d}_k = T_{k-1} \cdot \mathbf{d}, \; \mathbf{r}_k = \mathbf{r}_{k-1} + \mathbf{d}_k, \quad (2)$$

where I is the identity matrix, and the length unit is 3.8 Å.

Longer fragments will include more correlation than shorter fragments. However, the complexity that can be explored with the longer fragment lengths is limited severely by the relatively small number of known protein structures. Furthermore, a larger number of discrete states have to be determined for a longer segment. The minimal unit where the relative coordinates fix the angles and vice versa is the segment of four contiguous residues. Since our purpose is to obtain a concise coarse-grained description of protein tertiary structure, we shall concentrate mainly on the structure codes for the four residue unit.

2.2. The Mixture Model for the Angle Probability Distribution

The three pseudobond angles (θ, τ, θ') of the four-residue unit span the three-dimensional phase space. Our classifiers for conformational states are based on the following mixture model M: the probability distribution of 'points' $\mathbf{x} \equiv (\theta, \tau, \theta')$ is given by the mixture of several normal distributions

$$P(\mathbf{x}|M) = \sum_{i=1}^{c} \pi_i N(\mu_i, \mathbf{\Sigma}_i), \quad (3)$$

where c is the number of the normal distribution categories in the mixture, π_i the prior for category C_i, and $N(\mu, \mathbf{\Sigma})$ the normal distribution

$$N(\mu, \mathbf{\Sigma}) = (2\pi)^{-3/2} |\mathbf{\Sigma}|^{-1/2} \exp[-\tfrac{1}{2}(\mathbf{x} - \mu) \cdot \mathbf{\Sigma}^{-1} \cdot (\mathbf{x} - \mu)]. \quad (4)$$

Each normal distribution has six parameters for its symmetric covariance matrix $\mathbf{\Sigma}$ and three for its mean μ. Adding one more parameter of the prior for each category, the mixture model has $10c$ parameters for the total c categories. (The normalization $\sum_i \pi_i = 1$ reduces the number to $10c - 1$.) These categories will be translated as the structure codes.

To objectively determine the number c of categories and roughly estimate the means μ, we investigate density peaks in the phase space with the downhill simplex method of Nelder and Mead (1965). The method requires only function evaluations, not derivatives. It is not very efficient in terms of the number of function evaluations that it requires, but works well for our problem here. The value of the function for optimization at a given point is defined as the number of counts in the rectangular box centered at the point. The box size corresponds to the Parzon window width. A large box size has a low resolution, which helps us to focus on the main density peaks in the phase space, and to easily locate them near their real location. Reducing the box size we can see more peaks, which become less conspicuous and even disappear for larger box sizes. A box size which is too small makes local fluctuations visible and so is often misleading. Missing out any important modes will affect the model training and the efficacy of the generated structural codes.

We first search for maximal points of the one-dimensional marginal probability distributions of θ and τ, and then utilize them to generate a grid in the (θ, τ, θ') space to search for peaks in the space. We examine also density peaks in the five-dimensional phase space spanned by $(\theta_b, \tau_c, \theta_c, \tau_d, \theta_d)$ of the five-residue unit $abcde$ to investigate the effect of the angle correlation between two nearby quadra-peptides. A five-angle mode $(\theta_b, \tau_c, \theta_c, \tau_d, \theta_d)$ contains two three-angle modes $(\theta_b, \tau_c, \theta_c)$ and $(\theta_c, \tau_d, \theta_d)$. It is required that all the important three-angle modes implied by the main density peaks in the five-angle phase space must be included in the modes used for the construction of the mixture model.

The main purpose of our search for density peaks is to estimate the number c of categories and $\{\mu_i\}$ for each category. Once this has been done, we may start with some simple $\{\pi_i\}$ and $\{\mathbf{\Sigma}_i\}$, say $\pi_i = 1/c$ and a certain diagonal $\{\mathbf{\Sigma}_i\}$, and then update the mixture model by the Expectation-Maximization (EM) method as follows. For each point $\mathbf{x}_k = (\theta_{k-1}, \tau_k, \theta_k)$, we calculate the probability for the point to belong to the i-th category C_i according to the Bayes formula as

$$\begin{aligned} P(C_i|\mathbf{x}_k) &\propto \pi_i P(\mathbf{x}_k|C_i) \\ &\propto \pi_i |\mathbf{\Sigma}_i|^{-1/2} \exp[-\tfrac{1}{2}(\mathbf{x}_k - \mu_i)\cdot \mathbf{\Sigma}_i^{-1} \cdot (\mathbf{x}_k - \mu_i)], \end{aligned} \tag{5}$$

where we always shift τ_k to the interval $[\tau_{ci} - \pi, \tau_{ci} + \pi)$ centered at τ_{ci} (the τ-component of the mean μ_i) to avoid using the von Mises distribution. The probability $P(C_i|\mathbf{x}_k)$ satisfies the normalization condition $\sum_{i=1}^{c} P(C_i|\mathbf{x}_k) = 1$. The updated parameters for the mixture model are estimated by the EM method to be

$$n_i = \sum_k P(C_i|\mathbf{x}_k), \qquad \pi_i = n_i/n, \quad n = \sum_i n_i, \tag{6}$$

$$\mu_i = \frac{1}{n_i} \sum_k P(C_i|\mathbf{x}_k)\mathbf{x}_k, \tag{7}$$

$$\mathbf{\Sigma}_i = \sum_k P(C_i|\mathbf{x}_k)(\mathbf{x}_k - \mu_i)(\mathbf{x}_k - \mu_i)^T. \tag{8}$$

The EM updating is 'fuzzy', assigning a point $\mathbf{x}_k$ to all categories but with different probabilities. Thus, it is usually less sensitive to the initial conditions than a greedy algorithm.

Generally, the objective function for optimizing the mixture model is

$$\text{Prob}(\{\mathbf{x}_k\}) = \prod_k \sum_i P(\mathbf{x}_k, C_i) \propto \prod_k \sum_i P(C_i|\mathbf{x}_k). \tag{9}$$

However, when we convert point $\mathbf{x}_k$ to its structural code i^*, we use

$$i^* = \arg_i \max P(C_i|\mathbf{x}_k). \tag{10}$$

An alternative objective function would be

$$Q(\{\mathbf{x}_k\}) = \prod_k \max_i P(C_i|\mathbf{x}_k). \tag{11}$$

When starting with narrow distributions for Σ_i, a very high value of Q may be seen at the first step. However, by just one step of the EM iteration Q will drop significantly, and then increase at later steps. While $\text{Prob}(\{\mathbf{x}_k\})$ never decreases, Q will decrease after reaching its maximum. We may stop the model training before Q decreases again. Thus, the optimization here is a compromise between $\text{Prob}(\{\mathbf{x}_k\})$ and $Q(\{\mathbf{x}_k\})$. (Function Q does not directly correspond to a probabilistic model; a greedy updating with Q as the objective function would result in a loss of modes.)

2.3. Training of the Mixture Model

To establish the discrete structural states by training the mixture model we create a nonredundant set of 1544 non-membrane proteins from PDB_SELECT (Hobohm and Sander, 1994) with amino acid identity less than 25%, issued on 25 September 2001. The data of the three-dimensional structures for these proteins are taken from PDB. The secondary structures for these sequences are taken from the DSSP database (Kabsch and Sander, 1983). We consider the reduced three secondary structure states $\{h, e, c\}$ generated from the eight states of the DSSP by the coarse-graining $H, G, I \to h$, $E \to e$ and $X, T, S, B \to c$. The total number of contiguous fragments is 2248, which gives a total of 264,232 points in the three-angle phase space.

The model training involves a global optimization. The choice of a good initial trial plays an important role. Usually, an improper initial trial would not survive the later EM process of model training. Thus, we carefully explore the density distribution in the three-angle space.

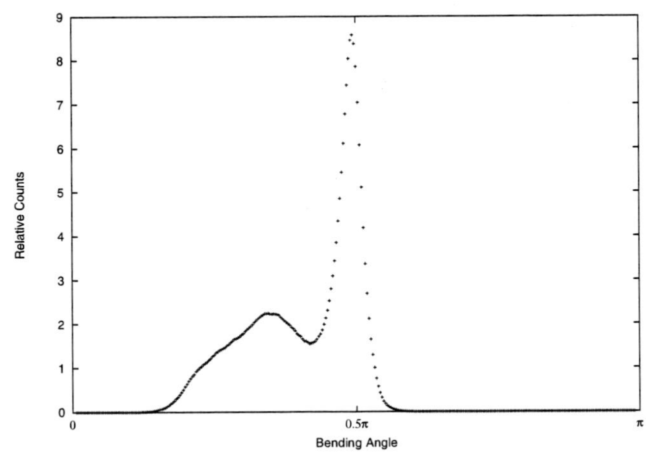

Figure 1. The 1D marginal distribution of the bending angle.

The marginal one-dimensional distribution of the pseudobond bending angle, as shown in figure 1, has two prominent peaks around $\theta = 1.10$ and 1.55 (radians). Non-zero θs are in the interval $[.4, 1.9]$. The marginal one-dimensional distribution of the torsion angle τ (figure 2) has one immediately noticeable peak at $\tau = 0.87$ (corresponding to the helix). Another peak at $\tau = -2.94$ is less prominent. There is a vague peak still recognizable around $\tau = -2.00$.

The 2D contour plot of θ-τ densities is shown in figure 3, where two peaks of the helix and sheet are clearly seen, and a few other peaks are recognizable. A grid generated with $\theta \in \{1.00, 1.55\}$ and $\tau \in \{-2.80, -2.05, -1.00, 0.00, 0.87\}$ is used to search a high dimensional phase space for density peaks by the downhill simplex method. In the box counting, the box size is taken from 0.1 to 0.2 for θ, and the width of τ is twice that of θ. The helices seen as a single peak in the three-angle phase space are clearly identified as several sub-peaks in the five-angle phase space. Combining the exploration of main peaks in the five-angle phase space of the quintapeptide, we identify 17 mode centers, which are then used as the main initial parameters to train the mixture model. Finally, the 17 structural states are obtained for the mixture model by the EM algorithm. They will be referred to as the conformational letters (CLs). The ultimate parameters for the model are listed in Table 1. Note that it is the entries of the inverse covariance matrix that are given. The determinant of the matrix is a measure of the divergence of the corresponding mode. The most sharp state is H, while the most vague state is Q, which occupies the least proportion of phase points.

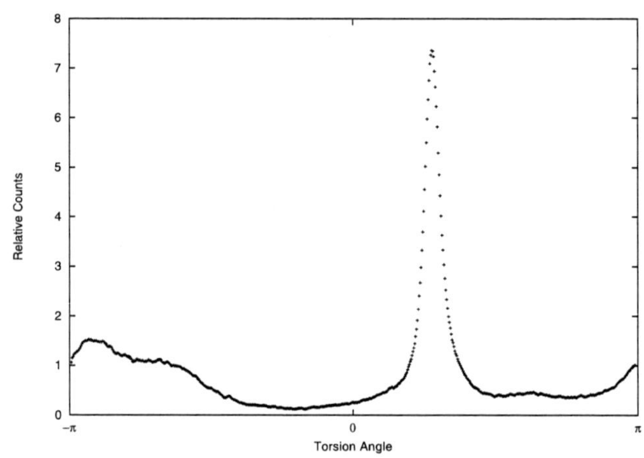

Figure 2. The 1D marginal distribution of the torsion angle.

Once we have the model trained, we may convert a structure to its conformational code sequence according to (10). Although no effect from the connection of states is directly considered, the model gains the advantage in being able to easily assign codes to short fragments. After converting all the structures in the training set to their conformational codes, we calculate the statistics of the 17 states directly from the annotated code sequences. The results are listed in Table 2. Although the frequencies f_i of the state occurrences are close to π_i, they are not the same. A tiny discrepancy is also observed between the observed means and μ_i. This is not so surprising. Both π_i and $P(\mathbf{x}_k|C_i)$ contribute to $P(C_i|\mathbf{x}_k)$. For a mode with a sharp distribution like H (or E), the factor $P(\mathbf{x}_k|H)$ surpasses those of other nearby modes, so mode H 'attracts' nearby points when we identify classes with Eq. (10). This also pushes the observed mean values of nearby modes away from their μ_i (along the direction away from μ_H). In fact, such a 'discrepancy' between the 'Baum-Welch' and 'Viterbi' schemes is rather general (Rabiner 1989). The representative shapes of the 17 conformational letters are shown in figure 4.

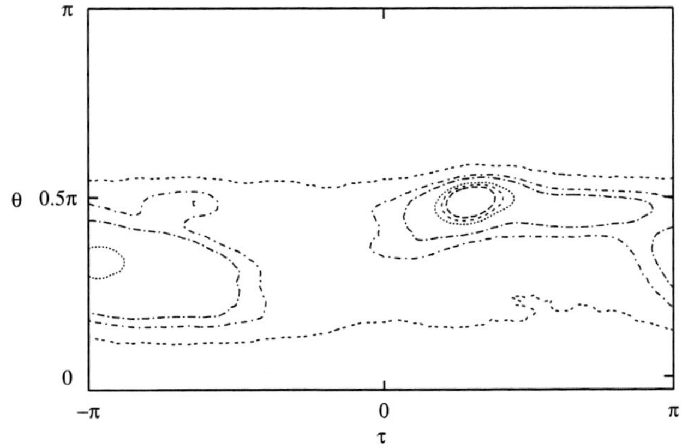

Figure 3. Contour plot of the density of bending-torsion (θ-τ) angles. The curves on the left side should be continuous with those on the right side. The plot is only schematic, and the levels of contours are not uniformly spaced.

Table 1. The 17 structural states from the mixture model. Angles are in units of radian.

	π	$\lvert\Sigma\rvert^{-1/2}$	μ			Σ^{-1}					
State			θ	τ	θ'	$\theta\theta$	$\tau\theta$	$\tau\tau$	$\theta'\theta$	$\theta'\tau$	$\theta'\theta'$
I	8.2	1881	1.52	0.83	1.52	275.4	-28.3	84.3	106.9	-46.1	214.4
J	7.3	1797	1.58	1.05	1.55	314.3	-10.3	46.0	37.8	-70.0	332.8
H	16.2	10425	1.55	0.88	1.55	706.6	-93.9	245.5	128.9	-171.8	786.1
K	5.9	254	1.48	0.70	1.43	73.8	-13.7	21.5	15.5	-25.3	75.7
F	4.9	105	1.09	-2.72	0.91	24.1	1.9	10.9	-11.2	-8.8	53.0
E	11.6	109	1.02	-2.98	0.95	34.3	4.2	15.2	-9.3	-22.5	56.8
C	7.5	100	1.01	-1.88	1.14	28.0	4.1	6.2	2.3	-5.1	69.4
D	5.4	78	0.79	-2.30	1.03	56.2	3.8	4.2	-10.8	-2.1	30.1
A	4.3	203	1.02	-2.00	1.55	30.5	9.1	8.7	6.0	5.7	228.6
B	3.9	66	1.06	-2.94	1.34	26.9	4.6	4.9	9.5	-5.0	54.3
G	5.6	133	1.49	2.09	1.05	163.9	0.6	3.8	2.0	-3.7	32.3
L	5.3	40	1.40	0.75	0.84	43.7	2.5	1.4	-7.0	-2.9	34.5
M	3.7	144	1.47	1.64	1.44	72.9	2.1	4.8	1.9	-7.9	72.9
N	3.1	74	1.12	0.14	1.49	25.3	3.2	3.1	9.9	0.9	83.0
O	2.1	247	1.54	-1.89	1.48	170.8	-0.7	3.7	-4.1	3.1	98.7
P	3.2	206	1.24	-2.98	1.49	48.0	8.2	7.3	-4.9	-6.6	155.6
Q	1.7	25	0.86	-0.37	1.01	28.4	1.5	1.2	3.4	0.1	19.5

2.4. The Conformational Alphabet and the Secondary Structure

Our 17 conformational states or letters of the structural alphabet describe the local structure of four-residue segments, and a code is assigned to the third residue of the unit of quadra-peptide. The total number of possible four-residue secondary structures is 37. The restriction of the minimal lengths 2 for e and 3 for h removes 44 quartets from the total $3^4 = 81$.

Table 2. The observed frequencies, means and covariances of the 17 structural states from the mixture model. Frequencies are in percentages, and angles are in radian units.

State	f	Mean			Covariance					
		θ	τ	θ'	$\theta\theta$	$\tau\theta$	$\tau\tau$	$\theta'\theta$	$\theta'\tau$	$\theta'\theta'$
I	7.50	1.50	0.81	1.50	0.005	-0.001	0.012	-0.004	0.002	0.007
J	6.29	1.59	1.12	1.56	0.003	-0.002	0.020	-0.001	0.005	0.004
H	19.53	1.55	0.88	1.55	0.001	0.000	0.004	-0.000	0.001	0.001
K	4.73	1.47	0.62	1.38	0.017	0.005	0.070	-0.002	0.020	0.021
F	3.12	1.16	-2.65	0.86	0.044	-0.003	0.044	0.014	0.009	0.017
E	14.11	1.02	-2.99	0.96	0.026	0.002	0.125	0.006	0.054	0.037
C	7.75	1.01	-1.80	1.15	0.035	-0.028	0.134	-0.005	0.011	0.013
D	4.61	0.72	-2.25	1.00	0.010	-0.005	0.155	0.001	-0.004	0.029
A	4.79	1.01	-1.97	1.55	0.041	-0.041	0.134	0.000	-0.004	0.004
B	3.11	1.00	-3.00	1.36	0.045	-0.058	0.221	-0.016	0.034	0.016
G	6.02	1.50	2.13	1.04	0.005	-0.001	0.249	-0.001	0.029	0.028
L	4.85	1.39	0.59	0.82	0.024	-0.053	0.739	-0.000	0.045	0.029
M	3.49	1.46	1.73	1.45	0.013	-0.002	0.170	-0.001	0.026	0.014
N	3.03	1.10	0.14	1.50	0.042	-0.049	0.326	-0.005	0.004	0.011
O	2.28	1.54	-1.88	1.48	0.005	0.000	0.251	0.000	-0.007	0.010
P	3.31	1.26	-3.03	1.50	0.020	-0.019	0.102	-0.001	0.001	0.005
Q	1.51	0.84	-0.13	0.97	0.032	-0.047	0.792	-0.009	0.009	0.048

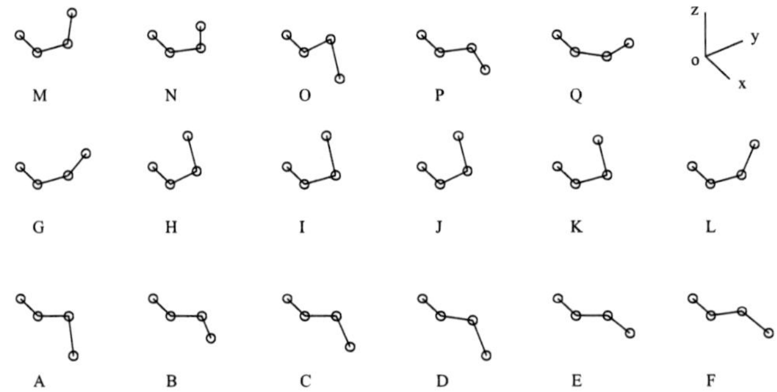

Figure 4. The representative shapes of the 17 conformational letters.

In order to make a detailed comparison between the secondary structures and the discrete conformational states, from the training set we extract a subset which contains 676 long fragments and 118,621 residues (hence 116,593 points in the three-angle phase space). Among the 37 four-residue secondary structures, *heeh* has no counts and so is omitted. Four more structures have their counts under 100. Table 3 shows the percentages of conformational letters in each secondary structure, while Table 4 shows the percentages of secondary structures in each conformational state.

Entries in Table 3 are normalized with respect to rows, while those in Table 4 to columns. It is clearly seen that there exists a correlation between the two types of local structure classifications. For example, from Table 3 we may attribute the *hhhh* mainly to H (60.1%), I (21.9%) and J (11.3%), and the *eeee* to E (48.3%) and D (17.2%). Among

Table 3. The percentages of conformational states in each secondary structure.

	I	*J*	*H*	*K*	*F*	*E*	*C*	*D*	*A*	*B*	*G*	*L*	*M*	*N*	*O*	*P*	*Q*	Counts
cccc	3	3	1	5	3	7	14	3	7	3	10	9	5	7	4	5	3	25090
ccce	0	1	0	3	3	6	15	2	5	4	24	9	4	4	4	4	3	3272
ccch	2	1	1	6	5	8	22	4	3	2	9	11	7	4	1	3	2	3028
ccee	0	0	0	0	7	20	17	7	0	1	15	24	0	0	0	0	2	4029
cchh	1	3	0	1	0	0	1	0	46	1	0	0	17	4	0	18	1	3664
ceec	0	0	0	0	6	36	26	14	0	8	0	2	0	0	0	0	3	620
ceee	0	0	0	0	7	43	12	22	0	4	0	2	0	0	0	2	3	3676
ceeh	0	0	0	0	3	11	49	19	0	3	0	0	0	0	1	1	7	51
chhh	21	38	28	4	0	0	0	0	0	0	0	0	4	2	0	0	0	4353
eccc	3	3	1	3	2	6	15	3	14	3	5	11	4	7	2	7	4	3007
ecce	3	1	1	6	1	4	5	1	1	1	1	4	1	26	35	0	1	492
ecch	1	1	0	5	1	6	18	5	4	1	9	24	5	6	0	3	2	258
ecee	1	0	0	0	5	18	16	10	1	2	3	33	1	0	0	2	3	80
echh	0	1	0	0	0	0	0	0	52	1	1	0	10	5	0	17	5	256
eecc	0	0	0	0	6	16	19	7	12	6	0	3	0	9	0	11	4	3807
eece	0	0	0	0	6	18	21	11	11	6	0	3	1	11	0	6	2	80
eech	0	0	0	0	4	15	25	16	2	5	0	3	0	6	0	10	10	256
eeec	0	0	0	0	7	36	19	14	1	9	0	6	0	0	0	1	3	3596
eeee	0	0	0	0	5	48	8	17	0	6	0	7	0	0	0	2	1	11418
eeeh	0	0	0	0	4	15	41	17	2	11	0	2	0	0	0	1	4	197
eehh	0	0	0	0	0	0	3	0	57	2	0	0	0	2	1	28	4	248
ehhh	13	43	25	3	0	0	0	0	0	0	0	0	6	5	1	0	0	248
hccc	4	5	2	4	2	4	7	1	4	2	14	6	6	5	18	6	1	3254
hcce	1	2	0	3	5	5	11	0	7	5	22	7	4	4	6	7	3	208
hcch	3	0	1	4	3	5	21	5	4	2	14	9	7	4	4	3	1	328
hcee	0	0	0	0	6	21	18	9	0	4	17	14	1	0	0	0	2	151
hchh	1	2	0	1	0	0	1	0	12	1	0	0	27	18	0	31	0	356
heec	0	0	0	0	5	50	15	10	0	10	0	5	0	0	0	5	0	20
heee	0	0	0	0	4	50	4	17	1	6	0	3	0	0	0	8	1	117
hhcc	9	15	11	9	1	0	0	0	1	0	15	10	9	1	10	2	0	3861
hhce	1	1	0	1	2	0	2	0	1	1	43	15	6	0	14	3	0	151
hhch	8	4	3	19	3	0	1	0	0	0	15	28	7	2	3	0	0	356
hhee	0	0	0	0	0	0	0	0	0	0	52	40	3	0	0	0	0	137
hhhc	23	21	25	20	0	0	0	0	0	0	0	2	2	3	1	0	0	4464
hhhe	29	30	15	11	0	0	0	0	0	0	0	4	4	2	2	0	0	137
hhhh	21	11	60	4	0	0	0	0	0	0	0	0	1	0	0	0	0	31327

Table 4. The percentages of secondary structures in each conformational state. Here, the last row 'Total' indicates the total counts of each letter in units of a hundred.

	I	*J*	*H*	*K*	*F*	*E*	*C*	*D*	*A*	*B*	*G*	*L*	*M*	*N*	*O*	*P*	*Q*
cccc	7	10	2	25	26	14	37	14	33	27	43	35	33	46	37	34	39
ccce	0	0	0	2	3	1	5	1	3	4	12	4	3	3	5	3	5
ccch	0	0	0	3	4	1	6	2	1	2	4	4	5	3	2	2	3
ccee	0	0	0	0	9	6	7	5	0	2	9	14	0	0	0	0	4
cchh	0	1	0	1	0	0	0	0	30	1	0	0	14	3	1	16	2
ceec	0	0	0	0	1	1	1	1	0	1	0	0	0	0	0	0	0
ceee	0	0	0	0	8	12	4	15	0	6	0	1	0	0	0	2	5
ceeh	0	0	0	0	0	0	0	0	0	0	0	0	0	0	0	0	0
chhh	8	19	5	3	0	0	0	0	0	0	0	0	4	3	0	0	0
eccc	0	1	0	2	2	1	4	1	7	3	2	4	2	5	2	5	5
ecce	0	0	0	0	0	0	0	0	0	0	0	0	0	3	6	0	0
ecch	0	0	0	0	0	0	0	0	0	0	0	0	0	0	0	0	0
ecee	0	0	0	0	0	0	0	0	0	0	0	0	0	0	0	0	0
echh	0	0	0	0	0	0	0	0	2	0	0	0	0	0	0	1	0
eecc	0	0	0	0	7	4	7	5	8	8	0	1	0	9	0	11	8
eece	0	0	0	0	0	0	0	0	0	0	0	0	0	0	0	0	0
eech	0	0	0	0	0	0	0	0	0	0	0	0	0	0	0	0	1
eeec	0	0	0	0	8	9	7	9	0	10	0	3	0	0	0	1	5
eeee	0	0	0	0	19	41	9	36	1	23	1	11	0	2	0	6	9
eeeh	0	0	0	0	0	0	0	0	0	0	0	0	0	0	0	0	0
eehh	0	0	0	0	0	0	0	0	2	0	0	0	0	0	0	1	0
ehhh	0	1	0	0	0	0	0	0	0	0	0	0	0	0	0	0	0
hccc	1	2	0	2	2	1	2	0	2	2	7	2	5	4	22	5	2
hcce	0	0	0	0	0	0	0	0	0	0	0	0	0	0	0	0	0
hcch	0	0	0	0	0	0	0	0	0	0	0	0	0	0	0	0	0
hcee	0	0	0	0	0	0	0	0	0	0	0	0	0	0	0	0	0
hchh	0	0	0	0	0	0	0	0	0	0	0	0	2	1	0	2	0
heec	0	0	0	0	0	0	0	0	0	0	0	0	0	0	0	0	0
heee	0	0	0	0	0	0	0	0	0	0	0	0	0	0	0	0	0
hhcc	3	7	1	7	1	0	0	0	1	0	9	5	8	1	14	2	0
hhce	0	0	0	0	0	0	0	0	0	0	1	0	0	0	0	0	0
hhch	0	0	0	1	0	0	0	0	0	0	0	1	0	0	0	0	0
hhee	0	0	0	0	0	0	0	0	0	0	1	0	0	0	0	0	0
hhhc	10	11	5	17	0	0	0	0	0	0	0	1	2	4	1	0	0
hhhe	0	0	0	0	0	0	0	0	0	0	0	0	0	0	0	0	0
hhhh	65	42	83	27	0	0	0	0	0	0	0	0	12	1	0	0	0
Total	104	82	224	50	32	131	98	53	55	29	62	68	44	37	26	40	21

the other 30 structures with counts more than 100, structure *eehh* is dominated by the single conformation A (58.0%), *echh* by A (53.1%), *heee* by E (50.4%), *cchh* by A (46.2%), *ceee* by E (43.4%), and *hhce* by G (43.0%). Structure *hhee* originates almost only from G (52.5%) and L (40.8%). Structure *cccc* shows rather uniform percentages in different conformational states as we would expect. From Table 4, H, I and J are associated to *hhhh*, which occupies 83.9% of H, 65.5% of I and 42.7% of J, while E and D are associated to *eeee*, which occupies 41.8% of E and 36.8% of D. Conformation K is quite evenly dominated by *cccc* (26.0%) and *hhhh* (27.6%). From this table it is also deduced that conformations L and G are responsible for the N-cap of β-strands, and P, Q, B, F and C for the C-cap of β-strands, while A, M, J and P give rise to the N-cap of helices, and O, K, G and M to the C-cap of helices. The mutual information between the conformational codes and the secondary structure states equals 0.731.

Table 5. The forward transition rates (multiplied by 100) between conformational letters.

	I	J	H	K	F	E	C	D	A	B	G	L	M	N	O	P	Q
I	28	13	26	12	0	0	0	0	0	0	8	3	6	0	4	0	0
J	19	21	23	12	0	0	0	0	0	0	11	3	5	0	5	0	0
H	11	7	74	2	0	0	0	0	0	0	2	0	1	0	3	0	0
K	8	6	4	23	2	3	2	0	5	2	8	14	11	4	4	6	0
F	0	0	0	0	3	32	16	16	14	8	0	1	0	5	0	1	5
E	0	0	0	0	6	44	12	15	5	7	0	1	0	3	0	4	2
C	0	0	0	3	7	22	22	0	17	5	1	3	1	10	0	8	1
D	0	0	0	0	7	44	14	8	10	6	0	2	0	3	0	4	2
A	12	26	19	8	1	0	0	0	0	0	13	10	5	0	2	2	0
B	2	2	1	5	6	12	5	0	2	4	9	28	8	4	2	10	1
G	0	0	0	1	7	21	28	7	11	7	0	2	1	6	0	7	2
L	0	0	0	0	2	20	17	22	12	9	0	1	0	5	0	2	9
M	14	10	8	11	2	3	2	0	2	2	10	13	9	3	3	8	0
N	3	5	2	4	3	2	2	0	1	1	35	8	13	3	13	4	0
O	1	1	0	2	2	2	1	0	0	3	56	6	17	2	4	4	0
P	12	18	7	7	2	1	1	0	0	0	12	23	9	1	4	3	0
Q	0	0	0	1	3	16	15	8	10	5	1	4	1	20	0	3	13

2.5. The Connection Effect of Sequential States

Using a zero-order Markov chain (M0) to model the 'hidden' states, our mixture model is closely related to the HMM which employs a first-order Markov chain (M1) for hidden states (Camproux *et al.*, 1999). Both adopt the Mahanalobis distance instead of the Euclidian distance. Compared with the M1-HMM, the mixture model does not include the connection effect of sequential states. The parameter number increases quadratically with the number of categories for a M1 model, while only linearly for a mixture model. We have to compromise between precision and correlation. A mixture model with finer categories is also promising. The connection effect can be recovered in later stages when dealing with strings of the conformational letters. (In fact, the correlation covered by a M1 model or code dimers is rather limited; for real applications longer strings have to be considered.) When considering dimers of the letters, any two sequential points $(\theta_{i-1}, \tau_i, \theta_i)$ and $(\theta_i, \tau_{i+1}, \theta_{i+1})$

share the common angle θ_i, which is a strong restriction. The effect of the connection of sequential structural states reflects the transition rates between structural states. These rates are determined by counting code pairs, and are listed in Table 5.

The entries of Table 5 show the forward transition rates as the conditional probability of the $(i+1)$-th site of a state chain at a given i-th site. Normalized according to the row, the table indicates where a row state would like to go. Extended states, e.g. H and E, are characterized by large diagonal elements, while transient states, e.g. A and G, have almost vanishing diagonal rates. From the table it can be seen that L and G have a high chance to transit to a β sheet, and A, M and P to a helix. A sheet has a high chance to go to A and B, and a helix to G, L and M. This is consistent with Table 4.

The 12 modes for the pentapeptide unit of Camproux et al. (1999) were found based on a M1-HMM. Instead of angles, they used a combination of four distances. Since only three of the four are independent their mode centers need not exactly correspond to a real conformation. However, we still can see the correspondence between their codes and ours: α_1-H, α_2-J, α'-K, α'_--O, α'_+-M, γ_1-N, γ_2-P, γ_β-Q, $\gamma_{\beta\alpha}$-A, $\gamma_{\alpha\beta}$-G, β_2-D, and β_1-E.

2.6. Structural Substitution Matrix

Sequence alignment is the main procedure of sequence comparison. Certain amino acid substitutions commonly occur in related proteins from different or the same species. Amino acid substitution matrices, extracted from our knowledge of most and least common changes in a large number of proteins, serve for the purpose of sequence alignment. The popular BLOSUM matrix of Henikoff and Henikoff (1992) is derived from a large set of conserved amino acid patterns without gaps representing various families. The frequency of amino acid substitutions in alignments is counted in sequence alignments. These frequencies are then divided by the expected frequency of finding the amino acids together in an alignment by chance. The ratio of the observed to the expected counts is an odds score. The BLOSUM entries are logarithms of the odds scores with the base 2 and multiplied by a scaling factor of 2.

To use our structural codes directly for the structural comparison, a score matrix similar to BLOSUM is desired. There is a database of aligned structures, the FSSP of Holm and Sander (1998a), which is based on exhaustive all-against-all 3D structure comparison of protein structures in the PDB. The proteins in the FSSP are divided into a representative set and sequence homologs of the representative set. The representative set contains no pair which has more than 25% sequence identity. In the version of Oct 2001, there are 2,860 sequence families representing 27,181 protein structures. A tree for the fold classification of the representative set is constructed by a hierarchical clustering method based on the structural similarities. Family indices of the FSSP are obtained by cutting the tree at levels of 2, 4, 8, 16, 32 and 64 standard deviations above the database average. We convert the structures of the representative set to their conformational code sequences.

All the pair alignments of the FSSP for structures with the same first three family indices in the representative set are collected for counting aligned pairs of conformational letters. The total number of structure pairs is 10047, involving 1657 of the 2860 proteins. The alignments consist of 175,723 fragment pairs and 1,284,750 code pairs. (Note that for each aligned fragment pair, the number of aligned residue duads is larger than that of code duads

Table 6. CLESUM: The conformation letter substitution matrix (in units of 0.05 bit).

J	37																
H	13	23															
I	16	18	23														
K	13	5	21	49													
N	-2	-34	-11	28	90												
Q	-44	-87	-62	-24	32	90											
L	-32	-62	-41	-1	8	26	74										
G	-21	-51	-34	-13	-8	8	29	69									
M	16	-4	1	12	7	-7	5	21	61								
B	-57	-96	-74	-50	-11	12	-12	13	-13	51							
P	-34	-60	-49	-36	-3	7	-12	5	8	42	66						
A	-23	-45	-31	-19	10	16	-11	-6	-2	20	35	73					
O	-24	-55	-34	5	15	-13	-4	-1	5	-12	4	25	104				
C	-43	-77	-56	-33	-5	29	0	-4	-12	7	4	13	3	53			
E	-93	-127	-108	-84	-43	-6	-21	-22	-47	15	-5	-25	-48	3	36		
F	-73	-107	-88	-69	-32	3	-16	-5	-33	7	0	-20	-30	20	26	50	
D	-88	-124	-105	-81	-44	14	-22	-31	-49	13	-10	-17	-42	21	22	21	52
	J	H	I	K	N	Q	L	G	M	B	P	A	O	C	E	F	D

by 3.) The derivation of the substitution matrix is similar to that for the BLOSUM. (Since the way to collect alignments is different, no clustering by identity rate is conducted. The requirement of first four identical indices would lead to less pairs of aligned structures, which are also closer to each other; this would correspond to a higher identity parameter of the BLOSUM. An alternative criterion for collecting alignments could be a Z-score cutoff.) We denote the count of code pairs C_i-C_j by N_{ij}. When a pair C_i-C_j is found, we increase both N_{ij} and N_{ji} by 1, no matter whether C_i and C_j are identical or not. The matrix elements are then calculated as

$$M(C_i, C_j) = 20 \log \frac{N \times N_{ij}}{N_i N_j}, \qquad \text{with } N_i = \sum_{j=1}^{17} N_{ij},\ N = \sum_{j=1}^{17} N_j, \tag{12}$$

where a scaling factor of 20 instead of 2 is used to show more detail. We call this substitution matrix for conformational letters CLESUM, which is shown in Table 6. Henikoff and Henikoff (1992) introduced for their BLOSUM the average mutual information per amino acid pair H, which is the Kullback-Leibler distance between the joint model of the alignment and the independent model. (A formal definition of the Kullback-Leibler distance will be given later.) The value of H for our CLESUM equals 1.03, which is close to that for BLOSUM83. (Note that the value of H is generally lower for an alphabet of smaller size.)

Although the history of using structural alphabets is long, the construction of substitution matrices occurred only quite recently, CLESUM being first constructed in 2004. Recently, a few research groups have developed several substitution matrices using various coding schemes for structural alphabets. For example, the coding of Godzik's group (Friedberg *et al.*, 2007) is based on a rigid fragment library, and their matrices were constructed using the multiple structure alignments (HOMSTRAD) of Mizuguchi *et al.* (1998) and alignments derived from multiple sequence alignments (BLOCKS). Tyagi *et al.* (2006)

published a matrix for their alphabet of 16 'protein blocks' representing eight dihedral angles of pentapeptides. Also, Chang *et al.* (2006) reported a coding based on the writhing number defined for a chain fragment and a matrix generated by using the CLUSTALW alignments of sequences derived from SCOP as lineages. Tung *et al.* (2007) derived a structural alphabet by clustering in the plot of two angles of pentapeptides, and its substitution matrix SASM. In another conformational substitution matrix in the literature (Martin, 2000) the codes were mainly for global orientation of secondary structure elements, and the crude substitution matrix was obtained empirically.

It should be mentioned that Matsuda et al. (1997) have also constructed a matrix for their uniform backbone encoding, whose enties are given by the cosine of angles between the normals of corresponding icosahedron surfaces for mapping the two-angle space. This matrix measures a geometrical similarity while CLESUM reflects an 'evolutionary' similarity, so more precisely it should be called a similarity matrix instead of a substitution matrix, reminiscent of the Grantham (1974) matrix for amino acids. There is a conceptual difference between the two kinds of matrices. For example, any two helices are geometrically very similar, but they need not be related in revolution. For CLESUM, scores between two frequent helical states are relatively low, which reduces the chance of accidental matching of two irrelevant helices.

2.7. Precision of the Conformational States

Based on conservative recurrent segments, we have discretized the combination of three psuedobond angles formed by quadrapeptides to convert the local geometry to 17 coarse-grained conformational letters according to a mixture model of the angle distribution.

The coarse-graining would introduce an error, so it is important to examine the precision of the codes. For this purpose we randomly pick 1,000 sample points for each letter, and calculate the distance root mean squared deviation (drms) for each of the total 499,500 pairs from their coordinates. The drms of structures a and b, without requiring a structure alignment, is defined as the averaged distance pair difference

$$\text{drms} = \left[\frac{2}{n(n-1)} \sum_{i=2}^{n} \sum_{j=1}^{i-1} (|\mathbf{r}_{ai} - \mathbf{r}_{aj}| - |\mathbf{r}_{bi} - \mathbf{r}_{bj}|)^2 \right]^{1/2}, \qquad (13)$$

where $\mathbf{r}_{ai}$ is the coordinate of atom i in structure a. The averaged coordinate pair difference, *i.e.* the coordinate root mean squared deviation (crms), is about 1.2 times of the drms (Park and Levitt, 1995).

The errors of the conformational codes are listed in Table 7. The most precise code H has an error 0.133 ± 0.060Å, while the vaguest code L has an error 0.604 ± 0.365Å. After averaging over the code relative frequencies, the mean error is found to be 0.330Å.

Many authors identified a structural alphabet as a set of cluster centers or centroids, which, as significant representative prototypes of conformations, provide a means of being able to reconstruct straightforwardly the tertiary structure of proteins. It is expected that concatenating these 'backbone building blocks' in an overlapping manner without introducing extra degrees of freedom should produce the backbone chain up to a certain precision. In fact, the coarse-grained description of a structural alphabet is by no means a

Table 7. Errors of the conformational codes.

Conformational code	I	J	H	K	F	E	C	D
Mean drms (Å)	0.244	0.246	0.133	0.452	0.398	0.307	0.392	0.262
Standard deviation (Å)	0.110	0.124	0.060	0.219	0.287	0.173	0.218	0.149
A	B	G	L	M	N	O	P	Q
0.347	0.322	0.390	0.604	0.481	0.551	0.538	0.252	0.506
0.163	0.197	0.192	0.365	0.231	0.321	0.318	0.134	0.287

stiff description. The distribution of conformations for a code is meaningful. The description of a single code may be rather vague. However, a string of codes implies a strong spatial constraint on the code conformations, leading to a much less vague description. For example, mode L located around $(\theta, \tau, \theta') = (1.392, 0.590, 0.817)$ has a large $\tau\tau$-covariance 0.739. However, L in dimer PL (with 1996 counts) is located around $(\theta, \tau, \theta') = (1.475, 0.191, 0.776)$, and has a smaller $\tau\tau$-covariance 0.368. Similarly, L in AL (with 1284 counts) at $(\theta, \tau, \theta') = (1.553, 0.261, 0.795)$ has $\tau\tau$-covariance 0.357. Not only is the covariance reduced, but the center location is also shifted. A further reduction of $\tau\tau$-covariance is observed in L of trimer PLD (with 573 counts) at $(1.463, -0.025, 0.669)$, whose $\tau\tau$-value is 0.311. The shift of centers of submodes makes a noticeable contribution to the mode covariance. Thus, instead of 'glueing' prototypes together to obtain conformations for long segments, a flexible way is to learn from conformations of sample strings in a known structure database. Furthermore, specifying an amino acid for a code often also refines the code conformations. For example, mode L associated with serine (1209 cases) has $\tau\tau$-covariance 0.568. Glycine is spatially most flexible. Indeed, L associated with it (2105 cases, around $(1.342, 0.156, 0.729)$) has an enormous $\tau\tau$-covariance 1.428. However, even for glycine, the reduction of covariance from assigning it to conformational states is observed in cases of N, O and Q, which are rather thickly populated by glycine. More about the relationship between amino acids and conformational states will be discussed in the next subsection.

2.8. Propensity of Amino Acids to Conformational States

From the correlation between the conformational letters and the secondary structures (Tables 3 and 4), and the known propensity of amino acids to secondary structures, it is not surprising that there exists a propensity of conformational letters to amino acids. The relationship between amino acids and codes can be quantified by means of the Kullback-Leibler divergence (also called KL-distance, information divergence, or relative entropy; Kullback and Leibler, 1951), which can be used as a measure for the distance between two probability distributions. The KL-distance of discrete distribution q from p is defined to be

$$d(p, q) = \sum_i p_i \log \frac{p_i}{q_i}, \tag{14}$$

which is asymmetric and always non-negative, and $d = 0$ if and only if the two distributions are identical. Regarding $d(p, q)$ as a function of p and expanding it around $\{p_i = q_i\}$, we have

$$d(p,q) \approx \sum_i \frac{(p_i - q_i)^2}{2q_i}, \tag{15}$$

which, under the null hypothesis $\{p_i\} = \{q_i\}$, is proportional to a χ^2 variable. Many of the other quantities of information theory can be interpreted as applications of the KL divergence to specific cases. For example, the mutual information

$$I(X;Y) = d(P(X,Y), P(X)P(Y)), \tag{16}$$

is the KL-distance of the product $P(X)P(Y)$ of the two marginal probability distributions from the joint probability distribution $P(X,Y)$. Intuitively, mutual information measures the information that X and Y share, and can be used to characterize both the relevance and redundancy of variables.

To measure how strongly a quadrapeptide conformational state is related to amino acids at different sites, we calculate the mutual information

$$I_\delta(C;a) = \sum_{\sigma,\alpha} p_\delta(\sigma,\alpha) \log \frac{p_\delta(\sigma,\alpha)}{q(\sigma)r(\alpha)}, \tag{17}$$

where $p_\delta(\sigma,\alpha)$ is the joint probability for amino acid α to occur in conformational state σ, and δ is the shift of site index of α with respect to that of σ. Note that, by convention, the conformational code of the quadrapeptide $a_{i-2}a_{i-1}a_ia_{i+1}$ is of site index i. In formula (17) the marginal probability contributions $q(\sigma)$ and $r(\alpha)$ are related to the joint probability as

$$q(\sigma) = \sum_\alpha p_\delta(\sigma,\alpha), \qquad r(\alpha) = \sum_\sigma p_\delta(\sigma,\alpha). \tag{18}$$

The values I for different δ are $I_{-2} = 0.026$, $I_{-1} = 0.091$, $I_0 = 0.113$, $I_1 = 0.052$ and $I_2 = 0.025$. The highest value of I_0 at $\delta = 0$ verifies that the convention for indexing conformational codes is reasonable.

The log-odds $\log[p(\alpha|C)/r(\alpha)]$, where $p(\alpha|C)$ is the conditional probability of amino acid α at state C and $r(\alpha)$ the background probability of α, may be taken as a propensity score of amino acid α to conformational state C. Again, the average score

$$\bar{R}_{c\delta} = \sum_\alpha p(\alpha|C) \log \frac{p(\alpha|C)}{r(\alpha)} \tag{19}$$

is a relative entropy, where subscript δ indicates the site shift between α and C. The calculated values of the average propensity scores at shift 0 and ± 1 are listed in Table 8. It is seen that nine conformational states have their maxima at $\delta = 0$, but the other eight at $\delta = -1$. Scores at $\delta = \pm 2$ are relatively small, and not shown. Generally, the difference between $p(\alpha|C)$ and $r(\alpha)$ is not so tremendous. To seek for strong correlation between sequence and structure, one has to look at code strings.

2.9. Reduction of Amino Acid Alphabet and Structural Alphabet

The counts of of a specific code string or residue string drop quickly with increasing length of the string. Besides regrouping 'similar' strings by means of substitution matrices, another way is to consider reduction of either or both of the amino acidic and structural alphabets.

Table 8. Average propensity scores of amino acids to different conformational states at shift 0 and ± 1.

Code	A	B	C	D	E	F	G	H	I
$\delta = -1$	0.24	0.04	0.04	0.11	0.10	0.03	0.27	0.05	0.03
$\delta = 0$	0.13	0.03	0.18	0.05	0.08	0.04	0.05	0.08	0.04
$\delta = +1$	0.11	0.03	0.04	0.04	0.02	0.01	0.08	0.09	0.04
Code		J	K	L	M	N	O	P	Q
$\delta = -1$		0.04	0.03	0.03	0.08	0.06	0.10	0.21	0.43
$\delta = 0$		0.08	0.04	0.10	0.04	0.36	0.89	0.12	0.29
$\delta = +1$		0.05	0.02	0.03	0.04	0.04	0.03	0.08	0.05

Let us examine the effect of the reduction of amino acid alphabet on the mutual information (with index shift $\delta = 0$)

$$I(C; a) = \sum_{\sigma,\alpha} p(\sigma, \alpha) \log \frac{p(\sigma, \alpha)}{q(\sigma) r(\alpha)}. \tag{20}$$

Clustering amino acids α_i and α_j together into $\alpha_{i\&j}$ leads to

$$r(\alpha_{i\&j}) = r(\alpha_i) + r(\alpha_j), \quad p(\sigma, \alpha_{i\&j}) = p(\sigma, \alpha_i) + p(\sigma, \alpha_j). \tag{21}$$

The difference between values of I after and before clustering is given by

$$\begin{aligned} &[p(\sigma, \alpha_i) + p(\sigma, \alpha_j)] \log \frac{p(\sigma, \alpha_i) + p(\sigma, \alpha_j)}{q(\sigma)[r(\alpha_i) + r(\alpha_j)]} \\ &-p(\sigma, \alpha_i) \log \frac{p(\sigma, \alpha_i)}{q(\sigma) r(\alpha_i)} - p(\sigma, \alpha_j) \log \frac{p(\sigma, \alpha_j)}{q(\sigma) r(\alpha_j)}, \end{aligned} \tag{22}$$

which, by introducing

$$x_i = \frac{p(\sigma, \alpha_i)}{r(\alpha_i)}, \quad \omega_i = \frac{r(\alpha_i)}{r(\alpha_i) + r(\alpha_j)}, \tag{23}$$

and their analogs x_j and ω_j, and defining $\langle F(x) \rangle = \omega_i F(x_i) + \omega_j F(x_j)$, is proportional to $f(\langle x \rangle) - \langle f(x) \rangle$ with $f(x) = x \log x$. From the Jenssen theorem (Rassias, 2000), for the convex function $x \log x$ here we have $f(\langle x \rangle) \leq \langle f(x) \rangle$, so I never increases after any step of clustering. That is, merging any two amino acids will result in a loss of mutual information. The above argument is also valid for the merging of conformational states. To make the loss of mutual information as small as possible, I should be maximized. Thus, mutual information I can be naturally chosen as the objective function for reducing either the amino acid alphabet, the structural alphabet, or both at the same time. We call this approach entropic clustering (Zheng, 2004).

The total number of ways to partition n objects into m classes is given by the Sterling number of the second kind (Duran and Odell, 1974)

$$S(n, m) = \frac{1}{m!} \sum_{i=0}^{m} (-1)^{m-i} \binom{m}{i} i^n. \tag{24}$$

It is easy to prove that for $m_2 > m_1$ the maximal I at m_2 is always greater than the maximal I at m_1. We may sketch the proof for $m_2 = 3$ and $m_1 = 2$. Assume that we have found two groups to reach the maximum I_1 for $m_1 = 2$. Splitting one group into two will never decrease I. If the value of I after splitting is I' and the maximum at $m_2 = 3$ is I_2, we have $I_2 \geq I' \geq I_1$. This argument is valid for any two m.

For the amino acid alphabet, $n = 20$. There are 190 ways to reduce the alphabet size by one, i.e. $m = 19$. The optimal scheme is to merge phenylalanine (F) and tyrosine (Y) together (with a dropping of I from 0.1128 to 0.1127). Generally, it is not practical to search for the optimal partition by enumeration. We conduct the Monte Carlo method to find the optimal grouping. It is found that for $m = 2$, glycine (G) and proline (P) are in one group, and all others in the other group. At $m = 3$ glycine (G) and proline (P) are each in one group, and all others still form one group. The first attractive grouping appears at $m = 4$, where the last group splits into ACDEHKNQRS and VFIWLMYT with a rough correspondence to hydrophobicity. For the conformational code alphabet, $n = 17$. The optimal scheme for reducing the alphabet size by one is to merge states `G` and `M` (with a dropping of I from 0.1128 to 0.1113). The Monte Carlo result for $m = 2$ gives `ACNOPQ` in one group. It is a little surprising that for $m = 3$ codes typically for helices and sheets are still in the same group, while the group `ACNOPQ` splits into `ACP` and `NOQ`.

For a periodic structure like a helix, residues on the hydrophobic and hydrophilic sides have different distributions. However, in the above analysis these distributions are mixed together, which weakens the correlation. For example, one might imagine that code `H` consists of several subclusters, say $\mathtt{H}_h$ and $\mathtt{H}_p$. Thus, grouping them into a single `H` would reduce I, *i.e.*, dividing `H` into $\mathtt{H}_h$ and $\mathtt{H}_p$ would make I increase. The mutual information between amino acids and conformational letters forms a basis for the structure prediction. Partition of highly populated patterns into subpatterns can enhance the mutual information. We may regard `PLD` as a subpattern of `L`, so using strings is an effective way for information enhancement. However, to maintain the statistical significance grouping similar strings together is necessary.

An important application of the alphabet reduction is to construct a joint substitution matrix for both amino acids and conformational letters, which will include both geometric and physico-chemical correlations. The CLESUM includes only the structural information. When we compare two structures we usually know also their amino acid sequences. Many papers considered a linear combination of structural alignment score and sequence alignment score. This is an approximation of independency. From the FSSP, which contains also the amino acidic information, it is possible to construct a substitution matrix in the joint space of the structure and sequence. However, such a matrix would have about $(20 \times 17)^2/2 \approx 6 \times 10^4$ parameters. When the structure is to be emphasized, we may use a reduced amino acid alphabet. For example, clustering 20 amino acids into two groups would reduce the parameter number to about 600. We should emphasize that for this purpose the average score H is the natural objective function for clustering. The extension of the above argument for mutual information I to score H is rather straightforward. Here, instead of the relationship between sequences and structures, the relationship between two structures is involved, *i.e.*, it is quantity H that we want. However, despite the advantage of using H as the objective function, realization of the clustering is much more complicated. For grouping of the 20 amino acids into two categories, the Monte Carlo method finds AVCFIWLMY and DEGHKNPQRST as the groups, which are denoted by h and p, respectively. The relative entropy increases from 0.715 to 0.760 nat. (Here, the grouping result is quite different from that based on I where the role played by various residues in local conformations is emphasized, not the correspondence of structures.)

Table 9. CLESUM-hh (lower left) and CLESUM-pp (upper right): the conformation letter substitution matrices for aligned amino acid types h-h and p-p (in units of 0.05 bit).

			J	H	I	K	N	Q	L	G	M	B	P	A	O	C	E	F	D	
J	40		45	19	24	19	3	-41	-27	-13	21	-49	-32	-21	-22	-37	-79	-62	-73	J
H	18	30		30	24	7	-31	-76	-60	-45	-5	-84	-57	-46	-55	-69	-113	-92	-108	H
I	20	26	30		31	27	-6	-57	-38	-28	3	-64	-44	-30	-31	-48	-93	-72	-89	I
K	18	14	28	59		57	39	-16	5	-8	20	-43	-33	-12	6	-24	-66	-52	-64	K
N	-2	-16	3	40	108		94	39	15	-5	11	-12	-1	14	16	0	-35	-25	-31	N
Q	-41	-110	-67	-19	27	109		95	33	14	-4	10	8	21	-14	33	3	13	20	Q
L	-33	-58	-37	10	8	30	86		76	35	12	-12	-8	-3	-4	7	-11	-5	-14	L
G	-22	-48	-35	-10	-11	2	40	80		80	33	7	9	-1	-4	1	-11	5	-17	G
M	24	7	9	16	10	-5	9	25	71		69	-6	13	4	9	-4	-32	-21	-34	M
B	-64	-99	-82	-61	-7	15	-3	31	-12	59		64	55	31	-4	13	20	14	25	B
P	-31	-57	-47	-39	-1	13	-21	4	13	45	76		74	38	7	9	3	7	0	P
A	-19	-36	-26	-25	13	11	-24	-10	-6	21	42	85		75	31	20	-16	-11	-4	A
O	-9	-37	-18	-14	25	29	-8	13	11	13	36	32	121		107	10	-37	-24	-32	O
C	-49	-87	-67	-45	-17	36	-8	-7	-24	10	7	17	14	62		60	12	28	33	C
E	-110	-138	-126	-98	-56	-5	-24	-22	-58	24	3	-23	-22	9	44		45	32	30	E
F	-92	-131	-105	-106	-46	3	-24	-4	-45	10	3	-24	5	29	34	61		58	30	F
D	-98	-138	-111	-95	-67	33	-20	-36	-66	14	-10	-22	-31	26	30	28	63		61	D
	J	H	I	K	N	Q	L	G	M	B	P	A	O	C	E	F	D			

Using the two reduced groups h and p of the 20 amino acids, we calculate the joint substitution matrix from the same FSSP alignments. The result is now given as three submatrices: CLESUM-hh, CLESUM-pp, and CLESUM-hp. The first two, shown in Table 9, correspond to the aligned amino acid types h-h and p-p, respectively. They are symmetric. The third, shown in Table 10, corresponds to the types h-p. This is asymmetric since the rows (h) and columns (p) correspond to different types. (The symmetry is recovered between CLESUM-ph and CLESUM-hp.) As we may expect, compared with the original CLESUM, elements of CLESUM-hh and CLESUM-pp generally become larger in absolute values, and those of CLESUM-hp show the opposite tendency. The tendency is stronger for letters dominated by helices or sheets.

3. Structure Comparison Using Conformational Letters

We have discretized the quadra-peptide conformations to convert the local geometry to 17 coarse-grained letters according to a mixture model of the angle distribution. The substitution matrix CLESUM extracted from the database FSSP of structure alignments contains the information of structure database statistics. For example, frequently appearing helix codes J, H, I have relatively small diagonal elements, which indicates the weak significance of their match. On the contrary, in geometric score matrices derived as local conformation distances for dynamic programming, the helix match score generally would be rather high, and matrices have to be calculated for each comparison.

Table 10. CLESUM-hp: Conformation letter substitution matrices for aligned amino acid type h-p (row-column) (in units of 0.05 bit).

	J	H	I	K	N	Q	L	G	M	B	P	A	O	C	E	F	D
J	27	10	13	11	-12	-50	-35	-20	11	-54	-34	-26	-35	-42	-81	-64	-84
H	3	14	9	-2	-48	-94	-65	-54	-13	-95	-62	-49	-62	-74	-116	-95	-114
I	6	10	13	12	-25	-66	-47	-35	-5	-75	-50	-37	-43	-57	-95	-79	-100
K	3	-1	13	36	4	-37	-8	-15	4	-53	-36	-24	4	-36	-80	-62	-79
N	-3	-25	-9	23	76	14	12	-17	12	-19	-6	7	9	-10	-44	-20	-30
Q	-47	-86	-64	-37	16	70	25	6	-1	-1	0	6	-21	30	-6	2	20
L	-38	-69	-44	-15	-10	11	65	16	-3	-19	-19	-19	-7	-4	-23	-15	-25
G	-31	-60	-40	-19	-8	5	17	52	9	5	4	-8	-4	-8	-24	-9	-32
M	6	-11	-7	4	-1	-13	-3	12	47	-21	1	-7	-3	-12	-43	-30	-45
B	-65	-112	-80	-48	-9	16	-13	-5	-17	34	21	2	-35	-3	12	4	11
P	-40	-67	-59	-40	-6	5	-8	0	4	35	53	26	-12	1	-1	5	-3
A	-28	-46	-30	-25	3	12	-10	-8	-4	18	32	66	0	7	-21	-18	-17
O	-8	-35	-21	8	15	-9	10	21	13	4	17	26	67	9	-19	2	-29
C	-50	-87	-65	-40	-13	15	-4	-5	-16	6	-1	2	-22	42	8	21	23
E	-103	-140	-119	-94	-46	-14	-26	-31	-54	-2	-22	-37	-69	-14	25	14	9
F	-84	-126	-108	-77	-41	-11	-24	-13	-41	-3	-16	-30	-55	0	18	33	9
D	-101	-139	-121	-93	-59	-5	-29	-37	-54	-1	-27	-31	-57	2	14	14	34

3.1. Structure Comparison by 1D Code Sequence Alignment

The conversion of a 3D structure of coordinates to its conformational codes requires little computation. Once we transform 3D structures to 1D sequences of letters, tools for analyzing ordinary sequences can be directly applied.

3.1.1. Structure Comparison via Global Alignment of Conformational Sequences

Conformational letters are a local descriptor. For a pair of highly similar structures, which have a high Z-score of the FSSP alignment, we expect that the global similarity is to a great extent determined simply by the local similarity. That is, the maximal sum of local similarity would imply optimal global similarity. In this case, the Needleman-Wunsch algorithm (Needleman and Wunsch, 1970) for global sequence alignment is applicable. We shall now examine the performance of the conformational alphabet derived above.

To show different levels of structural similarity, Holm and Sander (1998b) gave an example of the α/β-meander cluster with four members, whose PDB-IDs are 1urnA, 1ha1, 2bopA and 1mli. Here the fifth optional letter of a PDB-ID is a chain identifier. The structure of 1urnA was taken as the frame to superimpose the other structures. The structure 1ha1 shows the highest similarity to 1urnA with score $Z = 10.3$. Taking the scaling factor for the CLESUM to be 2 (instead of 20), and using -12 for the the gap-opening penalty and -4 for the gap extension, the global Needleman-Wunsch alignment of 1urnA and 1ha1 is shown in figure 5, where the amino acid sequences aligned according to the FSSP are also given. It is seen that, except for segment boundaries, the two alignments coincide.

```
1urnA avpetRPNHTIYINNLNEKIKKDELKKSLHAIFSRFGQILDILVSRS
1ha1     ahLTVKKIFVGGIKEDT    EEHHLRDYFEQYGKIEVIEIMTDRGS
1urnA   CCPMCEALEEEENGCPJGCCIHHHHHHHHIKMJILQEPLDEEEBGAIK
1ha1    ...BBEBGEDEENMFNMLFA....HHHHHKKMJJLCEBLDEBCECAKK

1urnA LKMRGQAFVIFKEVSSATNALRSMqGFPFYDKPMRIQYAKTDSDIIAKM
1ha1  GKKRGFAFVTFDDHDSVDKIVIQ kYHTVNGHNCEVRKAL
1urnA ...GNGEDBEEALAJHHHHHHIKKGNGCENOGCCEFECCALCCAHIJH
1ha1  AGCPOLEDEEEALBJ.HHHHIIJGALEEENOGBFDEECC.........
```

Figure 5. The alignment of 1urnA and 1ha1. The first two lines are their amino acid sequences aligned according to the FSSP, while the last two lines are the global Needleman-Wunsch alignment of the conformational code series. Lowercase letters of amino acids indicate structural nonequivalence.

The global alignment involves the spatial arragement of fragments. Even though a global sequence alignment algorithm is used, this does not guarantee that the alignment found corresponds to the optimal structure superposition. However, the CLESUM alignment is not affected by the domain move, so it is good for analyzing the structure evolution. For example, the first helix of 1ha1 is shorter than its counterpart in 1urnA by one turn. The FSSP aligns the N-cap (with codes FA) of the 1ha1 helix to the helix (with codes HH) of 1urnA, but the local structure FA is closer to CC (with positive scores) than to HH (with negative scores). The CLESUM alignment opens a gap for an extra helix turn of 1urnA.

The alignment of 1urnA and 2bopA in the FSSP is of $Z = 5.0$. The global CLESUM alignment and the FSSP alignment share a common alignment of four pieces with lengths 5, 14, 13, and 5. Another example we examined is the alignment of 1urnA and 1u2fA with $Z = 7.0$, which shows a shared alignment of 35 sites (in a single piece) between the FSSP and CLESUM alignments.

3.1.2. Structure Comparison via Local Alignment of Conformational Sequences

The lengths of proteins 1urnA and 2scuB are respectively 96 and 385. Their FSSP alignment has $Z = 2.1$. Due to the great difference in their lengths, the global alignment algorithm is not appropriate. Their conformational sequence alignment by the Smith-Waterman algorithm (Smith and Waterman, 1981) with CLESUM is shown in figure 6, together with the FSSP alignment. The common sites of the FSSP and CLESUM are marked with '*', being 22 in length.

This example shows that the global structure alignment and conformational sequence alignment usually tell us different aspects of the structural relation of two proteins. A gap in a conformational sequence alignment is generally a blocker to the global structure match. Hence, the sequence alignment as a whole does not automatically correspond to the structure alignment. However, a core of structure alignment often can be found from high score pairs of ungapped short aligned segments. In this example, for a window width of 20, the highest score 68 is found at the center of the block where the FSSP and CLESUM alignments coincide. Window widths can be chosen rather freely, say from 15 to 22. At width 18, the highest score 63 is also found in the same block (near the right end). Since the highest score pair need not belong to the global structure alignment, several pairs of top scores must be examined to find the optimal. Although the Smith-Waterman alignment has no direct use in structure alignment, its information about the arrangement of local similar structural elements could be useful in understanding the structure evolution, and helpful in finding candidates for structure neighbors.

```
1urnA   9 HTIYINNLNEKIKKDELKKSLhAIFSRFGQILDILVSRSLKMRGQA
2scuB  43 WVVKCQVHAG-----GRGKAGG------------------------
1urnA                    HHHHHIKMJILQEPLDEEEBGA...IKGNGE
2scuB                    JHJIHJJIIILQLQFDEEDELAGDNMJHJOM

1urnA 55 FVIFKEVSSATNALRSMQGFPFY-------DKPM-RIQYAKTDSDIIAK
2scuB 60 VKVVNSKEDIRAFAENWLGKRLVTYQTDANGQPVNQILVE--AATD---
         *********************
1urnA    DBEEALAJHHHHHHIKKGNGC..ENOGCCEFECCAL
2scuB    LEDFDLAJHHHJIHHHKMNGCEEEAIGCAIFQCEPL
```

Figure 6. Alignment of 1urnA and 2scuB. The first two lines are their amino acid sequences aligned according to the FSSP, while the last two lines are the local Smith-Waterman alignment of the conformational sequences. Only the relevant subsequence is shown.

3.1.3. Multiple alignment of conformational sequences

There are many tools for multiple sequence alignment. In contrast to ordinary amino acid sequence alignment, when we align two conformational sequences from N-termini to C-termini, an early error would be accumulated. An algorithm based on a maximal star tree is appropriate and convenient (Zheng, 2005a). The method takes a segment, say a 20-mer, from a sequence as a seed, then searches each of the other sequences for the 20-mers which are most similar to the seed. If the similarity is above an afore-given threshold, the 20-mer is regarded as a neighbor node of the seed (the center node), and an edge is linked between them. The CLESUM scores provide a similarity measure. Among all the star trees constructed by taking all 20-mers as seeds, the one that has the highest degree for its center node or the highest score sum with respect to its center is the maximal star tree. This provides a core of alignment as a pivot for determining trial transformations for superposition. Maximal stars need not be unique.

As an example, we consider 12 proteins from the superfamily 'Actin depolymerizing proteins' of the SCOP (Murzin *et al.*, 1995): 4 Cofilin-like with PDB-IDs 1ak6, 1cfyB, 1cnuA, 1f7s, and 8 Gelsolin-like consisting of 6 domains (27–152, 153–262, 263–383, 384–532, 533-628, and 629–755) of 1a0nA, 1svy and 2vik, which is the CL-GL ensemble of Dror *et al.* (2003b). Setting the similarity threshold to 40, we find the maximal star is centered at the 20-mer starting at 80 of 1cnuA with degree 11 and score sum 668. The seed, in the conformational letters, is EEBEBEAJGCAHIIHIIHHI. The loop AJGCA between a strand and a helix is at the middle of the 20-mer, contributing significantly to the score sum (since the A-A or G-G match score is rather high). Furthermore, the 20-mer window can be extended to the whole region from 78 to 84 covering 7 consecutive 20-mers. The score sums of the 7 stars are each above 604, and the degrees of the stars are each 11. For this 'easy' example, this core of alignment immediately determines the transformation for the correct global alignment. The correctly determined conformational block of multiple local structural alignment as a core of multi-structural alignment is sufficient for nailing down a transform accurate enough for determining structure superposition. It is clear that this method is also valid for detecting nontopological structural motifs where the order of segments in the structure alignment does not follow the same order in the sequences.

A global structure match may consist of several spread-out short segments. Generally, for ungapped neighbor searches, a smaller seed width, say 8 to 12, is more suitable for better sensitivity. The existence of nearby windows with high score sum will indicate that the aligned fragment is wider. Often several maximal stars can be found, which are not simply shifts of each other. Only some combinations from a few of them furnish a structure alignment scaffold. After obtaining maximal stars as possible correspondence relations, some crude criterion can be further used to identify consistent star pairs to extend the alignment.

The final objective of structure alignment is to find optimal global similarity in structures to be compared. The local conformational similarity is a necessary condition for structure alignment although it is not sufficient. In terms of conformational alphabets, substitution matrices like CLESUM provide a simple and accurate measure of local similarity for fast searching of high quality local matches. In contrast to the approximation of secondary structure elements by vectors, the similarity match extracted with CLESUM directly assigns the atomic correspondence to the structures under comparison. A significant local structure match may then serve as a structure alignment seed, based on which the transform for rigid superposition of the involved structures can be derived. Once the transformation for superposition is determined, more aligned atoms can be located for a global structure match.

3.2. Fast Pairwise Structure Alignment

The comparison of protein structures has been an extremely important problem in structural and evolutional biology. The detection of local or global structural similarity between a new protein and a protein with known function allows the prediction of the new protein's function. Since protein structures are better conserved than amino acid sequences, remote homology is detectable more reliably by comparing structures. Structural comparison methods are useful for organizing and classifying known structures, and for discovering structure patterns and their correlation with sequences.

The common goal of all pairwise structure alignment methods is to identify a set of residue pairs from each protein that are structurally similar, or to find the optimal correspondence between the atoms in the two molecular structures. An exhaustive search for such atomic correspondence is intractable, and various heuristics have been developed. For example, to lower the dimensionality of the problem, DALI identifies interaction patterns of fragment pairs (Holm and Sander, 1994; 1997), VAST describes secondary structure elements (SSEs) as a set of vectors (Madej *et al.*, 1995; Gibrat *et al.*, 1996) while CE designates short similar fragment pairs (SFPs) of local structural similarities (Shindyalov and Bourne, 1998). There are several excellent reviews, e.g. (Eidhammer *et al.*, 2000; Koehl, 2006).

For a given correspondence of two point sets, finding the best rigid transposition to superpose the correspondence sets can be easily done by using a closed-form solution based on the singular value decomposition (Kabsch, 1978; Umeyama, 1991). When the transformation between the two sets is given, the problem to find the correspondences (of ϵ-

congruence at the maximal or average error tolerance ϵ) is rather straightforward. However, when aligning two protein structures, at the beginning we know neither the transformation nor the correspondence. A few methods, like DALI and CE, directly search for a good alignment. Many methods start with an initial correspondence (seed matches), from which the optimal transformation for the correspondence is determined. The transformation is then used to update the correspondence. The procedure of progressively building up larger correspondence is iterated until the best correspondence is finally found. The methods vary in the way of seed finding and correspondence updating. There may be an optional follow-up refinement of alignments. Typical examples are STRUCTAL (Subbiah *et al.*, 1993) and ProSup (Lackner *et al.*, 2000).

Despite the existence of various pairwise structural alignment algorithms, fast, efficient and reliable tools for pairwise alignment are in ever increasing demand for analyzing the rapidly growing data of protein structures. We have developed an algorithm called CLePAPS for fast pairwise alignment of protein structures by fully using our conformational alphabet and its substitution matrix CLESUM. The flow chart of CLePAPS is shown in figure 7.

3.2.1. Finding Similar Fragment Pairs of High Scores

Many tools find SFPs of local similarity as preliminary correspondences. Local similarity is a necessary, but insufficient, condition for the global structural alignment. Structurally similar fragments found in a pair of proteins form the basic objects for further examination of their consistency in the spacial arrangement. Consistent pieces then may be joined to obtain a global alignment. Different methods use various criteria and strategies for seed matching, consistency checking and pieces merging. Generally, a stringent criterion for local similarity would create less objects of seed matches, and hence speed up the merging process. However, it would miss some substructures constituting the final global alignment. On the other hand, due to the insufficiencies of local similarity in the global alignment, too loose conditions of local similarity would overload the later filtering task. One has to balance sensitivity with specificity, and make a compromise between efficiency and accuracy.

Protein structural alignment involves the geometric representation of structures. In most cases, only the backbone of pseudobonds formed by C_α atoms are considered. Coordinates of C_α atoms, which change under translation and rotation in 3D space, are not geometric invariants. Distances used by DALI are the intrinsic property of a geometric object. The bending and torsion angles of pseudobonds, as the chain counterparts of curvature and torsion of a smooth curve, are also geometric invariants. In VAST, SSEs are replaced by the vectors of their axes. This vector representation speeds up the computation, but has low precision for structural elements.

Using a 3D structure coding of protein backbones chain of C_α pseudobonds, CLePAPS regards an SFP as an ungapped string pair with a high sum of pairwise CLESUM scores. CLePAPS searches for SFPs by simple string comparison.

Suppose that the pair of structures to be aligned is P and P' with P being the shorter. The coordinates $\{\mathbf{r}_i\}$ and $\{\mathbf{r}'_j\}$ of C_α atoms of the two proteins are converted to the sequences S and S' of conformational letters, respectively. Since each letter corresponds to a quadrapeptide unit, the length of S (S') is shorter than that of P (P') by 3. By convention,

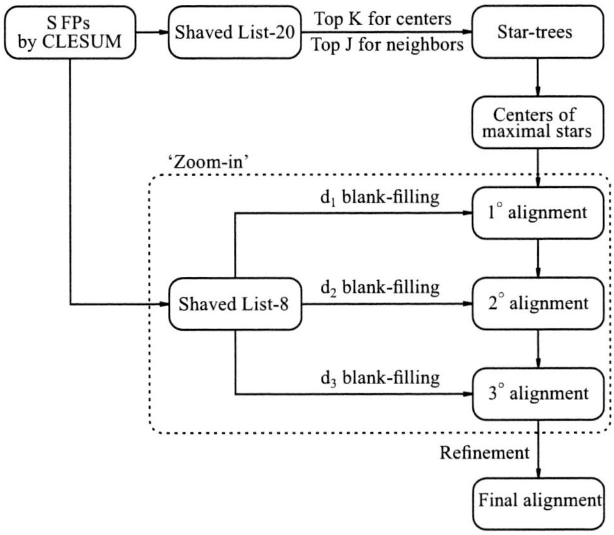

Figure 7. Flow chart of the CLePAPS algorithm.

we assign the first letter to the third residue, the second to the fourth and so on, until finally the last letter is assigned to the last residue but one.

Consider two fragments of the same length l, one of which starts at residue i of P and the other at j of P'. The local structural similarity of the fragment pair may be measured by

$$\sigma = \sum_{k=0}^{l-1} M(s_{i+k}, s'_{j+k}), \tag{25}$$

where $M(a, b)$ is the (a, b)-entry of the CLESUM, and s_i and s'_j are the conformational letters of corresponding residues. Note that a residue and its conformational letter are of the same site index. Setting a threshold T, if the pair score $\sigma \geq T$, we call the pair an SFP, which defines a correspondence (l residue duads). Searching for SFPs by string comparison is fast. Furthermore, compared with the usual definition of SFP by purely geometrical similarity, our definition also gains specificity. For example, two helices are geometrically similar, but their CLESUM score is relatively low since CLESUM has taken the abundance of helix sites into account.

Comparing each string of S of length l with those of S', we find all the SFPs. When an SFP contains residues of SSEs at its ends, shifts of the SFP often also have a high similarity score, hence they also form SFPs. To remove such redundancy, we keep only the one with the highest score among the nearby SFPs which are shifts of each other and which share common residue duads. A width w is set to restrict the maximum overlap for this 'shaving'. After shaving we have a reduced list of the representative SFPs. We sort the list in descending order of scores. Usually, a small l and a low T will result in a long list of SFPs.

For a given long enough SFP, we can find a rigid transformation to superpose its two members and make the spatial deviation of its duad C_α atoms very small. Since an SFP is determined only by local similarity, a superposition valid for one SFP need not be valid for another. We define the spatial distance or separation between two members of a certain SFP after a transformation by

$$\delta = \max_{(\mathbf{r}_i,\mathbf{r}'_j)\in\mathrm{SFP}} \{|x_i - x'_j|, |y_i - y'_j|, |z_i - z'_j|\}, \tag{26}$$

where $(\mathbf{r}_i, \mathbf{r}'_j)$ is a duad of the SFP after transformation, and (x, y, z) denotes the 3D coordinates of $\mathbf{r}$. A small separation δ implies a good superposition of the two SFP members. Taking highly similar SFPs as a pivot to determine the transformation for superposition, CLePAPS collects consistent SFPs under the guide of their similarity scores to extend the alignment.

3.2.2. The Greedy 'Zoom-In' Strategy

There is no clearly defined unique way to evaluate the quality of protein structure alignments. We adopt the standard of ProSup: the goal is to maximize the number N_e of structurally equivalent residues subject to a fixed Euclidean distance cutoff d_0 for judging correspondence between a residue pair and a minimal aligned segment size ρ. (In most cases, after filtering by a d_0 around 3 – 5Å, alignments from different methods agree very well with each other.)

To balance speed with accuracy, we generate two lists of SFPs, one for $l = 20$ with threshold $T_{20} = 350$, and the other for $l = 8$ with $T_8 = 0$. (The scaling factor of CLESUM here is 20.) Any two helices are locally similar. Length 20 will exclude many such purely local coincidence. Length 8 is necessary for including most significant aligned pieces. We denote them as List-20 and List-8, respectively. The two lists can be generated in a single run. We expect that a significant alignment should contain at least one SFP of length around 20. Initial primary correspondences will be taken from the top K, say top ten, SFPs of the sorted List-20. If the list size is less than 10, 'top ten' means all.

Once an SFP is chosen, the transformation optimal to the SFP may be used to superimpose the two proteins. The separation δ of any SFP under the transformation can then be calculated. Some SFPs are consistent with each other. That is, under the transformation optimal to one of the SFPs, which is referred to as the center SFP or simply the center, some other SFPs will have a small separation δ. By thinking in terms of graph theory, SFPs are nodes, and the center is the center node. At a fixed threshold d for δ, if the separation of an SFP is smaller than d, an edge is linked between the node of that SFP and the center. The center and its linked neighbors form a star tree or star. We define the size of a star as the total number of its nodes. Taking each of the top K SFPs as a center, we find its neighbor nodes in the top J SFPs of the sorted List-20, where $J \geq K$. The stars really used by CLePAPS are subject to a further restriction: for a given center, we search the sorted List-20 successively from the top for neighbors of the center, and add a new neighbor SFP only when it does not overlap with any existing neighbor SFPs. In this way, we obtain K restricted stars. We sort them first by their sizes, and then by similarity score σ of their centers in descending order. We remove the stars whose centers are a neighbor of the first star. Then, we examine the next star, and remove the stars associated with its neighbors, and so on, until all stars are examined. Only the centers of the retained stars will be taken as an initial alignment seed. The effect of this star removal is twofold: removing seed redundancy and selecting the seeds which better reflect the global consistency.

The extension of an initial seed alignment is mainly done by the following blank-filling of the SFPs from List-8 which are consistent with the seed. Blanks are residue positions not included in an existing correspondence. The transformation optimal to the seed SFP need not be globally optimal. We use a multi-step 'zoom-in' strategy, starting with a low precision to avoid local trapping. We first use a large cutoff d_1, say 8Å, as the consistency criterion. That is, we add only the SFPs with $\delta < d_1$ to the existing correspondence. The procedure of blank-filling is greedy. The SFPs with a higher σ have a priority to be filled. We examine the top half of the SFPs in the sorted List-8 one by one from the top. If none of the residues of an SFP is contained in the existing correspondence, we calculate its separation δ. If $\delta < d_1$, we add the SFP to the correspondence. When blank-filling is fulfilled, the transformation optimal to the enlarged correspondence is determined to update the superposition of the two proteins. In the next run of blank-filling, cutoff d_1 is reduced to d_2, and five-sixths of the SFPs in List-8 are examined. In a third run, d_2 is further reduced to d_3, and the whole list is examined. Usually, three runs of iteration are enough for obtaining a full alignment.

There are mainly two ways to update the correspondence. One is to keep the existing duads and add new ones. The other is to re-start with an empty correspondence and then fill in blanks with SFPs from List-8. The latter strategy is used in CLePAPS. In the final polishing stage, the SFPs which have only a limited overlap with the existing correspondence can also be used for blank-filling.

We speed up computation by means of marking. At the beginning, all SFPs in List-8 are identified as 'unmarked'. If an SFP has no two contiguous residue pairs whose coordinate differences are both smaller than d_i, it will be marked as 'invalid', and then would never be examined again.

3.2.3. Refinement by Elongation and Shrinking

After blank-filling we obtain an alignment usually as disjunct pieces. Due to the finite size of SFPs and the redundancy removal by shaving, it is possible that an aligned piece can be elongated near its ends. Assume that the residue duad (i, i') is at one end of an SFP. Thus, either $(i + 1, i' + 1)$ or $(i - 1, i' - 1)$ will be an outer residue duad. If the Euclidean distance of the pair is smaller than d_0 we elongate the aligned piece by joining the residue duad to it. More nearby residue pairs can be further examined for elongation. On the other hand, depending on the relative quantities of d_0 and d_3, the Euclidean distance between some residue duads on the aligned pieces would be greater than d_0. We remove such duads from the alignment (as a shrinkage of the SFPs). Although elongation and shrinking may be conducted subsequently for every complete run of blank-filling, CLePAPS conducts elongation and shrinking only after the final run.

A finetuning version of refinement is as follows. In the last run of blank-filling, checking of position confliction in protein P' is ignored, so one residue of protein P may correspond to several residues of protein P'. We sort in ascending order of the root mean square deviations (RMSDs) all the SFPs which are candidates for filling, then add these SFPs into an empty correspondence list one at a time and avoid any position confliction. We perform possible elongation of each filled SFP to enlarge the correspondence. For any unfilled gap in protein P, we search List-8 from the top. If an SFP which covers the gap is found, we calculate the Euclidian distance of residue duads whose positions coincide with the gap. When the distance of a duad is smaller than d_0, we check also the distances of nearby pairs by shifting protein P', and add the pair with the smallest distance to the correspondence.

A filter for a minimal aligned segment length ρ may be finally applied. A further iteration of transformation would additionally improve the quality of the alignment. Once a global alignment is accomplished, the total number N_e of equivalent residue duads and the RMSD of the alignment are calculated as quality indicators. Despite the star removal two alignments generated from two star centers may still be very similar. We compare entries of the rotation matrices. If the greatest difference between two corresponding matrix elements (or relative difference) is under ε, say 0.1, the two alignments are regarded as identical. A more careful criterion is the number of coincident residue duads, which is used by ProSup.

Structure comparison often yields several distinct alignments as multiple solutions. The existence of alternative alignments is mainly due to structure repeats at different levels ranging from secondary structure, supersecondary structure to domains. Another source is the domain move. CLePAPS often reports several alignments and ranks them according to their N_e.

3.2.4. The Fischer Benchmark Test

A well-known comprehensive test set for assessing the performance of fold recognition methods is the benchmark of Fischer *et al.* (1996), which contains 68 pairs of proteins. All pairs of the set are known to be structurally similar, but they have low sequence identity, ranging from 8% to 31% with an average of 18.6% and a standard deviation of 4.4. This set covers a wide range of protein families. We test the benchmark with CLePAPS.

One case from the benchmark is phycocyanin versus colicin A, which was thoroughly studied by CE for tuning its operation parameters. The two proteins have PDBIDs 1colA and 1cpcL. The former is classified as a 'membrane and cell surface protein', while the latter as an 'all alpha protein' by SCOP (Murzin *et al.*, 1995). Their lengths are 204 and 172, respectively. Coordinates of 1colA are determined only for 197 residues.

Table 11. Numbers of SFPs found at different lengths l and CLESUM score thresholds T, before and after shaving. N_i denotes the number of SFPs after shaving for the overlap range $i \times l/2$.

		before	after shaving					before	after shaving		
l	T	N_0	N_1	N_2	N_3	l	T	N_0	N_1	N_2	N_3
	500	21	7	6	6		200	513	239	200	195
20	400	164	45	42	41		150	4273	1512	1093	953
	350	381	93	87	85	8	100	7753	2462	1616	1245
	300	717	156	140	135		50	10350	3161	1960	1418
							0	13207	3942	2316	1547

A great proportion of both the protein structures is helices. It is expected that the number of SFPs should be large. Many SFPs were found at two different lengths l and different

CLESUM score thresholds T before and after shaving, and various overlaps for shaving were examined. As shown in Table 11, the number of SFPs drops significantly for long l and high T, and shaving dramatically reduces the number of SFPs. Finally, in CLePAPS the maximum overlap is set to be 20 for List-20 and 4 for List-8. For $T_{20} = 300$ the number of SFPs after shaving is reduced to about one fifth of the original SFPs, and for $T_8 = 50$ to one third.

We have examined the pair 1colA: 1cpcL and many other pairs to optimize the operation parameters of CLePAPS. The recommended values of these parameters are shown in Table 12.

Table 12. Recommended parameters of CLePAPS.

Symbol	Value	Meaning
l_l	20	length of long AFPs
T_l	300	similarity threshold for long AFPs
K	10	number of long AFPs used as seed candidates
J	50	number of long AFPs for building a star-tree
l_s	8	length of short AFPs for blank-filling
T_s	50	similarity threshold for short AFPs
ρ	4	minimum length of aligned fragments
d_0	5Å	distance cutoff for evaluating overall alignment
d	10Å	separation threshold for star construction
d_1	8Å	separation cutoff for blank-filling at first run
d_2	6Å	separation cutoff for blank-filling at second run
d_3	5Å	separation cutoff for blank-filling at third run
ε	0.1	maximal difference for rotational matrix entries of two 'identical' alignments

Although SFPs reflect mainly the local similarity, the construction of star trees as a partial global consistency check helps us to select seed SFPs for pivoting superposition. We rank seeds first according to the sizes of their center-stars, and then according to the similarity scores σ of the center seeds. In 68 protein pairs, the centers of 59 (87%) rank-1 star trees of List-20 lead to optimal alignments. The nine protein pairs whose optimal alignments do not correspond to a rank-1 star tree form three groups a, b and c. Four pairs in group a have structural repeats, and the rank-1 stars do correspond to one of the multiple best choices although their N_e are relatively lower. For the two pairs in group b, the alignments from the rank-1 stars agree with the optimal. For one pair of the two, 1dsbA: 2trxA, the rank-1 center SFP leads to only partial alignment due to local trapping. An extra 'zoom-in' step with cutoff 10Å is able to obtain the optimal alignment from that SFP. The other three of group c (1aep: 256bA, 1rcb: 2gmfA and 1bgeB: 2gmfA) belong to a 'four-helical up-and-down bundle'. The alignments of DALI consist of almost just helical regions; DALI and CE have nothing in common for the alignment between 1aep and 256bA. It seems that the evidence of local similarity to support the global alignments of these three pairs are not as strong as for other pairs in the benchmark. We see that the loss of sensitivity by keeping only the rank-1 star and obtaining alignment only from its center is rather limited.

Indeed, with the aid of the star tree construction, the SFPs in terms of CLESUM scores are efficient for generating protein structure alignment. The selection of a seed for initial correspondence plays an essential role in the quality of the final alignment. In most cases (41 of 68) the SFP of List-20 with the highest σ is the one from which the optimal alignment is derived. It appears that $K = 10$ is enough to include the seed leading to the optimal alignment. A large value of J for construction of a star tree is favorable for picking up a right seed. The size of a star tree is small for a small J. When the structures to be compared are large or contain repeats, usually a large J is required for star tree construction. For example, the pair 2sim: 1nsbA is a highly repeated fold '6-bladed beta-propeller', with component lengths 381 and 390, respectively. At $J = 30$, every star tree is of size 1, consisting of only the center itself. In CLePAPS we use $J = 50$; increasing J from 30 to 50 does not add much extra computational cost.

Since different criteria are used there is no simple direct comparison between the various methods. For example, a high RMSD would lead to a large number of equivalent residues. With an extra restriction in the minimal size for aligned segments, ProSup usually has a smaller N_e than others. It is difficult to compare N_e and RMSD directly in different methods. To make a close comparison with DALI, we derive a 'DALI kernel' alignment from the original DALI alignment as follows. We superimpose a given pair structure according to the DALI alignment, and calculate the distance of each correspondent residue duad. We remove residue duads with distances greater than d_k and aligned segments whose lengths are smaller than ρ. The remaining reduced correspondence is the DALI kernel of the original alignment at cutoff d_k. (We use $d_k = d_0 = 5\AA$.) The transformation optimal to the kernel is then determined, and N_e and RMSD of the kernel are calculated. It is seen that the performance of CLePAPS is comparable with DALI.

Ten protein pairs from the Fischer benchmark set of 68 pairs were regarded as 'difficult' for fold recognition, and treated as a test set by CE and ProSup. The comparison of CLePAPS with DALI, CE and ProSup is shown in Table 13. To make a direct comparison, as done in the above for DALI alignments, we also derive CE-kernel alignments from the original CE alignments. Generally, alignments of CLePAPS are comparable with those of other alignment tools.

The two proteins of pair 1fxiA: 1ubq are of lengths 96 and 76, respectively. There is only one member in List-20. This is rather risky, being prone to local trapping. It is easy for CLePAPS to give a warning. A simple way of rescue is to use a weaker criterion for SFPs, e.g. List-12 with $T_{12} = 180$ or a lower T_{20}. The alignment listed in Table 12 is obtained by using either $T_{20} = 200$ or $T_{12} = 180$. The CLePAPS alignment for the pair 1bgeB: 2gmfA has nothing in common with the first of DALI's three alignments. The List-20 of the protein pair has 31 members, but none coincides with any segment of the first DALI alignment. This means that the local similarity of the alignment is rather weak. The CLePAPS alignment for the pair is very similar to the second DALI alignment of $N_e = 94$ with RMSD 3.3Å. Of the total 82 aligned residue duads, 72 are identical with those of DALI.

Table 13. Comparison of structure alignments obtained by DALI, CE, ProSup and CLePAPS for 10 'difficult' cases from the Fischer benchmark. N_e: total number of equivalent residue duads; rmsd: RMSD in the unit of Å; CE-kern: CE-kernel; DALI-kern: DALI-kernel; IDA: number of residue duads which are identical to those of DALI.

Pair		CE N_e/rmsd	DALI N_e/rmsd	CE-kern N_e/rmsd	DALI-kern N_e/rmsd	ProSup N_e/rmsd	ProSup IDA	CLePAPS N_e/rmsd	CLePAPS IDA
1fxiA	1ubq	64/2.8	60/2.6	59/2.5	55/2.3	54/2.6	41	55/2.4	42
1ten	3hhrB	87/1.9	86/1.9	85/1.7	84/1.7	85/1.7	79	84/1.7	77
3hlaB	2rhe	85/3.5	75/3.0	71/3.0	63/2.3	71/2.7	37	65/2.3	57
2azaA	1paz	85/2.9	81/2.5	73/2.5	76/2.1	82/2.6	8	78/2.3	72
1cewI	1molA	81/2.3	81/2.3	78/2.0	78/1.9	76/1.9	68	78/2.0	75
1cid	2rhe	98/3.0	97/3.2	79/2.0	82/2.0	84/2.3	70	87/2.2	72
1crl	1ede	220/3.9	211/3.5	155/2.5	168/2.5	161/2.6	147	169/2.7	146
2sim	1nsbA	276/3.0	292/3.3	236/2.5	240/2.5	248/2.6	231	248/2.6	213
1bgeB	2gmfA	109/4.6	94/3.3	62/2.7	79/2.2	87/2.4	0	82/2.4	0
1tie	4fgf	117/3.0	114/3.1	99/2.3	97/2.2	101/2.4	48	100/2.3	94

3.2.5. Database Search with CLePAPS

The fischer benchmark is arranged in 'structure: probe' pairs. Only 56 of the 68 protein structures in the set are distinct. We take 4 proteins from the 68 probes of the benchmark as query structures: 2mtaC from Class α Fold Cytochrome, 1fxiA from $\alpha + \beta$ Ubiquitin-like, 1tie from β Trefoil, and 3chy from α/β Open sheet. Each query structure is aligned with each of the 56 target structures of the benchmark. The first eight structures found by CLePAPS as similar ones to these probes are shown in Table 14. In the benchmark, the structure counterparts of 2mtaC, 1fxiA, 1tie and 3chy are 1ycc, 1ubq, 4fgf and 2fox, respectively. It is seen that, indeed, 1ycc, 1ubq and 4fgf have the highest N_e among the similar structures found. However, 2fox appears rather behind. A close inspection reveals that the six structures with a higher N_e than 2fox are rather large, about three or five times longer than 3chy (except 3adk), and all have a domain or repeat of the same CATH topology 'Rossmann fold' as 3chy (Orengo *et al.*, 1997). Taking the lengths of the aligned proteins into account, the DALI Z-score (Holm and Sander, 1998b) assigns the second highest value to 2fox. The DALI Z-scores for alignments to 3chy are also listed in the table. All the listed structures found for 3chy are coincident with 3chy, at least at the CATH topology level.

According to the Fischer benchmark, four probe sequences 1tlk, 3cd4, 3hlaB and 1cid are associated with the structure 2rhe. We have also taken 2rhe as a query to search the 68 probes for similar folds, and the results are also listed in Table 14. As we may expect, indeed, the structures of the above four probes appear to be highly ranked among those similar to 2rhe. However, an additional one, 2fbjL, has the highest value of N_e. It is verified that the two domains of 2fbjL share the same SCOP superfamily or CATH homology with 2rhe, and has the highest DALI Z-score. Protein 1sacA is not classified as Immunoglobulin (IG) or IG-like, but it still shares the same CATH Architecture with 2rhe.

Table 14. Structures found by CLePAPS as highly similar to five queries. $Z_{\rm dali}$: the Z-score given by DALI; rmsd: RMSD in Åunits.

Query structure										
2mtaC		1fxiA		1tie		3chy			2rhe	
ID	N_e/rmsd	ID	N_e/rmsd	ID	N_e/rmsd	ID	N_e/rmsd	$Z_{\rm dali}$	ID	N_e/rmsd
1ycc	71/1.9	1ubq	50/2.3	4fgf	100/2.3	2liv	102/2.6	7.9	2fbjL	102/1.5
1tca	47/3.1	1rbp	48/3.0	5ptp	60/2.9	2minB	97/2.9	6.7	3cd4	89/1.6
1hbg	47/2.6	5ptp	44/3.2	9rnt	51/2.5	1tca	96/2.6	5.6	1cid	87/2.2
1atr	46/3.0	1aozA	39/3.1	1cauA	49/2.7	1ede	88/2.7	4.9	1tlk	86/1.7
2scpA	44/2.9	1nsbA	38/3.4	2plv1	49/2.8	3adk	88/3.0	4.3	1fc1A	76/2.6
2cpp	43/2.9	1atr	35/2.7	1rbp	45/2.5	6ldh	87/2.7	6.0	1ten	68/2.8
4cpv	42/3.2	1ifc	35/2.5	1fnb	45/2.9	2fox	86/2.8	6.8	1sacA	67/2.6
2mnr	41/2.8	1fbpA	35/2.8	1nsbA	45/3.4	1fnb	84/2.9	3.1	3hlaB	65/2.3

3.2.6. Multiple Solutions of Alignments

To be less greedy, CLePAPS generates several alignments from star centers of highly scored SFPs. Often there is one alignment which has a much higher N_e than others. (A better measure of significance for alignments is the P-value or Z-score.) There are situations where several meaningful alignments do exist. We see such examples in the Fischer benchmark.

Proteins 8i1b and 4fgf belong to SCOP fold beta-Trefoil, which exhibits a three fold rotational symmetry. Indeed, the alignment ranks 1 to 3 clearly demonstrate this symmetry. A more complicated example is the pair 2sim: 1nsbA. Both belong to SCOP fold 6-bladed beta-propeller, but in PFAM (Bateman *et al.*, 2004) classification 2sim has four repeats of PF02012 while 1nsbA belongs to PF00064. Multiple alignments reflect the rough symmetry. Another example is the pair 3rubL: 6xia, which contains a SCOP fold TIM beta/alpha-barrel. Among the multiple alignments, only one follows a sequential order, while all the others correspond to a circular permutation.

Proteins 1osa and 4cpv belong to all alpha SCOP fold EF hand-like. The former contains two EF hand domains, while the latter has a single dpmain. The top two alignments are mainly the two ways of alignment of the EF hand domains. Another similar example is the pair 3hhrB: 1ten of SCOP fold IG-like beta-sandwich.

Proteins 2gbp and 2liv both belong to SCOP class α/β superfamily Periplasmic binding protein-like I, consisting of two CATH 3-Layer ($\alpha\beta\alpha$) Sandwich domains, which will be denoted by I and II with I near the N-terminus. The two domains present a repeat of highly similar $\alpha\beta\alpha$ segments. For example, the SFP search at length 20 of 2gbp against itself discovers a region of contiguous SFPs with the correspondence of sites 23 – 60 to 163 – 200 around an SFP of $\sigma = 775$. The ranks 1 to 3 in CLePAPS alignments of 2gbp to 2liv are mainly the alignments of domain I to I, II to I, and II to II, respectively. The SFP search of 2gbp against 2liv ascertains a long piece of contiguous SFPs with a correspondence of sites 24 – 80 of 2gbp to 34 – 90 of 2liv around the SFP of the highest $\sigma = 803$ in List-20, a piece with a correspondence of 167 – 201 to 37 – 71 around an SFP of $\sigma = 709$, and another with a correspondence of 139 – 174 to 135 – 170 around an SFP of $\sigma = 566$. These roughly correspond to the cores of the three alignments. They are easily located by string comparison.

A relative move between the two domains with respect to the two proteins makes the RMSD for the alignment of the proteins as a whole rather high. This is seen in the DALI and CE alignments, which are of $N_e = 260$ with RMSD 6.8Å and $N_e = 252$ with RMSD 4.6Å, respectively.

It is well known that the 3D structures of two proteins may be surprisingly similar in secondary structure element packing while the sequential order of their SSEs is completely different. We do not see any clear example of such non-topological alignment other than a simple rotation in the Fischer benchmark. A good example of such protein pairs is SCOP fold SH3-like barrel 1ihwA: 1sso (of lengths 52 and 62, respectively; Jung and Lee, 2001). We represent an aligned fragment by the triplet $(a : a'; w)$, where w is the fragment length, and a and a' are starting positions in the two proteins. The alignment of 1ihwA: 1sso consists of 5 pieces:

$$(2 : 26; 10), (13 : 36; 10), (28 : 0; 7), (36 : 8; 11), (47 : 20; 5).$$

and has $N_e = 43$ with RMSD 2.21Å. The first two segments show that the N-terminus of 1ihwA aligns with the C-terminus of 1sso, while the other three segments are the alignment between the C-terminus of 1ihwA and the N-terminus of 1sso. Such alignments cannot be found by algorithms using dynamic programming.

3.3. BLOMAPS: a Fast Multiple Structure Alignment Tool

Multiple structure alignment carries significantly more information than pairwise alignment, and hence is a much more powerful tool for classifying proteins, detecting evolutionary relationship and common structural motifs, and assisting structure/function prediction.

Most existing methods of multiple structural alignment combine a pairwise alignment and some heuristic with a progressive-type layout to merge pairwise alignments into a multiple alignment (Ye and Janardan, 2004; Lupyan *et al.*, 2005). Besides the computaional cost, such pairwise-based methods have the limitation that the alignments that are optimal for the whole input set might be missed. There are a handful of truly multiple methods (Shatsky *et al.*, 2002; Benyamini *et al.*, 2003; Dror *et al.*, 2003a; 2003b; Ebert and Brutlag, 2006). Many multiple alignment tools often start with sets of structurally common fragments extracted from as many as possible input proteins, and then combine them into a global common substructure. For example, in doing this, MASS (Dror *et al.*, 2003a; 2003b) implements a two-level alignment, using both secondary structure and atomic representation.

We have described CLePAPS for fast pairwise alignment of protein structures using the conformational alphabet in last section. CLePAPS starts with similar fragment pairs (SFPs), whose two members are structurally similar and can be superimposed one on the other after transformation. CLePAPS is anchor-based, initiating alignments by superimposing the two structures based on a seed SFP. Its main difference from other algorithms is its way to define an SFP as a pair of strings with a high ungapped CLESUM score. Replacing local structural comparison of fragments by string comparison leads a significant gain in speed and specificity.

The common goal of all multiple structure alignment methods is to identify a set of residue 'columns' from each 'row' protein that are structurally similar, or to find an optimal correspondence among the atoms in these molecular structures. Here we describe an anchor-based tool called BLOMAPS for fast multiple structure alignment by means of conformational alphabet. The conformational alphabet representation of structures has been also used for multiple structure alignment as a multiple version of CE-type (Liu and Zheng, 2007). BLOMAPS is a multiple version of CLePAPS.

3.3.1. Finding Fragment Blocks of High CLESUM Similarity Scores

Different from pairwise structure alignment, when dealing with multiple alignment we face a problem of the 'combinatorial explosion'. The concept of similar fragment pair is now replaced by that of similar fragment block (SFB).

Considering an l long fragment taken as a seed from a 'pivot' protein which is chosen from a set of structures to be aligned, we search the structures other than the pivot for fragments similar to the seed as SFPs. For a given seed, several neighbors may be found in a single structure. Taking one neighbor from each structure which has neighbors of the seed, we may form a block of similar fragments with the seed being its center. If a seed fragment averagely has k neighbors in another strucutre, for a set of n structures there would be $k^{n-1}L$ SFBs, where L is the effective structure length. We have to think various greedy heuristics.

There is a block which consists of the seed and the neighbors which score the highest among neighbors in each structure possessing neighbors. We call such a block a 'highly similar fragment block' (HSFB). It could happen that a seed does not have any neighbor in some structures. The total number of fragments in an HSFB will be called the depth of the HSFB. Another characteristic of an HSFB is its total score Σ which is the sum of σ scores between the seed and its each neighbor in the HSFB, where σ is the CLESUM score for a string pair. We define the consensus letter of a set of conformational letters as the letter which belongs to the set and has the highest sum of CLESUM scores between itself and all letters in the set. For example, the consensus of $\{A, H, Q\}$ is A. The consensus of a HSFB is then defined as the string which consists of the consensus letters of columns when we align the 'row' strings of the HSFB. Thus, besides the width and positions of its member fragments, an HSFB has its depth, score and consensus.

We call a block of fragments which is extracted from the multiple alignment a multi-aligned block (MAB). The concept of SFB reflects only the local similarity, so need not correspond exactly to a MAB. A necessary condition to a MAB is that a MAB must be a block of similar fragments in the sense of CLESUM score. To find all possible SFBs, one need to examine each l long string from every conformational sequence. However, this is inefficient. For a set of closely related structures, we expect that there is a good chance to find certain members of MABs in some HSFBs. Furthermore, our greedy algorithm uses the shortest protein to create HSFBs. That is, all seed centers are extracted from this protein.

Usually, shifts of the center for an HSFB with a large depth and score would plausibly be also centers of HSFBs with a large depth and score. To remove this redundancy we sort HSFBs first in descending order of depths and then in that of scores. We create a 2D grid of atom indices with its rows corresponding to individual proteins. We fill the first HSFB in the grid, (i.e. mark its atom indices in the grid,) and then examine the second. If the overlapping positions between the two HSFBs are less than $\Gamma = \gamma l m$, where l is the width of the block, m the depth of the second HSFB and γ a parameter for redundancy, we fill in the grid the second HSFB. Otherwise, we skip it, and examine the third. When we examine a new HSFB, we calculate the number of its positions that overlap with the marked indices of the grid. Only when its overlapping proportion is less than γ fill we it in the grid. We go on the procedure until the last HSFB is examined. We then have a sorted list of HSFBs with redundancy removed.

3.3.2. Scaffold: The Vertical Equivalency and Horizontal Consistency

The correspondence of a multiple alignment defines an equivalence relation of residues among proteins. This will be called the 'vertical equivalency'. For any two structures in the multiple alignment, the transformation to superimpose the residue duads in a subset of the correspondence will also bring the residue duads in the complement set of the correspondence spatially close. This is the 'horizontal consistency'. The requirement of both the equivalency and consistency increases the difficulty of multiple alignment, but also reduces the chance of making a wrong alignment.

The multiple alignment algorithms which progressively merge pairwise alignments may be classified as horizontal-first. Our BLOMAPS may be regarded as vertical-first. To speed up, we align all structures against some template, and at the beginning the shortest protein is taken as the template to construct HSFBs. BLOMAPS then starts with an HSFB taken from the top K HSFBs as an 'anchor'. To choose a structure to be most representative among the structures, we update the template to the protein whose member in the anchor HSFB is closest to the block consensus (*i.e.* has the highest similarity score with the consensus). Since HSFBs are created using the shortest protein it may happen that the updated template, or the new pivot protein, does not have a fragment in an HSFB. In this case we use the consensus of the HSFB to search the pivot protein for possible neighbor fragments with a lower threshold T_{-}, and add the found optimal neighbor to the HSFB. All structures will be then aligned against this pivot protein. The transformation to superimpose the fragments of the anchor HSFB will also be used to superimpose the whole strutures where the fragments are.

After we have superimposed all the structures which have a fragment in the anchor HSFB, we examine the horizontal consistency as follows. We mark the anchor HSFB and its fragments 'colored', all other HSFBs and their fragments 'uncolored', and set up a counter for each HSFB other than the anchor. If an HSFB contains a fragment of the pivot protein, the fragment will be regarded as the center of the HSFB. We calculate separations between the center and other members of the HSFB. When we find a separation below a preset cutoff d_1 we advance the count of the counter for the HSFB by one, and mark both the fragment and the center colored. If an HSFB has at least three colored fragments, we mark the HSFB colored. Ignoring the uncolored HSFBs, we assign the total counts of colored fragments to the anchor HSFB. After we have calculated the total counts for all the top K anchor HSFBs, we select the one with the highest number of colored HSFBs to go on to next step.

We now have an anchor HSFB and several colored HSFBs which are spatially consistent with the anchor. Thus, the colored fragments of the pivot protein are supported both horizontally and vertically, and form a scaffold for multiple alignment. The anchor HSFB may have no neighbor from some structures. Such structures are regarded as unanchored proteins. No transformation can be determined based on the anchor HSFB to superimpose an unanchored protein on the pivot protein. Furthermore, it is not guaranteed that the anchor HSFB is exactly a part of a MAB. If an anchored protein has no fragment in a colored HSFB, we search the protein for fragments which do not overlap with any colored fragments and are similar to the center either by checking the similarity score first, or directly examining the separation with respect to the center fragment. If a separation is found being smaller than cutoff d_1 we color the corresponding fragment and add it to the HSFB. For each structure which has a fragment in the anchor HSFB, we count the number of its colored fragments. If the number is smaller than n_h we regard the protein also as unanchored, and otherwise as anchored.

So far the transformation to superimpose two structures is based on a single pair of fragments. Using the colored fragments, we may update the transformation based on more consistent pairs of fragments. That is, for an anchored protein, the transformation optimal for superimposing all its colored fragments on the pivot protein is the updated transformation for the two proteins. With the transformation updated, we use a width l' smaller than l and a more stringent cutoff d_2 to examine the consistency of colored fragments. If an l long colored fragment has at least l' residues whose coordinates deviate from their counterparts on the pivot protein within d_2, the fragment remains colored, and otherwise it is changed to uncolored.

After examining all l long colored fragments, we then conduct the following procedure of 'recruiting aligned fragment pairs (AFPs)' for every anchored protein. For a given anchored protein, the colored fragments and their partner fragments on the pivot define the primary AFPs. We mask them from the anchored and pivot proteins. We take each fragment of length l' from the unmasked region of the pivot, and search the unmasked region of the anchored protein for SFPs using threshold T'. We sort the found SFPs in descending order of scores. According to the order we examine separations of SFPs successively. Whenever a separation is found to be smaller than d_2 and the positions of the corresponding SFP do not conflict with the existing AFPs, we record the SFP as an AFP. After this recruiting we examine the equivalency of AFPs. If a fragment of the pivot protein is shared by at least n_v AFPs, the fragment will be assigned as a part of the scaffold. We then obtain an extended scaffold. AFPs map residues of proteins other than the pivot to those of the pivot protein, and define columns of residue correspondence. We construct the first average template by averaging transformed coordinates of atoms over individual columns. The template will be used for dealing with unanchored structures.

3.3.3. Dealing with Unanchored Structures

We now try to align unanchored proteins on the average template. The anchor HSFB does not provide any guidance for aligning an unanchored protein. However, the protein may still have members in other colored HSFBs. Any of such members can be used to generate a transformation for superimposing the protein on the template, and then examining the consistency of other fragments. For efficiency, we may sort the members of colored HSFBs according to their similarity scores, and then examine the top K. If an unanchored protein has no enough number of members which can be associated with colored HSFBs, we have to try to pairwisely align it on the template as follows.

Guided by the scaffold on the pivot protein, we take each fragment of length l from the scaffold as a seed, and search the unanchored protein for SFPs using threshold T. We sort the found SFPs in descending order of scores. We use the top K SFPs to generate transformations, and top $J(\geq K)$ for consistency checking. We pick up one SFP from the top K, replace the fragment of the pivot protein by its image on the template, and use the transformation determined from the pair of fragments to superimpose the unanchored protein on the template. We examine separations of the top J SFPs successively. We count the total number n_p of non-overlaping SFPs whose separations are below d_2. The SFP which has the largest n_p among the top K and its consistent SFPs will be further used to construct the updated transformation for superimposing the unanchored protein. The procedure of recruiting AFPs is then followed to extend the proportion aligned to the scaffold. We do this for every unanchored protein.

3.3.4. Refinement

After having succeeded in superimposing unanchored proteins on the template, we update the average template, and then examine the deviation between residue duads of AFPs and their flanking sites. We may elongate or shrink fragments according to deviation cutoff d_3, and hence update AFPs. The changed AFPs would lead to the updated average template. This is an iteration, and its convergence is usually rather fast.

Due to the greedy nature of the approach above, only patterns shared by the pivot protein have a chance to be discovered. Some patterns could be shared by a subset of structures, but absent from the pivot protein. They are 'missing motifs' to the pivot protein. Their information has to be extracted from the structures sharing them.

The patterns or motifs considered here are ungapped fragments which are structurally similar locally, and are arranged in space very alike globally. So, a necessary condition for such a motif is that it must be also a motif of conformational sequences. There are many methods for discovering motifs in a set of sequences. We use a simple center-star approach in the sense of strings instead of sequences (Zheng, 2005a).

We first consider the case where all structures share a common core of multiple alignment, but some subsets may share their own extra motifs. From a protein we collect strings of length l in the 'blank' region where no AFPs are located. Taking each string as a seed from every protein other than the pivot, we find all HSFBs using similarity threshold T, and sort them in descending order of the depth and then of the total score. The main difference here is that now seeds are taken from blank regions and from all proteins other than the pivot. Since we have superimposed every structure on the average template, we examine the deviation between any two fragments of an HSFB using cutoff d_2 if their positions do not conflict with any AFBs of early found motifs. If n_v pairs of fragments with deviation below d_2 are found, they are regarded as a motif, and we then try to recruit more AFPs by examining the deviation of SFPs to extend the depth of the motif.

An extraordinary large protein will have a large proportion of blank regions, most of which have no contribution to missing motifs, and hence make the above procedure inefficient. This inefficiency occurs also when the number of structures is large. In such cases, we propose another way to rescue missing motifs. We divide the space occupied by the structures after superimposition into uniform cubic cells of a finite size, say 6 Å. The number of different proteins which have their residues falling in a given cell is the depth of the cell. We discard cells whose depths are below a preset cutoff, then we sort the remaining cells in descending order of depth. Picking a cell of a large depth as a base, in each dimension we select from its two nearby cells the one with the higher depth to expand the base cell and double its size into a 'octad'. We mark the residues falling into the octad. In this octad, if we find a protein which has at least three marked residues with their indices being within l', we take the segment of the protein which covers the most marked residues as a seed, and examine all the segments which are from other proteins and have some marked residues by checking their separation from the seed. If an AFB is found, we discard the eight cells of the octad, and continue with the cell of the next highest depth until all cells are examined.

Table 15. Recommended parameters for BLOMAPS.

Symbol	Value	Meaning
l	12	width of wide similar fragment blocks (SFBs)
T	200	similarity threshold for wide HSFBs
K	5	number of wide HSFBs tested as anchors
l'	8	width of narrow SFPs for scaffold improving
T'	100	similarity threshold for narrow SFPs
γ	0.5	overlap proportion for removing redundancy of HSFBs
ρ	3	minimum length of aligned fragments
T_{-}	180	similarity threshold for adding fragments not in an HSFB
d_0	5Å	distance cutoff for evaluating overall alignment
d_1	12Å	separation threshold for consistency
d_2	8Å	separation bound for recruiting AFPs based on the pivot
d_3	5Å	separation cutoff for recruiting AFPs based on the template
n_h	3	minimal number of colored fragments for an anchored protein
n_v	3	minimal number of colored fragments for vertical equivalency

In the case when a common core is shared by only a subset of the input set, we divide the input set into two subsets: one with that core and the other without it. The algorithm first accomplish the alignment for the first subset, and then treats the second subset as a new input set.

The final alignment is given by the complete residue correspondence. A full column of the correspondence has residues from every protein of the structure set. The common core of the alignment is rigorously defined by all full colomns. We may also define a 'partial core' by introducing a parameter of proportion. For example, core-60 is given by columns covering sixty percent or more proteins of the structure set. Assume that a new motif is found on a protein other than the pivot. With the help of the common core, we can map the protein together with the missing motif on the template, and update the template by averaging coordinates over individual new columns. After having superimposed structures on the final average template, we can calculate the total squared deviation of aligned residues with respect to the template for full or partial core of the alignment, and then the RMSD for evaluation.

The recommended parameters for BLOMAPS are listed in Table 15. The overall algorithm of BLOMAPS is summarized below. The algorithm has been tested on 17 protein structure ensembles.

```
        * * * Overall algorithm of BLOMAPAS * * *
create HSFBs using the shortest protein as a template;
sort HSFBs;
derive redundancy-removed HFSBs;

for each HSFB in top K
  select the pivot protein based on the HSFB consensus;
  superimpose anchored proteins on the pivot;
  find consistent HSFBs;
select the best HSFB which admits  most consistent HSFBs;

build a primary scaffold from the consistent HSFBs;
update the transformation using the consistent HSFBs;
recruit aligned fragments;
improve the scaffold;
create average template;

for each unanchored protein
  if (members are found in colored HSFBs) find top K members;
  else search for fragments similar to the scaffold,
       and select top K;
  pairwisely align the protein on the template;

find missing motifs by registering atoms in spatial cells;
refine the average templet
```

4. Conclusion

By means of different local criteria, such as solvent accessibility, residue burial, and backbone geometry, protein fragments may be clustered and then represented by an alphabet. Protein structural alphabets are discretized states of backbone fragment conformations. As a coarse-grained description of tertiary structure, structure alphabets bridge the gap between the secondary and tertiary structures. The description of 3D segmental structural states by a few conformational letters aptly balances precision with simplicity. The substitution matrix CLESUM provides us with a proper measure of the similarity between these discrete states or letters. Once we transform 3D structures to 1D sequences of letters, tools for analyzing ordinary sequences can be applied with some minor modification (Friedberg *et al.*, 2007).

The representation of segmental conformations by letters is not rigid. The convenience to indicate a conformational cluster by its representative center conformation is sometimes misleading. The constraint imposed by nearby conformational letters and associated amino acids greatly reduces the vagueness of the conformation of a letter. To keep the size of alphabet as small as possible, we have considered quadrapeptides. The quadrapeptide segment is the minimal unit for one-to-one correspondence between pseudobond angles and atomic Euclidean coordinates. The correlation amongst conformational letters is far beyond the nearest neighbor. The advantage of a Markov model of the first order is limited. Since it involves many parameter numbers and is inconvenient for assigning codes to a short segment, we have decided to use a simple independent model (M0). The correlation can be treated in a later stage by considering strings.

Structural alphabets have been used in database searches by mimicking the BLAST search for amino acid sequences (Carpentier *et al.*, 2005; Tung *et al.*, 2007; Lo *et al.*, 2007), or by means of a suffix tree (Guyon *et al.*, 2004; Täubig *et al.*, 2006). It is worth noticing that a sequence of structure codes is quite different from that of amino acids at least in two aspects. Due to the locality nature of structural codes a gap in an alignment of code sequences may destroy the global consistency of fragment arrangement in the space, so its explanation requires prudence. Furthermore, the regular secondary structures create simple repetitive segments in code sequences.

A simple and direct use of structure alphabets is to provide signatures for classes in a structure classification like SCOP or CATH. The CLESUM with the conformational alphabet equips us with a convenient way to represent structure motifs and structure categories. A class at various level may be represented by several ungapped consensus strings of structural codes. The description is much finer than when only the regular secondary structure components are mentioned. Conservative loops with definite consensus strings are often observed. The profile HMMs might be less useful for structural code sequences due to the difficulty in explaining the gaps.

We have described in full length the application of our conformational letters in developing fast tools for pairwise and multiple structure alignments. The description with conformational letters is very suitable for the ϵ-congruent problem. Furthermore, the substitution matrix CLESUM extracted from the database FSSP of structure alignments contains information of structure database statistics. Thus, the similarity it measures is not only geometric, but also evolutionary. The definition of similar fragment pairs or blocks by CLESUM scores leads to fast searching for such pairs or blocks by string comparison. Furthermore, they are more significant than those derived purely by geometric similarity. Guided by similarity scores, only the top few structurally similar fragments need to be examined to determine the superposition for alignment, and hence a reliable greedy strategy becomes possible. We have also discussed substitution matrices in the joint sequence-structure space by alphabet reduction. We expect that the substitution matrix of structure codes together with two amino acid classes h and p will improve the performance of structure alignment tools.

The problem of evaluating the significance of a structural alignment by a P-value or Z-score is by no means simple. It has been demonstrated by Lackner *et al.* (2000) that optimizing the number of equivalent residues within a distance cutoff for residue equivalence provides a simple and intuitive measure of structure similarity. It is hard to define the random reference alignment. We have not discussed the problem of the statistical significance of a structure alignment, which deserves a separate chapter (Jia *et al.*, 2004).

We also have not discussed the prediction problem, which, of course, is extremely important. Local structure prediction, predicting the category of conformation for a fragment from its amino acid sequence, is a crucial step towards the goal of tertiary structure prediction. The basis of the prediction is the correlation between sequence and structure. It is known that the sequence-structure relationships have not always been strong; a local structure can usually be adopted by many different sequences. For example, the sequence profiles of the hydrophobic and hydrophilic sites of a helix are different. Mixing them together will dilute their information content, and hence diminish their prediction capacities. Bystroff and Baker (1998) have built a library of structure-sequence motifs, which are expected to correspond to functional units recurring in different protein contexts and to be found in different combinations in distantly related or functionally unrelated proteins. Tang *et al.* (2005) used the sum of ϕ-ψ and amino acid profile deviations as a distance measure for sequence-structure motifs. Recently, Sander *et al.* (2006) partitioned sequence space into decision regions to optimize the inference of local structure and create local sequence-structure prototypes. To identify the structural features that have strong sequence preferences we have to locate peaks of density distribution in the joint structure-sequence space. Previously, structure-based clustering was a duty much more cumbersome than sequence-based clustering, so one had to start with a sequence-based clustering then transfer back and forth between the structure and sequence subspaces. It will thus be interesting to see whether the library can be improved by clustering directly in the joint structure-sequence space with the help of conformational codes.

All the methods successful for secondary structure prediction can equally well be applied to the prediction of structure codes of a protein. Almost all statistical methods and machine learning techniques have been tried for protein secondary structure prediction. To extract information of secondary structure from sequence to the utmost extent, it is crucial to consider secondary structure segments instead of the structure class at a single residue (Crooks and Brenner, 2004). Long segments or strings have the potential to include more correlation, but the occurrence of long strings is usually rare. Similar strings have to be grouped together to maintain statistical significance. We have to balance correlation with resolution, and use a suitable string length. This is particularly true for structural alphabets, whose sizes are usually about a dozen. We have to group similar strings into 'superstrings' to keep their counts at a reasonable level. Generally, the relationship between sequence and structure is described by a position specific weight matrix of amino acids, where correlation between neighboring amino acids is neglected. However, if the sample size (or the number of occurrence) of a string is high, more than one such matrix can be used to cover the correlation among different positions in a structural string. This is splitting of a pattern for a single string into subpatterns. For example, the occurrence of helices and beta sheets is high, and the strong periodicity in the hydrophobicity of helices and sheets is well known (Eisenberg *et al.*, 1984). Ignoring such subpatterns and mixing different 'phases' would lead to the conclusion that amino acids on a protein chain are only weakly correlated (Crooks and Brenner, 2004). According to the discussion on entropic clustering, some entropy, often the average score, can be used as an objective function to optimize the scheme for grouping and splitting. After designing the states in terms of superstrings and subpatterns, hidden Markov models can be implemented for prediction of structural states (Zheng, 2005b).

It has been found that the accuracy of secondary structure prediction can be enhanced substantially by basing the prediction upon a multiple sequence alignment of homologous protein sequences rather than just a single sequence. This strategy can be equally well employed for prediction of structural codes. However, the prediction quality is limited by the size of the family, the divergence of structure across the family and the quality of the alignment.

Appendix

A1. Conversion from Coordinates to Angles

```
//convert coordinates r[N][3] to bending and torsion angles bt[N][2]
#define norm(x) sqrt(dot(x,x))
#define angle(x,y) ((z=dot(x,y)/norm(x)/norm(y))>1.?3.1416:acos(z))
float dot(float x[3], float y[3])
{ int i;
  float t=0.;
  for(i=0;i<3;i++) t+=x[i]*y[i];
  return(t);
}
void cross(float x[3], float y[3], float z[3])
{ int i;
    z[0]=x[1]*y[2]-x[2]*y[1];
    z[1]=x[2]*y[0]-x[0]*y[2];
    z[2]=x[0]*y[1]-x[1]*y[0];
}
void xyz2bt(int n, float r[N][3], float bt[N][2])
{ int i,j;
  float x[N][3],y[N][3],z;
  for(i=0;i<n-1;i++) for(j=0;j<3;j++)
    x[i][j]=r[i+1][j]-r[i][j];
  for(i=0;i<n-2;i++)
    cross(x[i],x[i+1],y[i]);
  for(i=0;i<n-2;i++)
    bt[i][0]=angle(x[i],x[i+1]);
  for(i=0;i<n-3;i++) {
    bt[i][1]=angle(y[i],y[i+1]);
    if(dot(y[i],x[i+2])<0.) bt[i][1]*=-1.; }
}
```

A2. From Angles to Conformational Letters

```
//convert angles to conformational letters
//pi: mode weight; sig: square root of |Sigma|^-1; mid: means (mu);
//var: inversed covariance matrices.
//(see Table 1)

float sig[17],mid[17][3],pi[17],var[17][3][3],bt[N][2];
float normal(int ic,int k)
{ int i,j;
  float y[3],z;
  y[0]=bt[k][0]-mid[ic][0];
  y[2]=bt[k+1][0]-mid[ic][2];
  y[1]=bt[k][1]-mid[ic][1];
  if(y[1]>3.1416) y[1]-=6.2832;
  else if(y[1]<-3.1416) y[j]+=6.2832;
  for(z=0.,i=0;i<3;i++) for j=0;j<=i;j++)
    z+=(i-j?1.:.5)*y[i]*var[ic][i][j]*y[j];
  return(exp(-z)*sig[ic]);
}
void bt2c(int n, char s[N])
{ i,j,ic;
```

```
    float q,qq;
    for(i=0;i<n-1;i++) {
      for(q=0.,ic=j=0;j<C;j++)
        if((qq=pi[j]*normal(j,i))>q) {
      q=qq; ic=j; }
      s[i]=ic+'A'; }
}
```

This work was supported in part by the National Basic Research Program of China (Grant No. 2007CB814800) and the National Natural Science Foundation of China.

References

Asai,K., Hazamizu,S., & Handa,K. (1993). Secondary structure prediction by hidden Markov model, *CABIOS* **9**, 141-146.

Bateman,A., Coin,L., Durbin,R., et al. (2004). The Pfam protein families database, *Nucleic Acids Res.*, **32**, D138-D141.

Benyamini,H., Gunasekaran,K., Wolfson,H., & Nussinov,R. (2003). Conservation and amyloid formation: A study of the gelsolin-like family, *Proteins*, **51**, 266-281.

Burge,C. & Karlin,S. (1997). Prediction of complete gene structures in human genomic DNA, *J. Mol. Evol.* **268**, 78-94.

Bystroff,C. & Baker,D. (1998). Prediction of local structure in proteins using a library of sequence-structure motifs, *J. Mol. Biol.* **281**, 565-577.

Camproux,A.C., Tuffery,P., Chevrolat,J.P., Boisvieux,J.F., & Hazout,S. (1999). Hidden Markov model approach for identifying the modular framework of the protein backbone, *Protein Eng.*, **12**, 1063-1073.

Carpentier,M., Brouillet,S. & Pothier,J. (2005). YAKUSA: a fast structural database scanning method. *Proteins*, **61**, 137-151.

Chang,P.L., Rinne,A.W., & Dewey,T.G. (2006). Structure alignment based on coding of local geometric measures, *BMC Bioinformatics,* **7:346**.

Crooks,G.E. & Brenner,S.E. (2004). Protein secondary structure: entropy, correlations and prediction, *Bioinformatics*, **20**, 1603-1611.

Dror,O., Benyamini,H., Nussinov,R., & Wolfson,H. (2003a). MASS: Multiple structural alignment by secondary structures, *Bioinformatics*, **19** suppl.1, i95-i104.

Dror,O., Benyamini,H., Nussinov,R., & Wolfson,H.(2003b). Multiple structural alignment by secondary structures: Algorithm and applications, *Protein Science*, **12**, 2492-2507.

Duran,B.S., & Odell,P.L. (1974). Cluster Analysis: A Survey, Springer Verlag, Berlin.

Ebert,J. & Brutlag,D. (2006). Development and validation of a consistency based multiple structure alignment algorithm, *Bioinformatics*, **22**, 1080-1087.

Edgoose,T., Allison,L. & Dowe,D.L. (1998). An MML classification of protein structure that knows about angles and sequences, Proc. 3rd Pacific Symposium on Biocomputing (PSB-98), Hawaii, USA, pp. 585-596.

Eidhammer,I., Jonassen,I., & Taylor,W.R. (2000). Structure comparison and structure patterns, *J. Comput. Biol.*, **7**, 685-716.

Eisenberg,D., Weiss,R.M., & Terwilliger,T.C. (1984). The hydrophobic moment detects periodicity in protein hydrophobicity, *Proc. Natl. Acad. Sci.* **81**, 140-144.

Fischer,D., Elofsson,A., Rice,D., & Eisenberg,D. (1996). Assessing the performance of fold recognition methods by means of a comprehensive benchmark, Proc. Pac. Symp. Biocomput., pp. 300-318.

Flocco,M.M., & Mowbray,S.L. (1995). C_α-based torsion angles: A simple tool to analyze protein conformational changes, *Protein Science* **4**, 2118-2122.

Friedberg,I., Harder,T., Kolodny,R., Sitbon,E., Li,Z., & Godzik,A. (2007). Using an alignment of fragment strings for comparing protein structures, *Bioinformatics*, **23**: e219-224.

Fujiwara,Y., Asogawa,M. & Konagaya,A. (1994). Stochastic motif extraction using hidden Markov model, Proc. ISMB94, pp.121-129.

Gibrat,J.F., Madej,T., & Bryant,S.H. (1996). Surprising similarities in structure comparison, *Curr. Opin. Struct. Biol.*, **6**, 377-385.

Grantham,R. (1974). Amino acid difference formula to help explain protein evolution, *Science*, **185**, 862-864.

Guyon,F., Camproux,A.C., Hochez,J., & Tuffery P. (2004). SA-Search: a web tool for protein structure mining based on a Structural Alphabet, *Nucleic Acids Res.*, **32**, W545-W548.

Henikoff,S., & Henikoff,J.G. (1992). Amino acid substitution matrices from protein blocks, *Proc. Natl. Acad. Sci.*, **89**, 10915-10919.

Hobohm,U. & Sander,C. (1994). Enlarged representative set of protein structures, *Protein Science*, **3**, 522-524.

Holm,L. & Sander,C. (1994). The FSSP database of structurally aligned protein fold families, *Nucleic Acid Res.*, **22**, 3600-3609.

Holm,L. & Sander,C. (1997). Dali/FSSP classification of three-dimensional protein folds, *Nucleic Acid Res.*, **25**: 231-234.

Holm,L., & Sander,C. (1998a). Touring protein fold space with Dali/FSSP, *Nucleic Acids Res.*, **26**, 316-319.

Holm,L., & Sander,C. (1998b). Dictionary of Recurrent Domains in Protein Structures, *Proteins*, **33**, 88-96.

Jia,Y., Dewey,T.G., Shindyalov,I.N., & Bourne,P.E. (2004). A New Scoring Function and Associated Statistical Significance for Structure Alignment by CE, *J. Comp. Biol.* **11**, 787-799.

Jung,J. & Lee,B. (2001). Circularly permuted proteins in the protein structure database, *Protein Sci.*, **10**, 1881-1886.

Kabsch,W. (1978). A discussion of the solution for the best rotation to related two sets of vectors, *Acta. Crystal.*, 34A, 827-828.

Kabsch,W. & Sander,C. (1983). Dictionary of protein secondary structure: Pattern recognition of hydrogen-bonded and geometrical features, *Biopolymers*, **22**, 2577-2637.

Koehl,P. (2006). Protein Structure Classification, in *Reviews in Computational Chemistry*, **V. 22**, K.B. Lipkowitz, T.R. Cundari, and VJ. Gillet (Eds.), pp. 1-55, Wiley-VCH, John Wiley & Sons, New York.

Krogh,A., Brown,M., Mian,I.S., Sjölander,K. & Haussler,D. (1994). Hidden Markov models in computational biology: Applications to protein modeling, *J. Mol. Biol.*, **235**, 1501-1531.

Kullback,S., & Leibler,R.A. (1951). On information and sufficiency, *Annals Math. Stat.*, **22**, 79-86.

Lackner,P., Koppensteiner,W.A., Sippl,M.J., & Domingues,F.S. (2000). ProSup: a refined tool for protein structure alignment, *Protein Eng.*, **13**, 745-752.

Liquori,A.M. (1969). The stereochemical code and the logic of a protein molecule. *Q. Rev. Biophys.*, **2**, 65-92.

Liu,X., & Zheng,W.M. (2007) CLEMAPS: Multiple Alignment of Protein Structures Based on Conformational Letters, *Proteins* (to appear).

Lo,W.C., Huang,P.J., Chang,C.H., & Lyu,P.C. (2007). Protein structural similarity search by Ramachandran codes *BMC Bioinformatics*, **8:307**.

Lupyan, D., Leo-Macias, A., & Ortiz, A.R. (2005). A new progressive-iterative algorithm for multiple structure alignment, *Bioinformatics*, **21**, 3255-3263.

Madej,T., Gibrat,J.F., & Bryant,S.H. (1995). Threading a database of protein cores, *Proteins*, **23**, 356-369.

Martin,A.C.R. (2000). The Ups and Downs of protein topology: Rapid comparison of protein structures, *Protein Eng.*, **13**, 829-837.

Matsuda,H., Taniguchi,F., & Hashimoto,A. (1997). An approach to detection of protein structural motifs using an encoding scheme of backbone conformations, Proc. Pac. Symp. Biocomput., pp. 280-291.

McCammon,J.A., Gelin,B.R., & Karplus,M. (1977). Dynamics of folded proteins, *Nature*, **267**, 585-590.

Micheletti,C., Seno,F., & Maritan,A. (2000). Recurrent oligomers in proteins: an optimal scheme reconciling accurate and concise backbone representations in automated folding and design studies, *Proteins*, **40**, 662-674.

Mizuguchi,K., Deane,C.M., Blundell,T.L., & Overington,J.P. (1998). HOMSTRAD: a database of protein structure alignments for homologous families, *Protein Science*, **7**, 2469-2471.

Murzin,A.G., Brenner,S.E. , Hubbard,T. , & Chothia,C. (1995). SCOP: A structural classification of proteins database for the investigation of sequences and structures, *J. Mol. Biol.*, **247**, 536-540.

Needleman,S.B., & Wunsch,C.D. (1970). A general method aplicable to the search for similarity in the amino acid sequence of two proteins. *J. Mol. Biol.*, **48**, 443-454.

Nelder, J.A., & Mead, R. (1965). A simplex method for function minimization, *Computer J.* **7**, 308-313.

Offmann,B., Tyagi,M., & de Brevern,A.G. (2007) Local Protein Structures, *Current Bioinformatics*, **2**, 165-202.

Orengo,C.A., Michie,A.D., Jones,S., Jones,D.T., Swindells,M.B., & Thornton,J.M. (1997). CATH: A hierarchic classification of protein domain structures, *Structure*, **5**, 1093-1108.

Park,B.H. & Levitt,M. (1995). The complexity and accuracy of discrete state models of protein structure, *J. Mol. Biol.*, **249**, 493-507.

Pauling, L., Corey, R.B. (1951). The structure of proteins: Two hydrogen-bonded helical cofigurations of the polypeptide chain, *Proc. Natl. Acad. Sci.*, **37**, 205-211.

Rabiner,L.R. (1989). A tutorial on hidden Markov model and selected applications in speech recognition, *Proc. IEEE*, **77**, 257-285.

Rassias,T.M. (2000). Survey on Classical Inequalities, Kluwer Academic, Dordrecht.

Robson,B., & Pain,R.H. (1974). Analysis of the code relating sequence to conformation in globular proteins: Development of a stereochemical alphabet on the basis of intra-residue information, *Biochem. J.*, **141**, 869-882.

Rooman,M.J., Kocher,J.P. & Wodak,S.J. (1991). Prediction of protein backbone conformation based on seven structure assignments: Influnce of local interactions, *J. Mol. Biol.*, **221**, 961-979.

Sander,O., Sommer,I., & Lengauer,T. (2006). Local protein structure prediction using discriminative models, *BMC Bioinformatics*, **7:14**, 1-13.

Shatsky,M., Nussinov,R., & Wolfson,H. (2002). MultiProt: A multiple protein structural alignment algorithm, in *Lecture notes in computer science 2452*, R.Guigo and D.Gusfield (Eds.), pp. 235-250, Springer Verlag.

Shindyalov,I.N., & Bourne,P.E. (1998). Protein structure alignment by incremental combinatorial extension (CE) of the optimal path, *Protein Eng.*, **11**, 739-747.

Smith,T.F., & Waterman,M.S. (1981). Identification of common molecular subsequences, *J. Mol. Biol.*, **147**, 195-197.

Subbiah,S., Laurents,D.V., & Levitt,M. (1993). Structural similarity of DNA-binding domains of bacteriophage repressors and the globin core, *Curr. Biol.*, **3**, 141-148.

Tang,T., Xu,J., & Li,M. (2005). Discovering sequence-structure motifs from protein segments and two applications, in Proc. Pac. Symp. Biocomput., R.B.Altman, T.A.Jung, T.E.Klein, A.K.Dunker, & L.Hunter (Eds.), pp. 370-81, World Scientific, Singapore.

Täubig,H., Buchner,A., & Griebsch,J. (2006). PAST: fast structure-based searching in the PDB, *Nucleic Acids Res.*, **34**, W20-W23.

Tung,C.T., Huang,J.W., & Yang,J.M. (2007). Kappa-alpha plot derived structural alphabet and BLOSUM-like substitution matrix for rapid search of protein structure database, *Genome Biology*, **8:R31**, 1-16.

Tyagi,M., Gowri,V.S., Srinivasan,N., de Brevern,A.G., & Offmann,B. (2006). A substitution matrix for structural alphabet based on structural alignment of homologous proteins and its applications, *Proteins*, **65**, 32-39.

Umeyama,S. (1991). Least-Squares Estimation of Transformation Parameters between 2-Point Patterns, IEEE Trans. Pattern Anal. & *Machine Intel.*, **13**, 376-380.

Ye,J. & Janardan,R. (2004). Approximate Multiple Protein Structure Alignment Using the Sum-of-Pairs Distance, *J. Comput. Biol.*, **11**, 986-1000.

Zheng,W.M. (2004). Clustering of amino acids for protein secondary structure prediction, *J. Bioinfor. Comp. Biol.*, **2**, 333-342; arxiv.org/pdf/q-bio/0310026.

Zheng,W.M. (2005a). Relation between weight matrix and substitution matrix: motif search by similarity, *Bioinformatics*, **21**, 938-943.

Zheng,W.M. (2005b). Protein secondary structure prediction by combining hidden Markov models and sliding window scores, *Int. J. Bioinfor. Res. Appl.*, **1**, 420-428.

Zheng,W.M. & Liu,X. (2005). A protein structural alphabet and its substitution matrix CLESUM, in *Lecture notes in Bioinformatics* **3680**, C. Priami & A. Zelikovsky (Eds.), pp. 59-67, Springer Verlag; arxiv.org/abs/q-bio/0412046.

In: Protein Conformation: New Research
Editor: Linda B. Roswell, pp. 51-124
ISBN: 978-1-60456-219-4

Chapter 2

The Proteomic Code: A Molecular Recognition Code for Proteins

Jan C. Biro [*]
Homulus Foundation, 88 Howard, #1205, San Francisco
CA 94105, USA

Abstract

The Proteomic Code is a set of rules by which information in genetic material is transferred into the physico-chemical properties of amino acids and determines how individual amino acids interact with each other during folding and in specific protein–protein interactions. The Proteomic Code is part of the redundant Genetic Code. The 25-year-old history of this concept is reviewed from the first independent suggestions by Biro and Mekler, through the works of Blalock, Root-Bernstein, Siemion, Miller and others, followed by the discovery of a Common Periodic Table of Codons and Nucleic Acids in 2003 and culminating in the recent conceptualization of partial complementary coding of interacting amino acids as well as the theory of the nucleic acid-assisted protein folding. A novel cloning method for the design and production of specific and with high affinity reacting proteins (SHARP) is presented. This method is based on the concept of proteomic codes and is suitable for large-scale, industrial production of specifically interacting peptides.

Introduction

Nucleic acids and proteins are the carriers of most (if not all) biological information. This information is complex, well organized in space and time. These two kinds of macromolecules have a polymer structure. Nucleic acids are built from four nucleotides and the proteins are built from 20 amino acids (as basic units). Both nucleic acids and proteins are able to interact with each other and in many cases these interactions are extremely strong (K_d ~10^9–10^{12} M) and extremely specific. The nature and origin of this specificity is well understood in the case of nucleic acid–nucleic acid (NA-NA) interactions (DNA–DNA,

[*] E-mail address: jan.biro@comcast.net

DNA–RNA, RNA–RNA), as is the complementarity of the Watson–Crick (W-C) base pairs. The specificity of NA-NA interactions is determined, undoubtedly, at the basic unit level where the individual bases have a prominent role.

Our most established view on the specificity of protein–protein (P-P) interactions is completely different [1]. In this case the individual amino acids in a particular protein together establish a large 3D structure. This structure has protrusions and cavities, charged and uncharged areas, hydrophobic and hydrophilic patches on its surface, which altogether form a complex 3D pattern of spatial and physico-chemical properties. Two proteins will specifically interact with each other if their complex 3D pattern of spatial and physico-chemical properties fits to each other as a mold to its template or a key to its lock. In this way the specificity of P-P interactions is determined at a level larger than one amino acid (Figure 1).

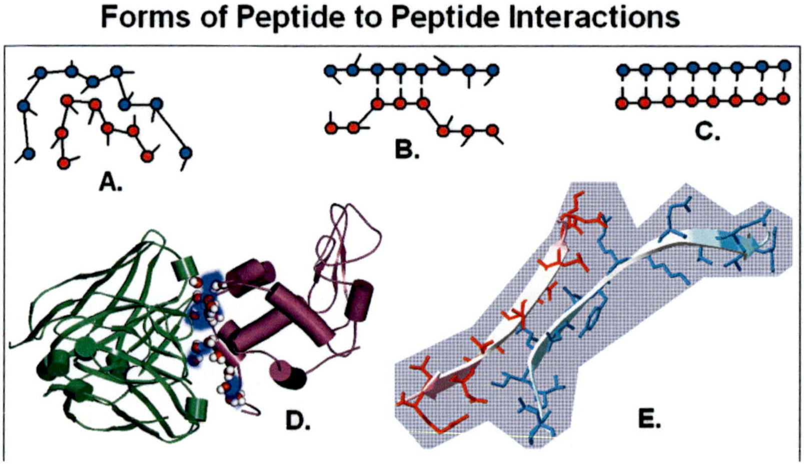

Figure 1. Forms of peptide to peptide interactions. The specificity of interactions between two peptides might be explained in two ways. First, many amino acids collectively form larger configurations (protrusions and cavities, charge and hydropathy fields) which fit each other (A and D). Second, the physico-chemical properties (size, charge, hydropathy of individual amino acids fit each other like "lock and key" (C and E). There are even intermediate forms (B).

The nature of specific nucleic acid–protein (NA-P) interactions is less understood. It is suggested that some bases together form some kind of 3D structure which fits to the 3D structure of a protein (in the case of single-stranded nucleic acids). Alternatively, double-stranded nucleic acid provides a pattern of atoms in the grooves of the double strands, with an atomic pattern that is in some way specifically recognized by nucleo-proteins [2].

Regulatory proteins are known to recognize specific DNA sequences directly through atomic contacts between protein and DNA, and/or indirectly through the conformational properties of the DNA.

There has been ongoing intellectual effort for the last 30 years to explain the nature of specific P-P interactions at the residue unit (individual amino acid) level. This view states that there are individual amino acids which preferentially co-locate with each other in specific P-P contacts and form amino acid pairs which are physico-chemically more compatible with each other than any other amino acid pairs. These physico-chemically highly compatible amino acid pairs are complementary to each other in analogy to the W-C base pair complementarity.

The comprehensive rules describing the origin and nature of amino acid complementarity is called the Proteomic Code.

The History of the Proteomic Code

People from the Past

This is my very subjective selection of scientists, for whom I have great respect and I think they contributed – in one or another way – to the development of the Proteomic Code.

Linus Pauling is regarded as "the greatest chemist who ever lived". *The Nature of the Chemical bond* is fundamental to the understanding of any biological interaction [3]. His works on the protein structure are classics [4]. His unconfirmed DNA model, in contrast to the established model, gives some theoretical ideas on how specific nucleic acid–protein interaction might happen [5,6].

Carl R Woese is famous for defining the Archaea, the third life form on Earth (in addition to bacteria and eucarya). Less known is his "RNA world" hypothesis. This is a theory which proposes that a world filled with RNA (ribonucleic acid)-based life predates current DNA (deoxyribonucleic acid)-based life. RNA, which can store information like DNA *and* catalyze reactions like proteins (enzymes), may have supported cellular or pre-cellular life. Some theories as to the origin of life present RNA-based catalysis and information storage as the first step in the evolution of cellular life.

The RNA world is proposed to have evolved into the DNA and protein world of today. DNA, through its greater chemical stability, took over the role of data storage while proteins, which are more flexible in catalysis through the great variety of amino acids, became the specialized catalytic molecules. The RNA world hypothesis suggests that messenger RNA (mRNA), the intermediate in protein production from a DNA sequence, is the evolutionary remnant of the "RNA world" [7].

His concept of a common origin of nucleic acid and protein "worlds" is very much compatible with the foundation of the Proteomic Code.

Margaret O Dayhoff is the mother of bioinformatics. She was the first who collected and edited the *Atlas of Protein Sequence and Structure* [8] and later introduced statistical methods into protein sequence analyses. Her work was a huge asset and inspiration to my first suggestion of the Proteomic Code [9,10,11].

George Gamow was a theoretical physicist and cosmologist and spent only a few years in Cambridge, UK, but he was there when the structure of DNA was discovered in 1954. He developed the first genetic code, which was not only an elegant solution for the information transfer from DNA to proteins, but at the same time explained how DNA might specifically interact with proteins [12,13,14,15,16,17]. In his mind, the codons were mirror images of the coded amino acids and they had a very intimate relationship with each other. His genetic code proved to be wrong and the nature of specific nucleic acid–protein interactions is still not known, but he remains a strong inspiration (Figure 2) [18,19].

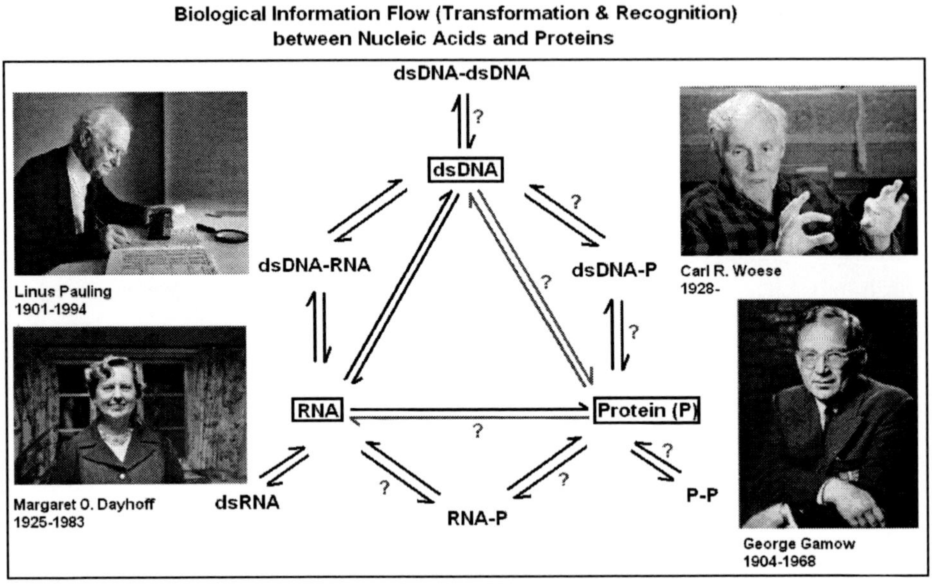

Figure 2. Biological information flow (transformation and recognition) between nucleic acids and proteins. All biological information is stored in nucleic acids (DNA/RNA) and much in proteins (P). The information transfer and interactions between nucleic acids and the formation of double-stranded (ds) forms are well known and understood. However, the exact nature of P-P and P-nucleic acid interactions is still obscure. The works of these four scientists played important roles in much that we know about these information transfers and interactions (subjectively chosen by the author of this article).

First Generation Models for the Proteomic Code

The first generation models (up to 2006) of the novel Proteomic Code are based on perfect codon complementarity coding of interacting amino acid pairs.

Mekler

Mekler described an idea of sense and complementary peptides which may be able to interact specifically, mediated by specific through-space, pair wise interactions between amino acid residues [20]. He suggested that amino acids of specifically interacting proteins, in their specifically interacting domains, are composed of two parallel sequences of amino acid pairs that are spatially complementary to each other, similar to the Watson–Crick base pairs in nucleic acids. The protein nucleic acid analogy in his theory continued and he proposed that these spatially complementary amino acids are coded by reverse-complementary codons (translational reading in the 5′→3′ direction).

It is possible to segregate 64 (the number of different codons, including three stop codons) of all the possible putative amino acid pairs (20×20/2=200) into three non-overlapping groups [21].

Biro

I was also inspired by the complementarity of nucleic acids and developed a theory of complementary coding of specifically interacting amino acids [9,10,11]. I had no knowledge of the publications of Mekler or Idlis (published in two Russian papers). I was also convinced that amino acid pairs coded by complementary codons (no matter if in the same 5′→3′/5′→3′ or opposite 5′→3′/3′→5′ orientations) are somehow special and suggested that these pairs of amino acids might be responsible for the specific intra- and intermolecular peptide interactions.

I developed a method for pair wise computer searching of protein sequences for complementary amino acids and found that these specially coded amino acid pairs are statistically overrepresented in those proteins known to interact with each other. In addition, I was able to find short complementary amino acid sequences within the same protein sequences and concluded that these might play a role in the formation or stabilization of 3D protein structures (Figure 3). Molecular modeling showed the size compatibility of complementary amino acids and that they might form a 5–7 atoms long bridge between the alpha C atoms of amino acids. It was a rather ambitious theory at a time when the antisense DNA sequences were called nonsense, and it was an even more ambitious method when computers were programmed by punch-cards and the protein databases were based on Dayhoff's three volumes of protein sequences [8].

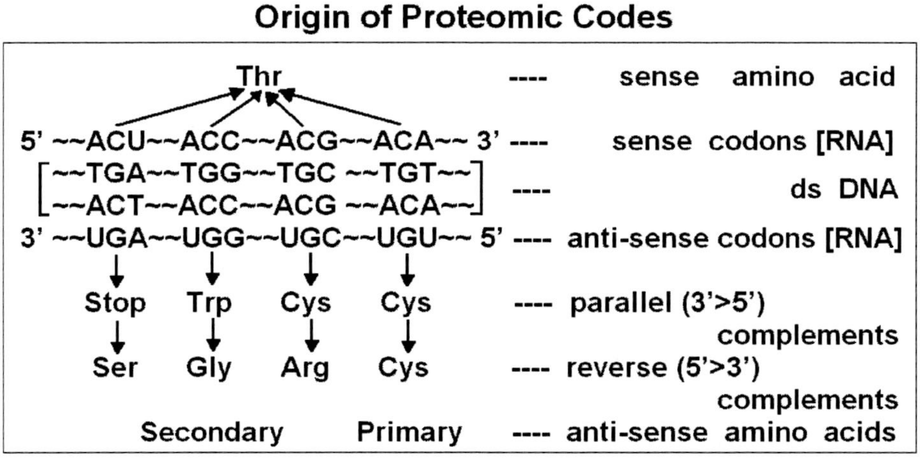

Figure 3. Origin of the Proteomic Code. Threonine (Thr) is coded by 4 different synonymous codons. Complementary triplets are coding different amino acids in parallel (3′→5′) and anti-parallel (5′→3′) readings. Amino acids coded by symmetrical codons are called "primary" and others "secondary" anti-sense amino acids (modified from [9]).

Blalock-Smith

This theory is called the *molecular recognition theory*. Synonymous names are *hydropathy complementarity* or *anti-complementarity theory* and was based on the observation [22] that codons for hydrophilic and hydrophobic amino acids are generally complemented by codons for hydrophobic and hydrophilic amino acids, respectively. This is the case even when the complementary codons are read in the 3′→5′’ direction. Peptides, specified by complementary RNAs, bind to each other with specificity and high affinity [23,24]. The theory turned out to be very fruitful in neuro-endocrine and immune research [25,26].

A very important observation is that antibodies against complementary antibodies also specifically interact with each other. Bost and Blalock [27] synthesized two complementary oligopeptides (i.e. peptides translated from complementary mRNAs, in opposing directions). The two peptides, Leu-Glu-Arg-Ile-Leu-Leu (LERILL), and its complementary peptide, Glu-Leu-Cys-Asp-Asp-Asp (ELCDDD), specifically recognized each other in radioimmunoassay. Antibodies were produced against both peptides. Both antibodies specifically recognized their own antigen. Using radioimmunoassays, anti-ELCDDD antibodies were shown to interact with ^{125}I-labeled anti-LERILL antibodies but not with ^{125}I-labeled control antibodies. More importantly, the interaction of the two antibodies could be blocked using either peptide antigen, but not with control peptides. Furthermore, ^{125}I-labeled anti-LERILL binding to LERILL could be blocked with anti-ELCDDD antibody and vice versa. It was concluded therefore that antibody/antibody binding occurred at or near the antigen combining site, demonstrating that this was an idiotypic/anti-idiotypic interaction.

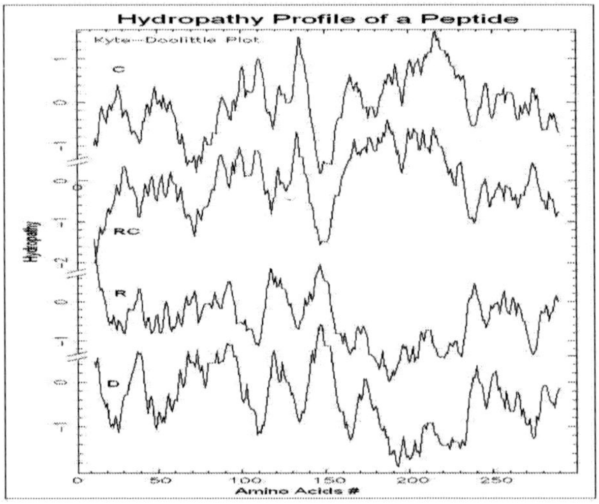

Figure 4. Hydropathy profile of a protein. An artificially constructed nucleic acid sequence was randomized and translated into the four possible directions (D, direct; RC, reverse-complement; R, reverse; C, complement) sequences. The D sequence was designed to contain equal numbers of the 20 amino acids. The hydropathy profile was determined by an online tool (http://www.tcdb.org/progs/hydropathy.php).

This experiment clearly showed the existence (and functioning) of an intricate network of complementary peptides and interactions. Much effort is being made to master this network and use it in protein purification, binding assays, medical diagnosis and therapy.

Recently, Blalock [28]{Blalock JE: Department of Physiology and Biophysics, The University of Alabama at Birmingham. *http://www.physiology. uab.edu/Blalock.htm# Research*} has emphasized that nucleic acids encode amino acid sequences in a binary fashion with regard to hydropathy and that the exact pattern of polar and non-polar amino acids, rather than the precise identity of particular R groups, is an important driver for protein shape and interactions. The perfect codon complementarity behind coding of interacting amino acids is no longer an absolute requirement for his theory.

It is easy to see that there is similarity between hydropathy profiles of the four possible translations of a nucleic acid (Figure 4), but the validity of hydrophobe–hydrophyl interactions remains unanswered.

Root-Bernstein

Another amino acid pairing hypothesis was presented by Root-Bernstein [29,30]]. He focused on whether it was possible to build amino acid pairs meeting standard criteria for bonding. He concluded that it was possible only in 26 cases (of 210 pairs). Of these 26, 14 were found to be genetically encoded by perfect complementary codons (read in the same orientation (5′→3′/3′→5′) while in 12 cases mismatch was found at the wobble position of pairing codons.

Siemion

There is a regular connection between activation energies (measured as enthalpies (ΔH^{++}) and entropies (ΔS^{++}) of activation for the reaction of 18 *N*''-hydroxysuccinimide esters of N-protected proteinaceous amino acids with *p*-anisidine) and the genetic code [31,32,33]

This periodic change of amino acid reactivity within the genetic code led him to suggest a peptide–anti-peptide pairing. This is rather similar to Root-Bernstein's hypothesis.

Miller

Practical use is the best test of a theory. Technologies based on interacting proteins have a significant market in different branches of biochemistry, as well as in medical diagnostics and therapy. The Genetic Therapies Centre (GTC) *(http://www.gtc.ch.ic.ac.uk/about_us/ overview.shtml* at the Imperial College London (*http://www.imperial.ac.uk/P2801.htm*) was founded in 2001 with major financial support from a Japanese company, the Mitsubishi Chemical Corporation, and the UK charity, the Wolfson Foundation) is one of the first academic centers that are openly investing in Proteomic Code-based technologies. With the clear intention for their science "to be used in the marketplace", Andrew Miller, the first director of GTC and co-founder of its first spin-off company, Proteom Ltd (http://www.prosarix.com/default.htm), is making major contributions to this field [34,35,36,37,38].

However, Miller and his colleagues came to realize that the amino acid pairs provided by perfectly complementary codons are not always the best pairs, and deviations from the original design sometimes significantly improved the quality of a protein–protein interaction.

Therefore the current view of Miller is that there are "strategic pairs of amino acid residues that form part of a new, through-space two-dimensional amino acid interaction code (Proteomic Code). The proteomic code and derivatives thereof could represent a new molecular recognition code relating the 1D world of genes with the 3D world of protein structure and function, a code that could shortcut and obviate the need for extensive research into the proteome to give form and function to currently available genomic information (i.e., true functional genomics)" (http://www.gtc.ch.ic.ac.uk/research_interests/code.shtml).

The Proteomic Code and the 3D Structure of Proteins

It is widely accepted that the 3D structure of proteins plays a significant role in their specific interactions and function. The opposite is less obvious, namely that specific and individual amino acid pairs or the sequence of these pairs might determine the foldings of proteins. Complementarity at the amino acid level in the proteins and the corresponding internal complementarity within the coding mRNA (the Proteomic Code) raises the intriguing possibility that some protein folding information is present in the nucleic acids (in addition or within the known and redundant genetic code). A higher propensity of complementarily coded amino acids was observed in real protein sequences than in translations of corresponding random nucleotide sequences [9,10,11]. The internal amino acid complementarity allows the polypeptides coded by complementary codons to retain the secondary structure patterns of the translated strand (mRNA). Thus, genetic code redundancy could be related to evolutionary pressure towards retention of protein structural information in complementary codons and nucleic acid subsequences [39,40,41,42,43,44].

Experimental Evidence

Experiments based on the idea of a Proteomic Code usually start with a well-known receptor-ligand type protein interaction. A short sequence is selected (often <10 amino acids long) which is known or suspected to be involved in the direct contact between the proteins in question (P-P/r). A complementary oligopeptide sequence is derived using the known mRNA sequence of the selected protein epitope, making a reverse complementation of the sequence, translating it and synthesizing it.

The flow of the experiments looks as follows:

(a) choose an interesting peptide
(b) select a short, "promising" oligo-peptide epitope (P);
(c) find the true mRNA of P;
(d) reverse-complement this mRNA;
(e) translate the reverse-complemented mRNA into complementary peptide (P/c);
(f) test P–P/c interaction (affinity, specificity);
(g) use P/c for finding P-like sequences (for histochemistry, affinity purification);
(h) use P/c to generate antibodies (P/c_ab);
(i) test P/c_ab for its interaction with the P-receptor (P/r) and use it for example labeling or affinity purification of P/r;

(j) use P_ab (as well as antibodies to P, P_ab) to find and characterize idiopathic (P_ab–P/c_ab) antibody reactions.

A nice feature of Proteomic-Code based technology is that the amino acid complementarity (information mirroring) does not stop with the P–P/c interaction but continues and involves even the antibodies generated against the original interacting domains; even P_ab–P/c_ab, i.e., antibodies against interacting proteins, will themselves contain interacting domains. They are idiotypes.

Peptides and interactions involved in Proteomic Code-based experiments are summarized in Figure 5.

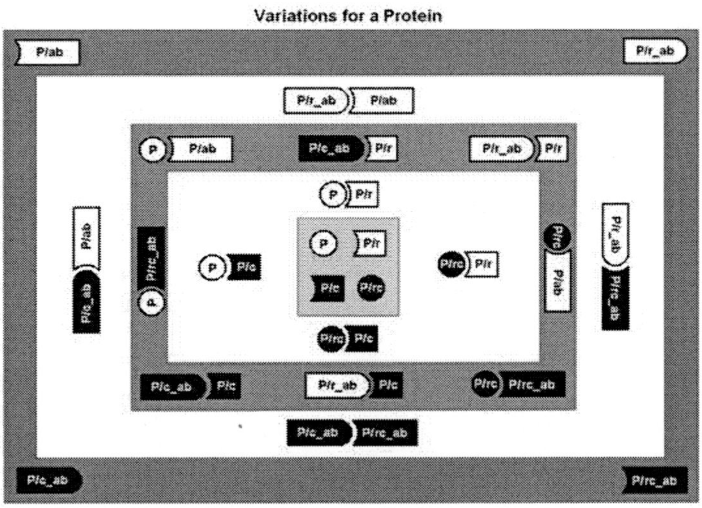

Figure 5. Variations for a protein. Experiments regarding the Proteomic Code are usually designed for the peptides and peptide interactions depicted in this figure. A peptide (P) naturally interacts with its receptor (P/r). Antibodies against this protein (P/ab) and its receptor (P/r_ab) might also be naturally present *in vivo* as part of the immune surveillance or artificially arise. The Proteomic Code gives a method to design artificial oligopeptides (P/c and P/rc) with the potency to interact with the receptor and its ligand. P and P/c as well as Pr and P/rc are expressed from complementary nucleic acid sequences. It is possible to raise antibodies against P/c (P/c_ab) and P/rc (P/rc_ab).

A very nice example of this technology and its potential is given by Bost and Blalock [27] (described above), It is reviewed by Heal et al. [37] and McGuian [45]. Table 1 presents a number of these kinds of experiments.

Some experiments or types of experiments require further attention.

Table I

Experiments related to the Proteomic Code

Target Proteins				Antibodies				ig.-Rec. typ				Ag-Ab type								Ab-Ab type				Comments	Authors	Year	Ref.
p.	P/r	P/c	P/rc	P/ab	P/r_ab	P/c_ab	Prc_ab	P | P/r	P | P/c	Prc | P/r	P/rc | P/c	P | P/ab	P/r | P/r_ab	P/c_ab | P/c	P/rc | Prc_ab	P | P/rc_ab	P/r | P/c_ab	P/c | P/r_ab	P/rc | P/ab	P/ab | P/r_ab	P/ab | P/c_ab	P/r_ab | P/rc_ab	P/c_ab | P/rc_ab				
protein (P2)		x				x			x								x							treatment of allergic neuritis	Araga et al.	1999	1
acetylcholine receptor	x		x		x		x															x		treatment of myastenia gravis	Araga et al.	1993	2
prolactin		x							x																Bajpai et al.	1991	3
endothelin A	x		x							x														antisense homology boxes (AHB)	Baranyi et al.	1995	4
anaphylatoxin (C5a)	x	x						x	x															antisense homology boxes (AHB)	Baranyi et al.	1996	5
interleukin-18		x							x															designe of "receptor-surrogates", lead,	Bhakoo et al.	2004	6
ACTH		x							x																Blalock & Bost	1986	7
trypsin-TMOF	x	x						x	x															Kd~microM	Borovsky et al.	1994	8
Antibodies (Anty-idiotypic		x		x		x						x		x							x			classical	Bost & Blalock	1989	9
ACTH		x							x																Bost et al.	1985	10
endorphin (gamma)		x							x																Bost et al.	1985	11
fibronectin	x	x			x			x	x				x					x						affinity purification of receptor	Brentani et al.	1988	12
substance P	x	x				x		x	x															receptor histochemy	Bret-Dibat et al.	1994	13
Somatostatin		x				x			x								x							receptor identification	Campbell-Thompson et al.	1993	14
enkephalin [Met]	x	x	x			x		x						x			x							opioid pep agonist	Carr et al.	1989	15
endorphin (Gamma)		x				x			x								x							affinity purification of opiate receptor	Carr et al.	1986	16
Laminin	X	X		X				x	x			x													Castronovo et al.	1991	17
fibrinogen (a_IIb_b3)	x		x							x														fibrinogen agonist	Derrick et al.	1997	18
ca-binding site	x		x							x														Ca-mimetic biol effect	Dillon et al.	1991	19
Angiotensin IIa	x	x				x		x	x								x							affinity purification of receptor	Elton et al.	1988	20
tumor necrosis factor		x							x															Kd 10 nM	Fassina et al.	1992	21
Interleukin-2	X	X						x	x															kd micromolar	Fassina et al.	1995	22
interleukin 1 beta		x							x															micromolar range, affinity purification	Fassina & Cassani	1992	23
Endithelin (Big)		x							x																Fassina et al.	1992	24
Endithelin (Big)		x							x															Kd ~ microMn (monomer) 1/100x (multimer)	Fassina et al.	1992	25
c-Raf		x							x															affinity purification of c-raf	Fassina et al.	1989	26
vasopressin [Arg8]	x	x	x					x	x		x													10(-3) to 10(-6) M., affinity purification	Fassina et al.	1989	27
anaphylatoxin (C5a)		x						x	x															shock prevention	Fujita et al.	2004	28
Fibrinogen		x						x	x																Gartner et al.	1991	29
fibrinogen		x							x															inhibirion of platelet aggregation	Gartner & Taylor	1991	30
cystatin C | C4		x							x																Ghiso et al.	1990	31
angiogenin	x		x							x														Kd 44 nM, inhibition of neovascularization	Gho & Chae	1997	32
angiogenin	x		x					x	x															blocks tumor-induced angiogenesis	Gho et al.	1997	33
interleukin 1beta	x	x							x															no interaction with X-ray crystallography	Heal et al.	1999	34
angiotensin II		x							x															Kd ~ 5 microM and 70 nM	Holsworth et al.	1994	35
HIV (T20)		x																						HIV inhibition	Imai et al.	2000	36
interferon-gamma		x		x		x						x		x			x								Johnson et al.	1982	37
antibody (idiotypic,T15)		x							x																Kang et al.	1988	38
vasopressin [Arg8]	x	x				x			0								0							negative results	Kelly et al.	1990	39
vasopressin [Arg8]	x	x						x	x																Knigge et al.	1988	40
Insulin	x	x				x		x	x					x			x							Kd of 3 nM	Knutson	1988	41
amyloid beta		x							x															Prevention of aggregate formation	Kwak et al.	2006	42
vasopressin [Arg8]	x	x						x	x															affinity purification of receptor	Lu et al.	1991	43
HIV proteins		x				x																		HIV diagnostic & vaccine development	Ludwig et al.	2006	44
interferon-beta		x							x															Kd=1.89 x 10(-4) - 1.22 x 10(-5) M, biosensors	Luo et al.	2007	45
Prion protein		x				x			x								x							rceptor recognition	Martins et al.	1997	46
Gastrin		x				x		x	x								x							gastrin antagonism	McGuigan & Campbell	1992	47
Protein	x	x				x		x	x								x							receptor identification	McGuigan JE.	1994	48
angiotensin II		x						x	x															inhibition of biol action	Moore et al.	1989	49
LHRH	X	X				X		x	x			x												LHRH antagonists	Mulchahey et al.	1986	50
anaphylatoxin (C5a)		x							x															prevention of anaphylactic shock	Okada et al.	2007	51
La/SSB		x		x		x						x		x							x			autoimmunity	Papamattheou et al.	2004	52
substance P	x	x				x		x	x															receptor purification	Pascual et al.	1989	53
fibronectin		x				x																			Pasqualini et al.	1989	54
chemoattractant		x							x															biol. Inhibitor	Pfister et al.	1989	55
LHRH		x							x															Kd mM	Root-Bernstein & Westall	1986	56
La/SSB		x		x		x						x		x							x			autoimmunity	Routsias et al.	2003	57
La/SSB		x		x		x						x		x							x			autoimmunity	Routsias et al.	2002	58
La/SSB		x		x		x						x		x							x			autoimmunity	Sakarellos-Daitsiotis et al.	2004	59
synthase (nitric oxide)		x							x																Sautebin et al.	2000	60
interferon-beta		x							x															affinity purification	Scapol et al.	1992	61
endorphin (beta)	x	x				x		x	x								x								Shahabi et al.	1992	62
ribonuclease S		x							x															10(-6) M	Shai et al.	1989	63
ribonuclease		x							x															not saturable binding	Shai et al.	1987	64
Endothelin A	x	x	x					x	x	x														inhibitor	Wu et al.	1997	65

The *antisense homology box*, a new motif within proteins that encodes biologically active peptides, was defined by Baranyi an coworkers, around 1995. They used a bioinformatical method for a genome-wide search of peptides coded by complementary exon sequences. They found that amphiphilic peptides, approximately 15 amino acids in length, and their corresponding antisense peptides exist within protein molecules. These regions (termed antisense homology boxes) are separated by approximately 50 amino acids. They concluded that because many sense–antisense peptide pairs have been reported to recognize and bind to each other, antisense homology boxes may be involved in folding, chaperoning and oligomer formation of proteins. The frequency of peptides in antisense homology boxes was 4.2 times higher than expected from random sequences ($p<0.001$) [46].

They successfully confirmed their suggestion by experimental results. The antisense homology box-derived peptide CALSVDRYRAVASW, a fragment of human endothelin A receptor, proved to be a specific inhibitor of endothelin peptide (ET-1) in a smooth muscle relaxation assay. The peptide was also able to block endotoxin-induced shock in rats. The finding of endothelin receptor inhibitor among antisense homology box-derived peptides indicates that searching proteins for this new motif may be useful in finding biologically active peptides [47,48,49].

A bioinformatics experiment, similar to Baranyi's, was performed by Segerstéen et al. [50]. The hypothesis that nucleic acids, which code specifically interacting receptor and ligand proteins, contain complementary sequences was tested. Human insulin mRNA (HSINSU) contained 16 sequences which were 23.8±1.4 nucleotides long and were complementary to the insulin receptor mRNA (HSIRPR, 74.8±1.9% complementary matches, $p<0.001$ compared to randomly occurring matches). However, when examining 10 different nucleic acids (coding proteins not interacting with the insulin receptor), 81 additional sequences were found which were also complementary to HSIRPR. Although the finding of short complementary sequences was statistically highly significant, we concluded that this is not specific for nucleic acid coding of specifically interacting proteins.

There are two kinds of antisense technologies based on the complementarity of nucleic acids: (a) when the production of a protein is inhibited by an oligonucleotide sequence which is complementary to its mRNA; this is a pre-translational modification and it usually requires transfer of nucleic acids into the cells; (b) when the biological effect of an already complete protein is inhibited by another protein which is translated from its complementary mRNA; this is a post-translational modification and does not block the synthesis of a protein.

Many experiments (Table 1) indicate that antisense proteins inhibit the biological effects of a protein. This suggests the possibility of *antisense protein therapy*. The P–P/c reaction is in many respects similar to the antigen–antibody reaction, therefore the potential of antisense protein therapy is expected to be similar to the potential of antibody therapy (passive immunization against proteinaceous toxins, such as bacterial toxins, venoms, etc.). However, antisense peptides are much smaller than antibodies (MW as little as ~1000 Da compared to IgG ~155 kDa). This means that antisense proteins are easy to manufacture in vitro; antibodies are produced in living animals (with non-human species characteristics). However, the small size is expected to have the disadvantage of a higher K_d and a shorter biological half-life.

Immunization with complementary peptides gains antibodies (P/c_ab) as any other protein. These antibodies contain a domain that is similar to the original protein (P) and specifically bind to the receptor of the original protein (P/r). This property is effectively used for affinity purification or immuno-staining of receptors. The P/c_ab is able to mimic or antagonize the in vivo effect of P by binding to its receptor. This property has the desired potential to treat protein-related disease, such as many pituitary gland-related diseases. A vision might be to treat, for example, pituitary dwarfism, with immunization to growth hormone complementary peptide (GH/c) or Type I diabetes with immunization to insulin/c peptide.

Reverse But not Complementary Sequences

The biochemical process of transcription and translation is unidirectional, 5′→3′ and reversion does not exist. However, there are many examples of sequences which are present in the genome (in addition to direct reading) even in reverse orientation, and if expressed (in the usual 5′→3′ direction) they produce mRNA and proteins that are virtually reversely transcribed and reversely translated.

An interesting observation is that direct and reverse proteins often have very similar binding properties and related biological effects even if their sequence homology is very low (<20%). For example, growth hormone-releasing hormone (GHRH) and the reverse GNRH specifically bind to the GHRH receptor on rat pituitary cells and to polyclonal anti-GHRH antibody in ELISA and RIA procedures although they share only 17% sequence similarity and they are antagonists in in vitro stimulation of GH RNA synthesis and in vitro and in vivo GH release from pituitary cells [51].

The same phenomenon is observed in complementary sequences. A peptide expressed by complementary mRNA often specifically interacts with proteins expressed by the direct mRNA and it does not matter if they are read in the same or opposite directions. A possible explanation is that many codons are actually symmetrical and have the same meaning in both directions of reading. The physico-chemical properties of amino acids is preferentially determined by the 2nd (central) codon letter [52] and in that way the physico-chemical pattern of direct and reverse sequences remains the same. In addition, I found that protein structural information is also carried by the 2nd codon letters [53].

Controversies Regarding the Original Proteomic Codes

All proteomic codes before 2006 required perfect complementarity, even if it was noticed that the "biophysical and biological properties of complementary peptides can be improved in a rational and logical manner where appropriate" [36] .

- Expression of the antisense DNA strand was simply not accepted before large scale genome sequences confirmed that genes are about equally distributed on both strands of DNA in every organisms containing dsDNA.
- Spatial complementarity is difficult to imagine between longer amino acid sequences, because the natural, internal folding of proteins will prohibit it in most cases.
- The usual is that residues with the same polarity show attraction to each other, because hydrophobes like the hydrophobe environment as well as lipohobes prefer lipohobe neighbors. Amphipathic interactions seems to be artificial for most chemists.
- Only complementary (but not reversed) sequences were also found effective. This requires 3′→5′ translation, which is normally prohibited.
- Inconsistency of the results; it works for some proteins but not for others; necessity to improve results, e.g., "M-I pair mutagenesis" [36].
- The protein 3D structure and interactions are thought to be arranged on a larger scale than individual amino acids.

- The number of possible amino acid pairs is 20×20/2=200. The number of perfect codons is 64, i.e., about a third of the number expected. This means that two-thirds of amino acid pairs are impossible to code by perfectly complementary codons.
  are these amino acid pairs not derived from complementary codons at all?
  are these amino acid pairs derived from imperfectly complementary codons?

Development of the Second Generation Proteomic Code

What did we learn about the Proteomic Code during its first 25 years (1981–2006)? My first and most important lesson is that I realize how terribly wrong it was (and is) to believe in scientific dogmas, like sense vs nonsense DNA strands. It is almost unbelievable today that many of us were able to see difference between two perfectly symmetrical and structurally identical strands.

We were able to provide multiple independent and convincing evidence that the concept of the Proteomic Code is valid. At the same time we had to understand that the first concepts – based on perfect complementarity of codons behind interacting amino acids – are imperfect.

There is protein folding information in the nucleic acids – in addition or within the redundant genetic code – but it is unclear how is it expressed and interpreted to form the 3D protein structure.

A major physico-chemical property, the hydropathy of amino acids is coded by the codons. Proteins translated from direct and reverse as well as from complement and reverse-complement strands results in the same hydropathic profile. This is possible only if the amino acid hydropathy is related to the second, central codon letter.

There is a clear indication that some biological information exists in multiple complementary (mirror) copies: DNA–DNA/c→RNA–RNA/c→protein–protein/c→IgG–IgG/c.

Some theoretical considerations and research that led to the suggestion of the 2nd generation Proteomic Codes are now reviewed.

Construction of a Common Periodic Table of Codons and Amino Acids

The Proteomic Code revitalizes a very old dilemma and dispute represented by Carl Woese and Francis Crick about the origin of the genetic code. Is there any logical connection between any properties of an amino acid on the one hand and any properties of its genetic code on the other?

Carl Woese [54] argued that there was stereochemical matching, i.e., affinity, between amino acids and certain triplet sequences. He therefore proposed that the genetic code developed in a way that was very closely connected to the development of the amino acid repertoire, and that this close biochemical connection is a fundamental of specific protein–nucleic acid interactions.

Table II. Common Periodic Table of Codons and Amino Acids

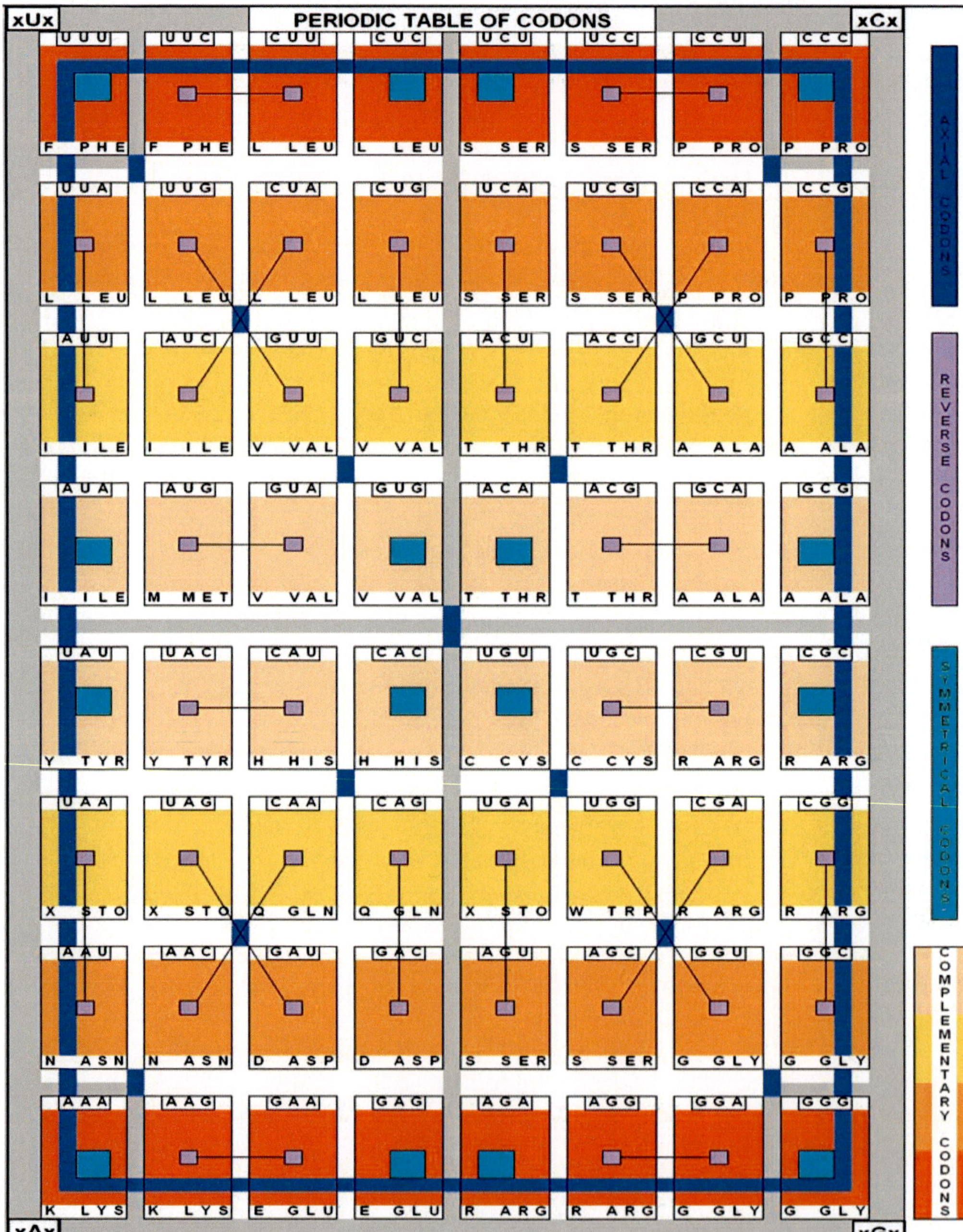

Table II. Continued

Table III. Effect of a Single Codon Residue on the Structure of the Amino Acids

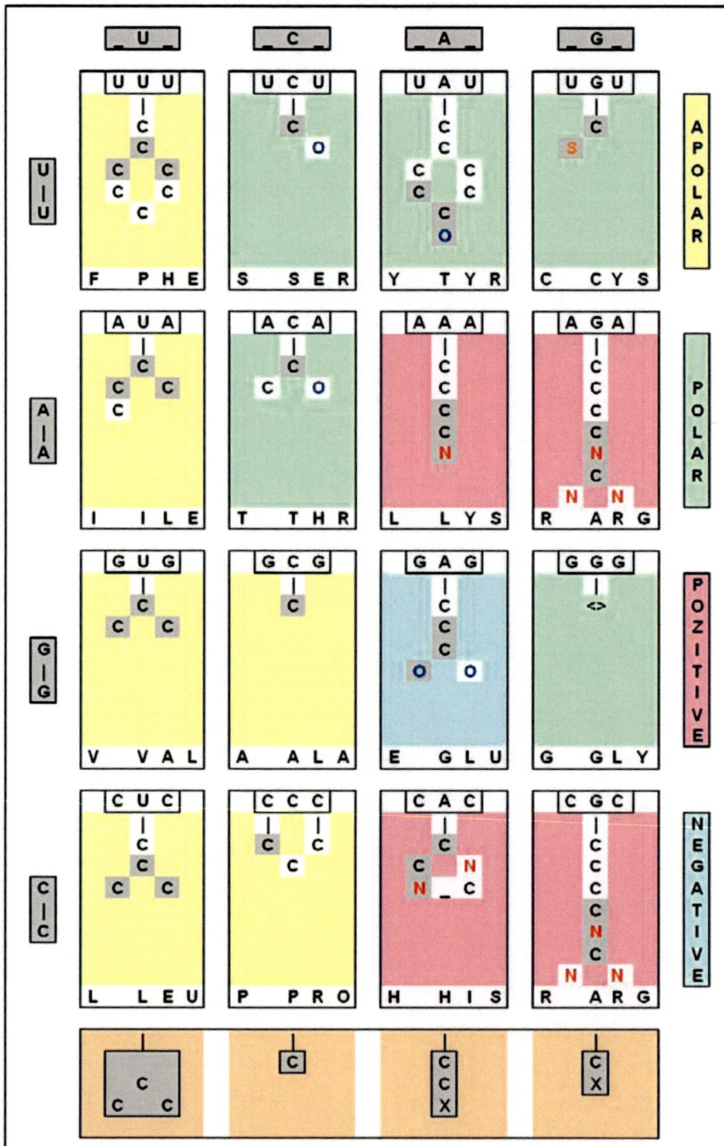

Crick [55] considered that the basis of the code might be a "frozen accident", with no underlying chemical rationale. He argued that the canonical genetic code evolved from a simpler primordial form that encoded fewer amino acids. The most influential form of this idea, "code co-evolution," proposed that the genetic code co-evolved with the invention of biosynthetic pathways for new amino acids [56].

A periodic table of codons has been designed where the codons are in regular locations. The table has four fields (16 places in each) one with each of the four nucleotides (A, U, G, C) in the central codon position. Thus, AAA (lysine), UUU (phenylalanine), GGG (glycine) and CCC (proline) are positioned in the corners of the fields as the main codons (and amino acids). They are connected to each other by six axes. The resulting nucleic acid periodic table shows perfect axial symmetry for codons. The corresponding amino acid table also displaces periodicity regarding the biochemical properties (charge and hydropathy) of the 20 amino acids, and the position of the stop signals. Table 2 emphasizes the importance of the central nucleotide in the codons, and predicts that purines control the charge while pyrimidines determine the polarity of the amino acids.

In addition to this correlation between the codon sequence and the physico-chemical properties of the amino acids, there is a correlation between the central residue and the chemical structure of the amino acids. A central uridine correlates with the functional group –$C(C)_2$–; a central cytosine correlates with a single carbon atom, in C_1 position; a central adenine coincides with the functional groups –CC=N and –CC=O, and finally a central guanine coincides with the functional groups –CS, –C=O, and C=N, and with the absence of a side chain (glycine). (Table III)

I interpret these results as a clear cut answer for the Woese vs Crick dilemma: there is a connection between the codon structure and the properties of the coded amino acids. The second (central) codon base is the most important determinant of the amino acid property. It explains why the reading orientation of translation has so little effect on the hydropathy profile of the translated peptides. Note, that 24 of 32 codons (U or C in the central position) code apolar (hydrophobe) amino acids, while only 1 of 32 codons (A or G in the central position) codes non-apolar (non-hydrophobe, charged or hydrophilic) amino acids. It explains why complementary amino acid sequences have opposite hydropathy, even if the binary hydropathy profile is the same.

The Physico-Chemical Compatibility of Amino Acids in the Proteomic Code

The complementary coding of two amino acids is not a guarantee per se for the special co-location (or interaction) of these amino acids within the same or between two different peptides. Some kind of physico-chemical attraction is also necessary. The most fundamental properties to consider are, of course, the size, charge and hydropathy. Mekler and myself suggested size compatibility [9,10,11,20], obviously under the influence of the known size complementarity of the Watson–Crick base pairs. Blalock emphasizes the importance of hydropathy, better say amphipathy (which makes some scientists immediately antipathic). Hydrophobe residues like other hydrophobe residues and hydrophilic residues like hydrophilic residues. Hydrophyl likes hydrophobe like water likes oil.

Amino Acid Co-locations

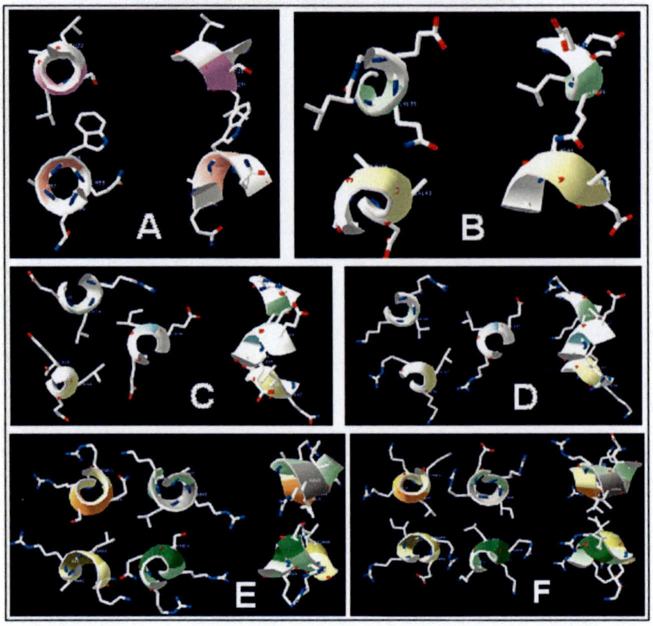

Figure 6. Amino acid co-locations. Randomly selected amino acid contacts from real proteins. The interactions between amino acid residues from 2 (A, B) 3 (C, D) and 4 (E, F) parallel alpha helices are perpendicular to the peptide backbones (helices). The orientation of residues show large variation; some are located side-by-side, others are end-to-end.

Visual studies on the 3D structures of proteins give some ideas how interacting interfaces look (Figure 6):

- the interacting (collocating) sequences are short (1–10 amino acid long);
- the interacting (collocating) sequences are not continuous; there are many mismatches;
- the orientation of the collocating residues is often not the same (not parallel);
- the contact between collocating residues might be side-to-side or top-to-top.

This is clearly a different picture than the base-pair interactions in a dsDNA spiral. Alpha-helices and beta-sheets are regular structures, which make their amino acid residues periodically ordered. Many residues are parallel to each other and W-C-like interactions are not impossible. But is it really the explanation for specific residue interactions?

SeqX

The interacting residues of protein and nucleic acid sequences are close to each other; they are co-located. Structure databases (e.g., Protein Data Bank, PDB and Nucleic Acid Data Bank, NDB) contain all the information about these co-locations; however it is not an easy task to penetrate this complex information. We developed a JAVA tool, called SeqX, for this purpose [57].

The SeqX tool is useful to detect, analyze and visualize residue co-locations in protein and nucleic acid structures. The user:

(a) selects a structure from PDB;
(b) chooses an atom that is commonly present in every residue of the nucleic acid and/or protein structure(s);
(c) defines a distance from these atoms (3–15 Å).

The SeqX tool then detects every residue that is located within the defined distances from the defined "backbone" atom(s); provides a dot-plot-like visualization (residues contact map), and calculates the frequency of every possible residue pair (residue contact table) in the observed structure. It is possible to exclude ±1–10 neighbor residues in the same polymeric chain from detection, which greatly improves the specificity of detections (up to 60% when tested on dsDNA). Results obtained on protein structures show highly significant correlations with results obtained from the literature (p<0.0001, n=210, four different subsets). The co-location frequency of physico-chemically compatible amino acids is significantly higher than is calculated and expected in random protein sequences (p<0.0001, n=80) (Figure 7).

These results gave a preliminary confirmation of our expectation that physico-chemical compatibility exists between co-locating amino acid pairs. Our findings do not support any significant dominance of amphipathic residue interactions in the examined structure.

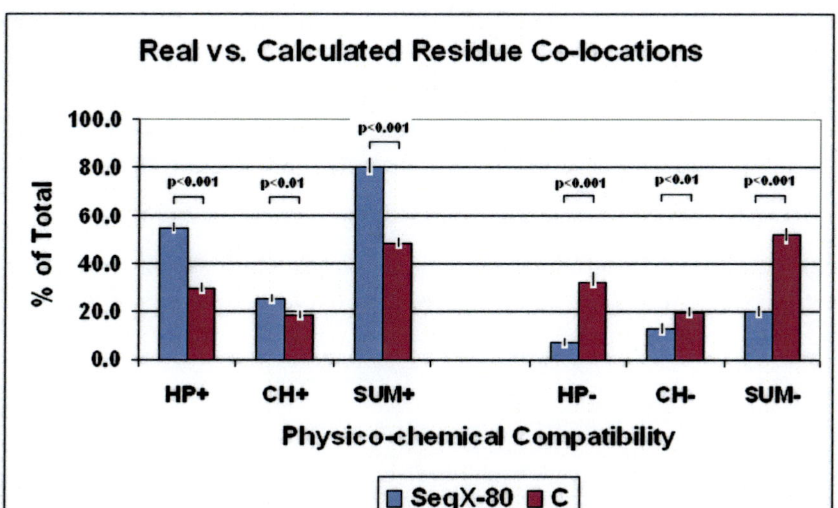

Figure 7. Real vs calculated residue co-locations (from [57] {. The relative frequency of real residue co-locations was determined by SeqX in 80 different protein structures and compared to the relative frequency of calculated co-locations in artificial, random protein sequences (C). The 200 possible residue pairs provided by the 20 amino acids were grouped into 4 subgroups regarding their physico-chemical compatibility to each other, i.e., favored (+) and un-favored (–) regarding hydrophobicity and charge. (HP+, hydrophobe–hydrophobe and lipophobe–lipophobe; HP–, hydrophobe–lipophobe; CH+, positive–negative and hydrophobe–charged; CH–: positive–positive and negative–negative and lipophobe–charged interactions). The bars represent the mean±SEM (n=80 for real structures and n=10 for artificial sequences). Student's t-test was applied to evaluate the results.

Availability and Requirements

Available from http://janbiro.com/Downloads.html SeqX, Java J2SE Runtime Environment 5.0 (available from [see Additional file 1] http://www.sun.com) and at least a 1 GHz processor and with a minimum 256 Mb RAM. Source codes are available from the authors.

Amino Acid Size, Charge, Hydropathy Indices and Matrices for Protein Structure Analysis

It was necessary to have a closer look at the physico-chemical compatibility of co-locating amino acids [58].

We indexed the 200 possible amino acid pairs for their compatibility regarding the three major physico-chemical properties – size, charge and hydrophobicity – and constructed size, charge and hydropathy compatibility indices (SCI, CCI, HCI) and matrices (SCM, CCM, HCM). Each index characterized the expected strength of interaction (compatibility) of two amino acids by numbers from 1 (not compatible) to 20 (highly compatible). We found statistically significant positive correlations between these indices and the propensity for amino acid co-locations in real protein structures (a sample containing total 34,630 co-locations in 80 different protein structures): for HCI, $p<0.01$, $n=400$ in 10 subgroups; for SCI, $p<1.3E–08$, $n=400$ in 10 subgroups; for CCI, $p<0.01$, $n=175$). Size compatibility between residues (well known to exist in nucleic acids) is a novel observation for proteins (Figure 8).

We tried to predict or reconstruct simple 2D representations of 3D structures from the sequence using these matrices by applying a dot-plot-like method. The location and pattern of the most compatible subsequences was very similar or identical when the three fundamentally different matrices were used, which indicates the consistency of physico-chemical compatibility. However, it was not sufficient to choose one preferred configuration between the many possible predicted options (Figure 9).

Indexing of amino acids for major physico-chemical properties is a powerful approach to understanding and assisting protein design. However, it is probably insufficient itself for complete ab initio structure prediction.

Anfinsen's Thermodynamic Principle and the Proteomic Code

The existence of physico-chemical compatibility of co-locating amino acids already on the residue level is, of course, a necessary support for the Proteomic Code. At the same time, it raises the possibility that the protein structure might be predicted from the primary amino acid sequence (de novo, ab initio prediction) and the location of physico-chemically compatible amino acid residues in the sequence. This idea is in line with a dominating statement about protein folding. Anfinsen's thermodynamic principle states that all information necessary to form a 3D protein structure is present in the protein sequence [60].

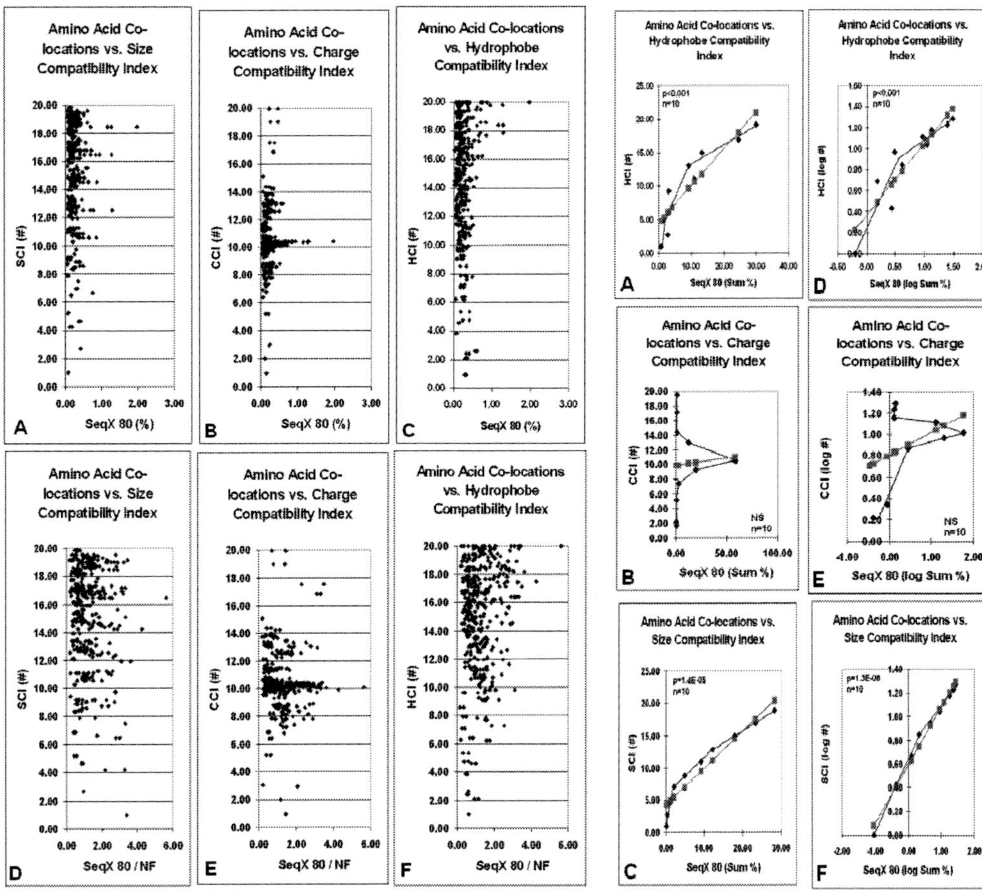

Figure 8. (a) Amino acid co-locations vs size, charge, and hydrophobe compatibility indexes (modified from [59]. Average propensity of the 400 different amino acid co-locations in 80 different protein structures (SeqX 80) are plotted against size, charge and hydrophobe compatibility indexes (SCI, CCI, HCI). The original "row" values are indicated in (A–C). The SeqX 80 values were corrected by the co-location values, which are expected only by chance in proteins where the amino acid frequency follows the natural codon frequency (NF) (D–F). (b) Amino acid co-locations vs size, charge, and hydrophobe compatibility indexes in major subgroups (modified from [59]. Data presented in Figure 1 were divided into subgroups and summed (Sum). The group averages are connected by the blue lines while the pink symbols and lines indicate the calculated linear regression.

Attempts were made to use the three different matrices in a dot plot to predict the place and extent of the most likely residue co-locations. This semi-quantitative method indicated that the three very different matrices located very similar residues and subsequences as potential co-location places. No single diagonal line was seen in the dot-plot matrices, which is the expected signature of sequence similarity (or compatibility in our case). Instead, block-like areas indicated the place and extent of predicted sequence compatibilities. It was not possible to reconstruct a real map of any protein 2D structure (Figure 9) [61].

This experience with the indexes provides arguments for as well as against Anfinsen's theorem. The clear-cut action of basic physico-chemical laws at residue level is well in line with the lowest free energy requirement of the law of entropy. Furthermore, this obvious presence of physico-chemical compatibility is easy to understand, even from an evolutionary perspective. In evolution, sequence changes more rapidly than structure; however, many sequence changes are compensatory and preserve local physico-chemical characteristics. For example, if, in a given sequence, an amino acid side chain is particularly bulky with respect to the average at a given position, this might have been compensated in evolution by a particularly small side chain in a neighboring position, to preserve the general structural motif. Similar constraints might hold for other physico-chemical quantities such as amino acid charge or hydrogen bonding capacity [62].

We were not able to reconstruct any structure using our indexes. There are massive arguments against Anfinsen's principle:

1) The connection between primary, secondary and tertiary structure is not strong, i.e., in evolution, sequence changes more rapidly than structure. Structure is often conserved in proteins with similar function even when sequence similarity is already lost (low structure specificity to define a sequence). Identical or similar sequences often result in different structures (low sequence specificity to define a structure).
2) An unfolded protein has a vast number of accessible conformations, particularly in its side chains of residues. Entropy is related to the number of accessible conformations. This problem is known as the Levinthal paradox [63].
3) The energy profile characteristics of native and designed proteins are different. Native proteins usually show a unique and less stable profile, while designed proteins show lower structural specificity (many different possible structures) but high stability [64].
4) The entropy minimum is a statistical minimum. The conformation entropy change of the whole molecule is the sum of local (residue level) conformation entropy changes and it permits the co-existence of many different local conformation variations. It is doubtful whether structural variability (heterogeneity, instability) is compatible with the function (homogeneity, stability) of a biologically active molecule.

The present experiments do not decide the "fate" of Anfinsen's dogma; however, they show that the number of possible co-locating places is too large, and searching this space poses a daunting optimization problem. It is not realistic to expect the ab initio prediction of only one single structure from one primary protein sequence. The development of a prediction tool for protein structure (like an mfold for nucleic acids [65]{4 Zuker M. 2003}), which provides only a few hundred most likely (thermodynamically most optimal) structure suggestions per protein sequence, seems to be closer. It is likely that SCM, CCI and HCM (or similar matrices) will be essential elements of these tools.

Additional folding information might be necessary (in addition to that carried in the protein primary sequence) to be able to create a unique protein structure. Such information is suspected to be present in the redundant genetic code [66,67,68] and chaperons [39,70,71].

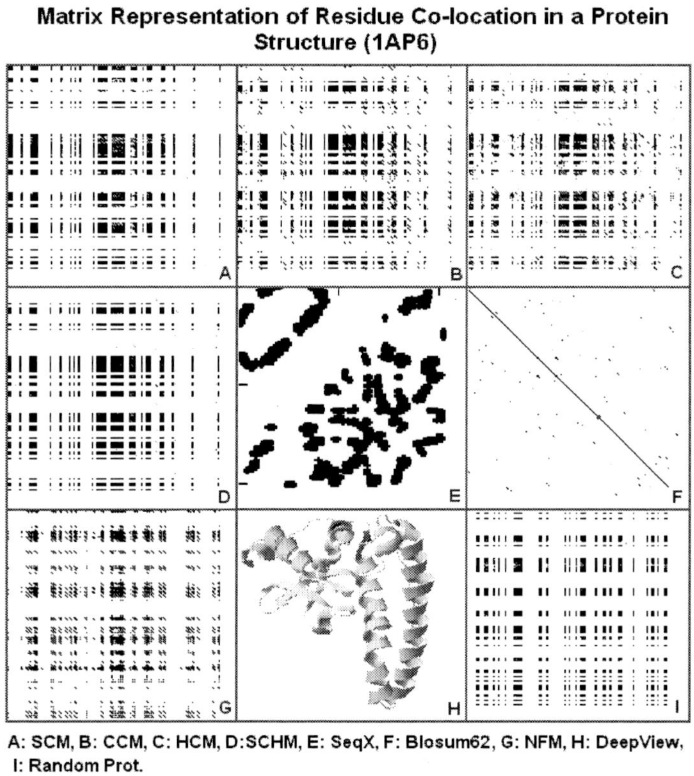

Figure 9. Matrix representation of residue co-locations in a protein structure (1AP6) (modified from [58.]). A protein sequence (1AP6) was compared to itself with DOTLET using different matrices, SCM (A), CCM (B), HCM (C), the combined SCHM (D) and NFM (G) and Blosum62 (F). Comparison of randomized 1AP6 using SCHM is seen in (I). The 2D (SeqX Residue Contact Map) and 3D (DeepView/Swiss-PDB Viewer) views of the structure are illustrated in (E) and (H). The black/gray parts of the dot-plot matrices indicate the respective compatible residues, except the Blosum62 comparison (F), where the diagonal line indicates the usual sequence similarity. The dot-plot parameters are otherwise the same for all matrices.

Protein Structure and the Functional Asymmetry of the Codons

The consequence of codon redundancy is that the third codon base is ambiguously defined; the same amino acid might be coded by more than one codon differing at the wobble base positions. These codons are called synonymous codons.

The Common Periodic Table of Codons and Amino Acids revealed another difference between codon bases, namely, that the second codon base is most significant in determining the physico-chemical properties of amino acids.

I wanted to find more evidence supporting the connection between amino acid (protein) and codon (nucleic acid structure). All these facts provide vital evidence for the Proteomic Code.

The protein folding problem has been one of the grand challenges in computational molecular biology. The problem is to predict the native three-dimensional structure of a protein from its amino acid sequence. It is widely believed that the amino acid sequence contains all the necessary information to make up the correct three-dimensional structure, since protein folding is apparently thermodynamically determined; i.e., given a proper environment, a protein will fold up spontaneously. This is called Anfinsen's thermodynamic principle [7].

The thermodynamic principle has been confirmed many times on many different kinds of proteins in vitro. Critics says that the in vivo chemical conditions are different from those in vitro, the correct folding is determined by interactions with other molecules (chaperons, hormones, substrate, etc.) and protein folding is much more complex than re-naturation of de-natured poly amino acids. The fact that many naturally occurring proteins fold reliably and quickly to their native state, despite the astronomical number of possible configurations, has come to be known as Levinthal's Paradox [8].

Anfinsen's principle was formulated in the 1960s using purely chemical experiments and a lot of intuition. Today, we have a lot of sequences and structures available to establish a logical and understandable link between sequence, structure and function. But it is still not possible to correctly predict the structure (or a range of possible structures) purely from the sequence, ab initio and in silico [9].

There are two potential, external sources of additional and specific protein folding information: (a) the chaperons (other proteins that assist in the folding of proteins and nucleic acids [71]; and (b) the protein coding nucleic acid sequences themselves (which are templates of the protein syntheses, but are not defined as chaperons).

The idea that the nucleotide sequence itself could modulate translation and hence affect co-translational folding and assembly of proteins has been investigated in a number of studies [72,73]. Studies on the relationships between synonymous codon usage and protein secondary structural units are especially popular [68,74,75]. The genetic code is redundant (61 codons code 20 amino acids) and as many as 6 synonymous codons can code the same amino acid (Arg, Leu, Ser). The "wobble" base has no effect on the meaning of most codons but still the codon usage (wobble usage) is not randomly defined [76,77] and there are well known, stable species-specific differences in the codon usage. It seems to be logical to search for some meaning (biological purpose) of the wobble bases and try to associate them with protein folding.

Another observation concerning the code redundancy dilemma is that there is a widespread selection (preference) for local RNA secondary structure in protein coding regions [78]. A given protein can be encoded by a large number of distinct mRNA species, potentially allowing mRNAs to simultaneously optimize desirable RNA structural features in addition to their protein coding function. The immediate question is whether there is some logical connection between the possible, optimal RNA structures and the possible, optimal biologically active protein structures.

Single-stranded RNA molecules can form local secondary structures through the interactions of complementary segments. W-C base pair formation lowers the average free energy, dG, of the RNA and the magnitude of change is proportional to the number of base pair formations. Therefore the free folding energy (FFE) is used to characterize the local complementarity of nucleic acids [78]. The free folding energy is defined as $\mathrm{FFE}=(\mathrm{d}G_{\mathrm{shuffled}} - \mathrm{d}G_{\mathrm{native}})/L\times 100$, where L is the length of the nucleic acid, i.e., free energy difference between native and shuffle (randomized) nucleic acids per 100 nucleotides. Higher positive values indicate stronger bias towards secondary structure in the native mRNA, and negative values indicate bias against secondary structure in the native mRNA.

We used a nucleic acid secondary structure predicting tool, mfold [65] to obtain d*G* values and the lowest d*G* was used to calculate the FFE. mfold also provided the folding energy dot-plots, which are very useful to visualize the energetically most favored structures in a 2D matrix.

A series of JAVA tools were used: SeqX to visualize the protein structures in 2D as amino acid residue contact maps [57]; SeqForm for selection of sequence residues in predefined phases (every third in our case) [79]; SeqPlot for further visualization and statistical analyses of the dot-plot views [80]; Dotlet as a standard dot-plot viewer [81]. Structural data were downloaded from PDB [82], NDB [83], and from a wobble base oriented database called Integrated Sequence–Structure Database (ISSD) [84].

Structures were generally randomly selected regarding species and biological function (a few exceptions are mentioned below). Care was taken to avoid very similar structures in the selections. A propensity for alpha helices was monitored during selection and structures with very high and very low alpha helix content were also selected to make sure of a wide range of structural representation.

Linear regression analyses and Student's *t*-tests were used for statistical analyses of the results.

Observations were made on human peptide hormone structures. This group of proteins is very well defined and annotated, the intron–exon boundaries are known and even intron data are easily accessible. The coding sequences were phase separated by SeqForm into three subsequences, each containing only the 1st, 2nd or 3rd letters of the codons. Similar phase separation was made for intronic sequences immediately before and after the exon. There are, of course, no known codons in the intronic sequences, therefore we continued the same phase that we applied for the exon, assuming that this kind of selection is correct and maintained the name of the phase denotation even for non-coding regions. Subsequences corresponding to the 1st and 3rd codon letters in the coding regions had significantly higher FFEs than subsequences corresponding to the 2nd codon letters. No such difference was seen in non-coding regions (Figure 10).

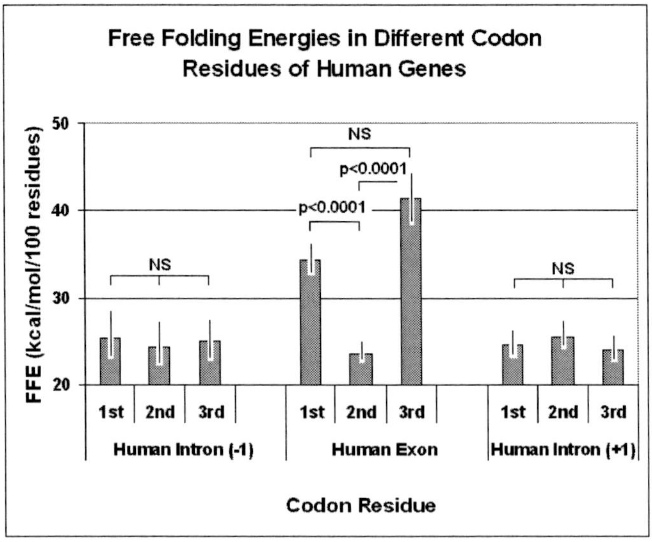

Figure 10. Free folding energies (FFE) in different codon residues of human genes. The coding sequences (exons) of 18 human hormone genes and the preceding (–1) and following (+1) sequences (introns) were phase separated into three subsequences each corresponding to the 1st, 2nd and 3rd codon positions in the coding sequence. The d*G* values were determined by mfold and the FFE was calculated. Each bar represents the mean±SEM, n=18.

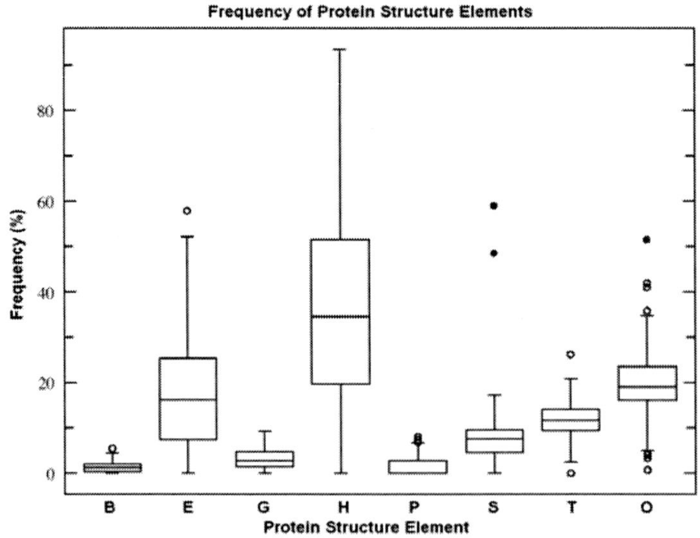

Figure 11. Frequency of protein structure elements. Box plot representation of protein secondary structure elements in 81 structures. L=317±20 (mean±SEM, n=81). Secondary structure codes: H, alpha helix; B, residue in isolated beta bridge; E, extended strand, participates in beta ladder; G, 3-helix (3/10 helix); I, 5 helix (pi helix); P, polyproline type II helix (left-handed); T, hydrogen bonded turn; S, bend.

In a larger selection of 81 different protein structures, the corresponding protein and coding sequences were used to extend the observations. These 81 proteins represented different (randomly selected) species and different (also randomly selected) protein functions and therefore the results might be regarded as more generally valid. The propensity of different secondary structure elements was recorded (as annotated in different databases) (Figure 11).

The proportion of alpha helices varied from 0 to 90% in the 81 proteins and showed a significant negative correlation to the proportion of beta sheets (Figures 12 and 13).

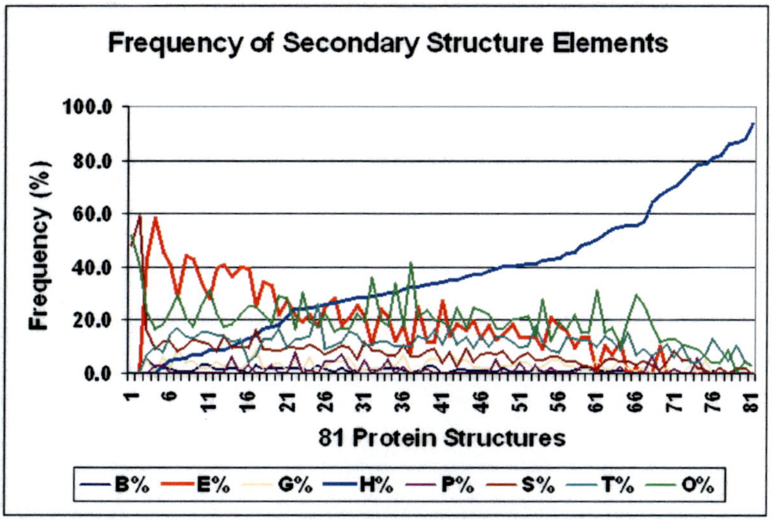

Figure 12. Frequency of secondary structure elements. The propensity of different structural elements in 81 different proteins is shown. L=317±20 (mean±SEM, n=81). Secondary structure codes: H, alpha helix; B, residue in isolated beta bridge; E, extended strand, participates in beta ladder; G, 3-helix (3/10 helix); I, 5 helix (pi helix); P, polyproline type II helix (left-handed); T, hydrogen bonded turn; S, bend.

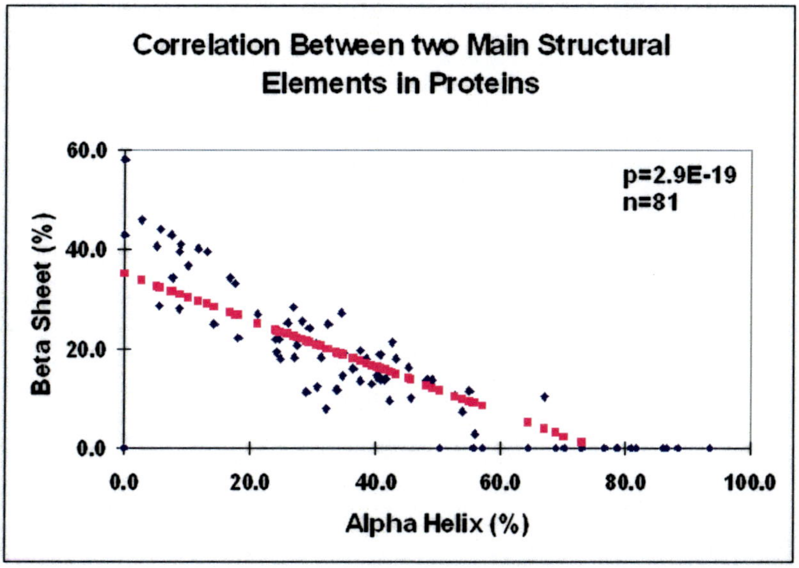

Figure 13. Correlation between two main structural elements in proteins. Data were taken from Figure 3 (H, alpha helix; E, beta sheet).

The original observation made on human hormone proteins, that significantly more free folding energy is associated with the 1st and 3rd codon residues than with the 2nd was confirmed on a larger and more heterogeneous protein selection. A significant difference showed up even between the 1st and 3rd residues in this larger selection (Figure 14).

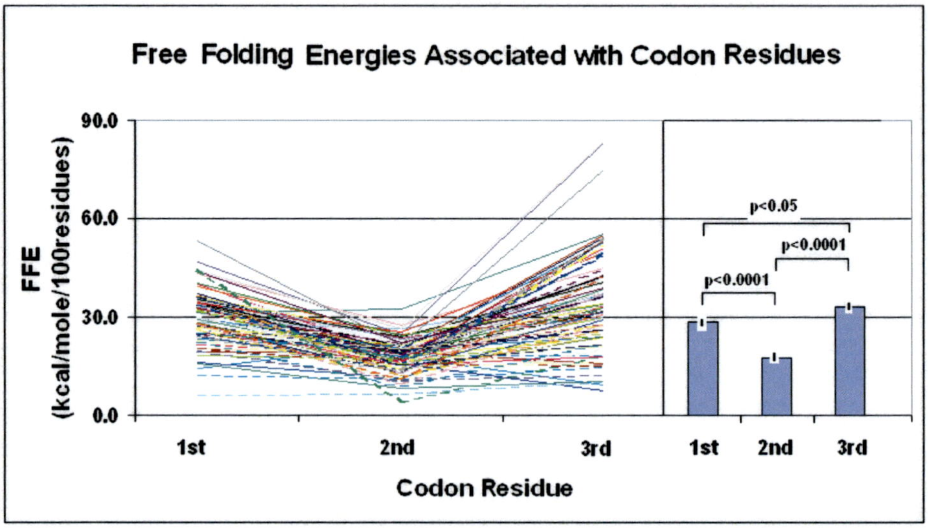

Figure 14. Free folding energies associated with codon residues (Free folding energies (FFE) were determined in phase-selected subsequences of 81 different protein coding nucleic acids. The lines indicate individual values (left part of the figure), while the bars (right part of the figure) indicate the mean±SEM (n=81).

There is a correlation between the protein structure and the FFE associated with codon residues. The correlation is negative between FFE associated with the 2nd (middle) codon residues and the alpha helix content of the protein structure. The correlation is especially significant when the FFE ratios are compared to the helix/sheet ratios (Figures 15 and 16).

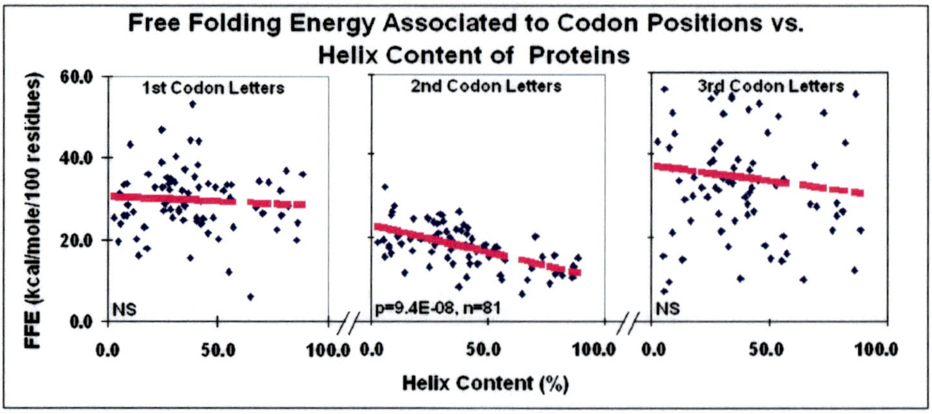

Figure 15. Free folding energy associated with codon positions vs helix content of proteins. Linear regression analyses; pink symbols represent the linear regression line.

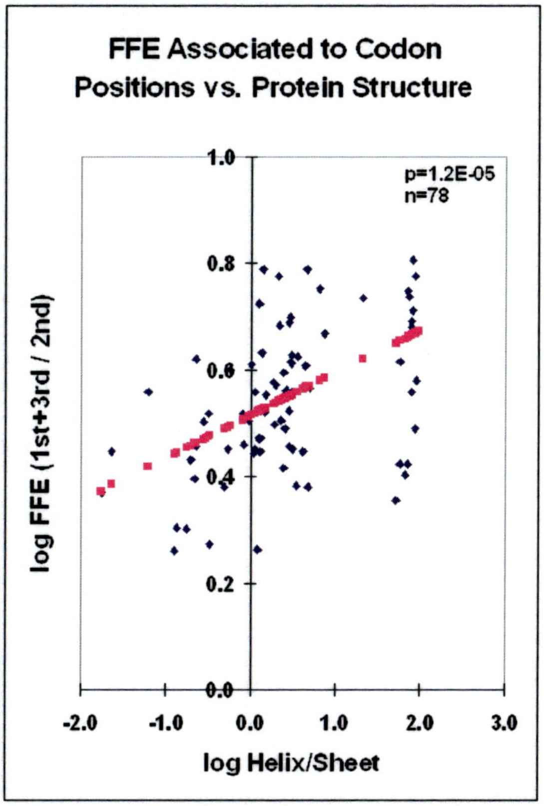

Figure 16. FFE associated with codon positions vs protein structure. Same data as in Figure 6 after calculating ratios and log transformation. Linear regression analyses; pink symbols represent the linear regression line.

The alpha helix is the most abundant structure element in proteins. It shows negative correlation to the frequency of the second most prominent protein structure, the beta sheet. The propensity of some amino acids and the major physico-chemical characteristics (charge and polarity) shows significant correlation (positive or negative) to this structural feature. We include statistical analyses of alpha helix content and other protein characteristics to show the complexity behind the term "alpha helix" and to show the insecurity in interpreting any correlation to this structural feature (Figures 17 and 18). Detailed analyses of these data are outwith the scope of this review.

Higher FFE in subsequences of 1st and 3rd codon residues than in the 2nd indicates the presence of a larger number of complementary bases at the right positions of these subsequences. However, this might be the case only because the first and last codons form simpler subsequences and contain longer repeats of the same nucleotide than the 2nd codons. This would not be surprising for the 3rd (wobble) base but would not be expected for the 1st residue, even though it is known that the central codon letters are the most important to distinguish between amino acids (as shown in the in the Common Periodic Table of Codons and Amino Acids [52]. It is more significant to see that the FFEs in 1st and 3rd residues are additive and together they represent the entire FFE of the intact mRNA (Figure 19).

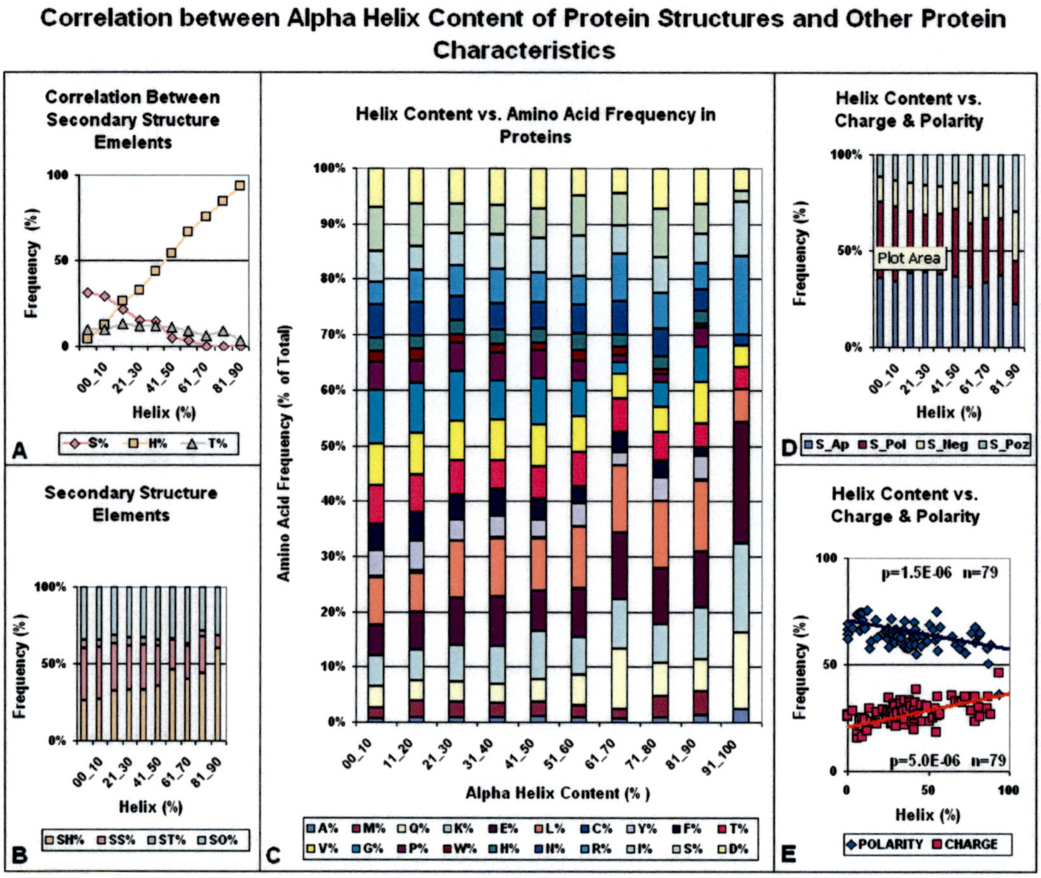

Figure 17. Correlation between alpha helix content of protein structure and other protein characteristics. The alpha helix content of 80 protein structures was compared to the frequency of other major structural elements (A,B), the frequency of individual amino acids (C) and the frequency of charged and hydrophobe residues (D,E). (A) The correlation between helix (H), beta sheet (S) and turn (T); (B) the proportions between the sum of helices (SH), beta strands (SS), turns (ST) and all other structural elements (TO). (D) The proportion between the sums of apolar (S_Ap), polar (S_Pol), negatively charged (S_Neg) and positively charged (S_Poz) amino acids. (E) The linear regression analyses correlation between helix content and the percentage of polar+apolar (Polarity) and positively+negatively charged (Charge) residues.

Correlation Between Frequency of Individual Amino Acids and the Main Secondary Structure Elements in Proteins

	A%	C%	D%	E%	F%	G%	H%	I%	K%	L%	M%	N%	P%	Q%	R%	S%	T%	V%	W%	Y%
HELIX (%)	3.E-04	1.E-04	2.E-01	1.E-05	2.E-06	2.E-06	4.E-01	2.E-01	1.E-04	9.E-04	2.E-02	1.E-01	2.E-06	1.E-04	8.E-02	2.E-01	3.E-03	4.E-03	1.E-02	6.E-02
HELIX CORRELATION	POS	NEG	NS	POS	NEG	NEG	NS	NS	POS	POS	POS	NS	NEG	POS	NS	NS	NEG	NEG	NEG	NS
SHEET (%)	2.E-03	6.E-05	6.E-02	3.E-05	5.E-06	6.E-06	5.E-01	3.E-01	1.E-03	5.E-05	4.E-02	6.E-02	9.E-02	5.E-03	4.E-02	5.E-02	8.E-04	8.E-03	2.E-01	4.E-02
SHEET CORRELATION	NEG	POS	NS	NEG	POS	POS	NS	NS	NEG	NEG	NEG	NS	NS	NEG	NEG	NS	POS	POS	NS	POS
TURN (%)	9.E-01	2.E-01	1.E-01	2.E-01	1.E-01	2.E-03	5.E-01	3.E-01	7.E-02	4.E-02	6.E-01	7.E-01	1.E-03	1.E-01	1.E-01	3.E-01	9.E-02	9.E-02	1.E-01	5.E-01
TURN CORRELATION	NS	NS	NS	NS	NS	POS	NS	NS	NS	NEG	NS	NS	POS	NS	NS	NS	NS	NS	NS	NS

A-Y: One letter code of amino acid, POS: positive correlation, NEG: negativer correlation, NS: not significant, linear regression analyses, n=79

Figure 18. Correlation between frequency of individual amino acids and the main secondary structure elements in proteins. See text for explanation.

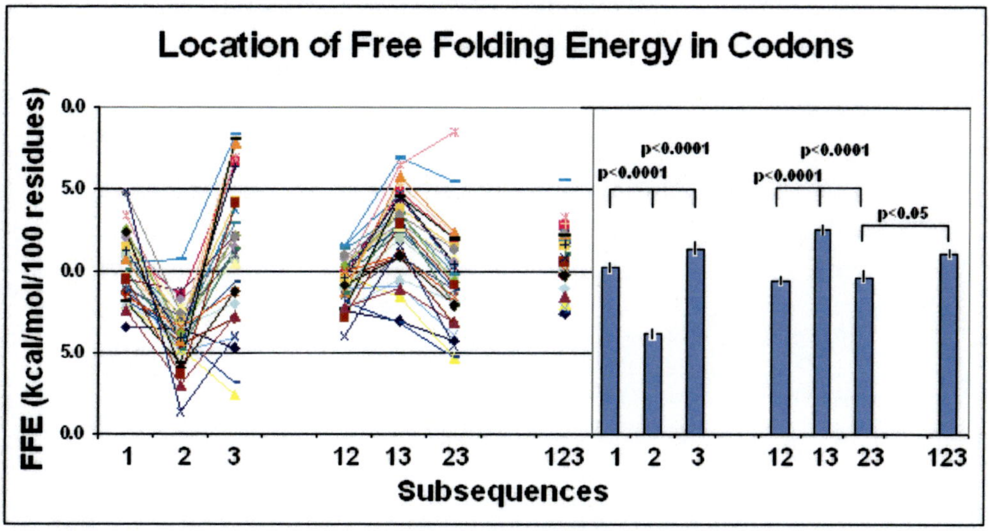

Figure 19. Location of free folding energy in codons. Free folding energies (FFE) were determined in phase-selected subsequences of 31 different protein coding nucleic acids. The original intact RNA contained the intact three-letter codons (123). Subsequences were constructed by periodical removal of one letter from the codon and maintaining the other two (12, 13, 23) or removing two letters and maintaining only one (1, 2, 3). The lines indicate individual values (left), while the bars (right) indicate the mean±SEM (n=31).

Higher FFE at the 1st and 3rd codon positions than at 2nd indicates that the number of complementary bases (a-t and g-t) is higher in the 1st and 3rd subsequences than in the second. This is possible only if more complementers are in 1-1, 1-3, 3-1, 3-3 position pairs than in 1-2, 2-1, 2-3, 3-2 position pairs. We wanted to know whether the 1-1, 3-3 (complement) or the 1-3, 3-1 (reverse-complement) pairing is more predominant.

The length of phase-separated nucleic acid subsequences (l) is a third of the original coding sequence (L). The number of different residues (a, t, g, and c) varies at different codon positions (1, 2, 3).

$$a1+u1+g1+c1=a2+t2+g2+c2=a3+t3+g3+c3=l=L/3$$

The highest number of complementary pairs might occur in the 1st subsequence if

$$a1=t1, \quad g1=c1 \quad \text{and} \quad a1/t1=g1/c1=1$$

If, for example, a1>t1, g1=c1 an excess of unpaired a1 occurs and a1/t1>g1/c1=1 and the possible FFE in subsequence 1 will be less. Following the same logic for other pairs in other subsequences we can conclude that any deviation from a/t=g/c=1 is suboptimal regarding the FFE. Counting the different residue ratios and combinations indicates that the optima are obtained if the residues in the first position form W-C pairs with residues at the third positions (1–3) and vice versa (3–1). This is consistent with the expectation that mRNA will form local loops, in which the direction of more or less double stranded sequences is reversed and (partially) complemented (Figure 20).

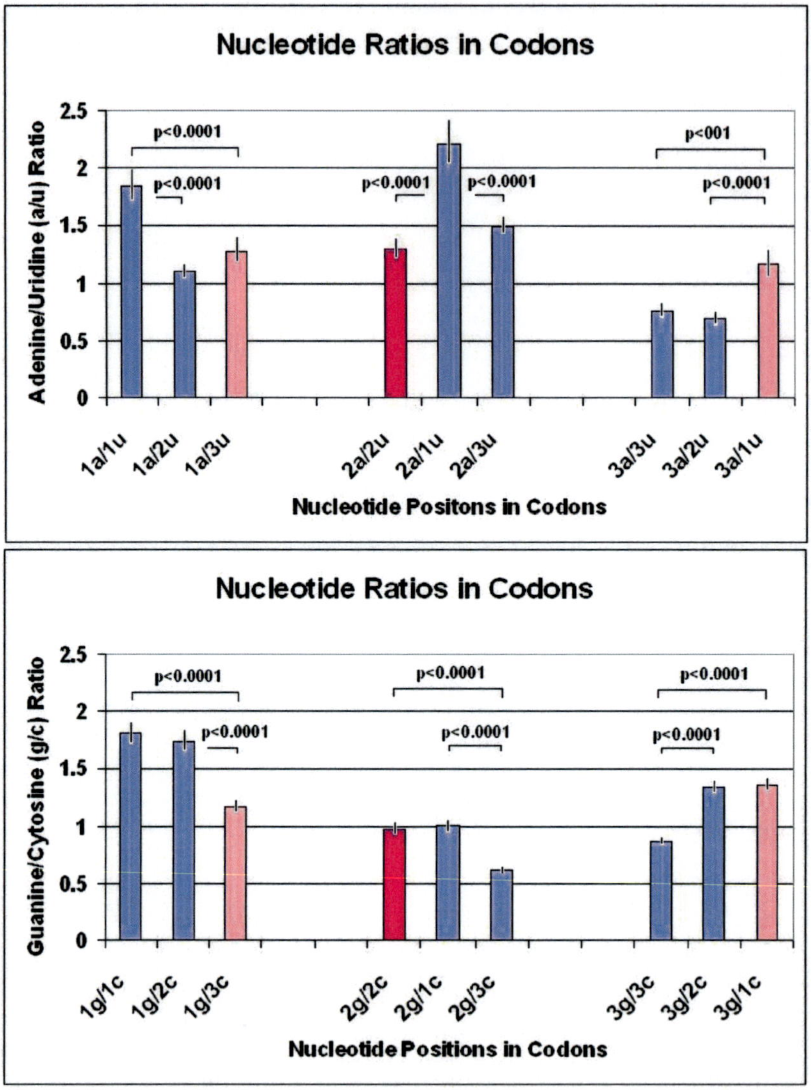

Figure 20. Nucleotide ratios in codons. The number of the 4 different nucleotide bases was counted at the 1st, 2nd and 3rd codon positions in 30 different protein coding RNA sequences. The ratio of the Watson–Crick pairs at different codon positions are indicated by bars (±SEM, n=30). Ideally, the ratio of complementary base pairs is ~1.0. This ideal situation was mostly satisfied when one of the complementary bases was located at codon position 1 with the other at codon position 3 (pink) or both complements at codon position 2 (violet).

Comparison of the Protein and mRNA Secondary Structures

The partial (suboptimal) reverse complementarity of codon-related positions in nucleic acids suggested some similarity between protein structures and the possible structures of the coding sequences. This possibility was examined by visual comparison of 16 randomly selected protein residue contact maps and the energy dot-plots of the corresponding RNAs. We could see similarities between the two different kinds of maps (Figure 21). However, this type of comparison is not quantitative and statistical evaluation is not directly possible.

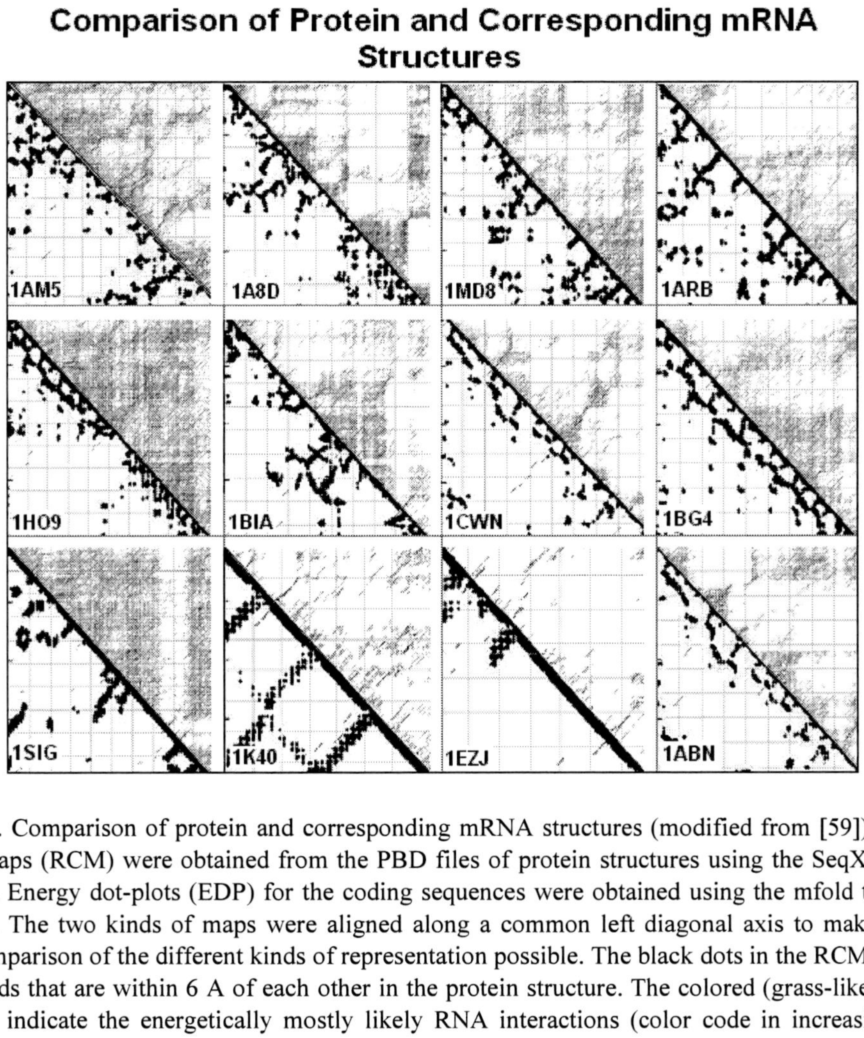

Figure 21. Comparison of protein and corresponding mRNA structures (modified from [59]). Residue contact maps (RCM) were obtained from the PBD files of protein structures using the SeqX tool (left triangles). Energy dot-plots (EDP) for the coding sequences were obtained using the mfold tool (right triangles). The two kinds of maps were aligned along a common left diagonal axis to make an easy visual comparison of the different kinds of representation possible. The black dots in the RCMs indicate amino acids that are within 6 A of each other in the protein structure. The colored (grass-like) areas in the EDPs indicate the energetically mostly likely RNA interactions (color code in increasing order: yellow, green red, black).

Another similar, but still not quantitative, comparison of protein and coding structures was performed on four proteins that are known to have very similar 3D structures but their primary structure (the sequence) is less than 30% similar, as well as the sequence of their mRNA. These four proteins are examples of the fact that the tertiary structure of proteins is much more conserved than the amino acid sequence. We asked the question whether this is true for the RNA structures and sequence. We found that there are signs of conservation even of the RNA secondary structure (as indicated by the energy dot-plots) and there are similarities between the protein and nucleic acid structures (Figure 22).

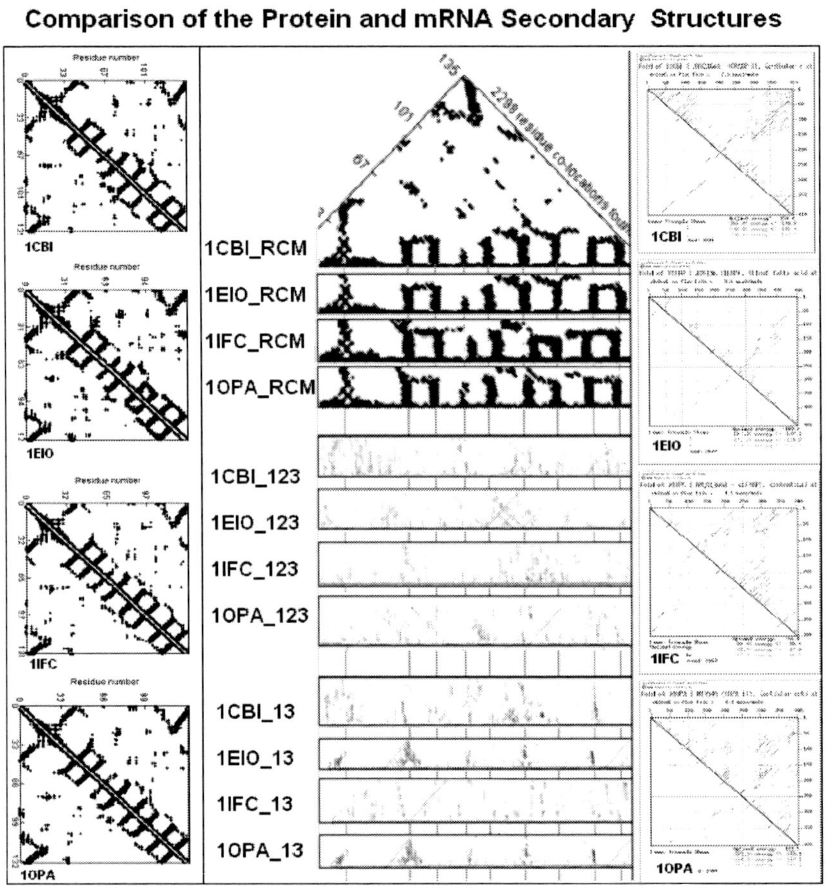

Figure 22. Comparison of the protein and mRNA secondary structures (modified from [59]). Residue contact maps (RCM) were obtained from the PBD files of 4 protein structures (1CBI, 1EIO, 1IFC, 1OPA) using the SeqX tool (left column). Energy dot-plots (EDP) for the coding sequences were obtained using the mfold tool (right column). The left diagonal portion of these two kinds of maps was compared in the central part of the figure. Blue horizontal lines in the background correspond to the main amino acid co-location sites in the RCM. Intact RNA (123) as well as subsequences containing only the 1st and 3rd codon letter (13) are compared. The black dots in the RCMs indicate amino acids that are within 6 A of each other in the protein structure. The colored (grass-like) areas in the EDPs indicate the energetically most likely RNA interactions (color code in increasing order: yellow, green red, black).

The similarity between mRNA and the coded protein secondary structure is an unexpected, novel observation. The 21/64 redundancy of the genetic code gives a 441/4.096 codon pair redundancy for every amino acid pair. It means that every amino acid pair might be coded by ~9 different codon pairs (some complementary but most not). The similarity between protein and their mRNA structures indicates extensive complementary coding of co-locating amino acids. The possible number of codon variations and possible nucleic acid structures behind a protein sequence and structure is very large (Figure 23) as well as the corresponding folding energies (dG, the stability of the mRNA).

Independency of Sequence and Structure of a Protein and its mRNA

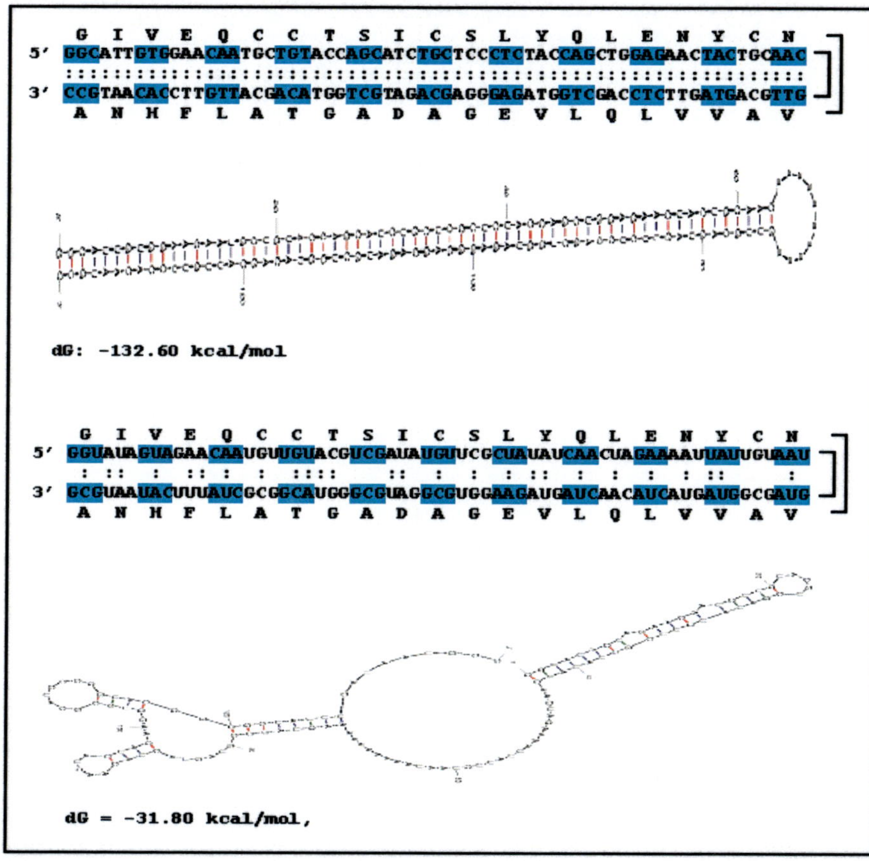

Figure 23. Independency of sequence and structure of a protein and its mRNA. The amino acids in a U-shaped protein structure are coded by complementary codons (rule PC1). The nucleic acid structure is uniform and the folding energy is –132 kcal/mol. Exactly the same amino acid sequence might be coded by non- or only partially complementary codons, which will fundamentally alter the mRNA structure and increase the folding energy to –31 kcal/mol (less stable). The nucleic acid structures were generated by mfold and dGs were calculated by the same program.

Complementary Codes vs Amino Acid co-Locations

Comparisons of the protein residue contact map with the nucleic acid folding maps suggest similarities between the 3D structures of these different kinds of molecules. However, this is a semi-quantitative method.

A more direct statistical support might be obtained by analyzing and comparing residue co-locations in these structures. Assume that the structural unit of mRNA is a tri-nucleotide (codon) and the structural unit of the protein is the amino acid. The codon may form a secondary structure by interacting with other codons according to the W-C base complementary rules, and contribute to the formation of a local double helix. The 5′-A1U2G3-3′ sequence (Met, M codon) forms a perfect double string with the 3′-U3A2C1-5′ sequence (His, H codon, reverse and complementary reading). Suboptimal complexes are 5′-

A1X2G3-3′ partially complemented by 3′-U3X2C1-5′ (AAG, Lys; AUG, Met; AGG, Arg; ACG, Pro; and CAU, His; CUU, Leu; CGU, Arg; CCU, Pro, respectively).

Our experiments with FFE indicate that local nucleic acid structures are formed under this suboptimal condition, i.e., when the 1st and 3rd codon residues are complementary but the 2nd is not. If this is the case, and there is a connection between nucleic acid and protein 3D structure, one might expect that the 4 amino acids coded by 5′-A1X2G3-3′ codons will preferentially co-locate with another 4 amino acids coded by 3′-U3X2C1-5′ codons. We have constructed 8 different complementary codon combinations and found that the codons of co-locating amino acids are often complementary at the 1st and 3rd positions and follow the D-1X3/RC-3X1 formula but not the 7 other formulas (Figures 24 and 25).

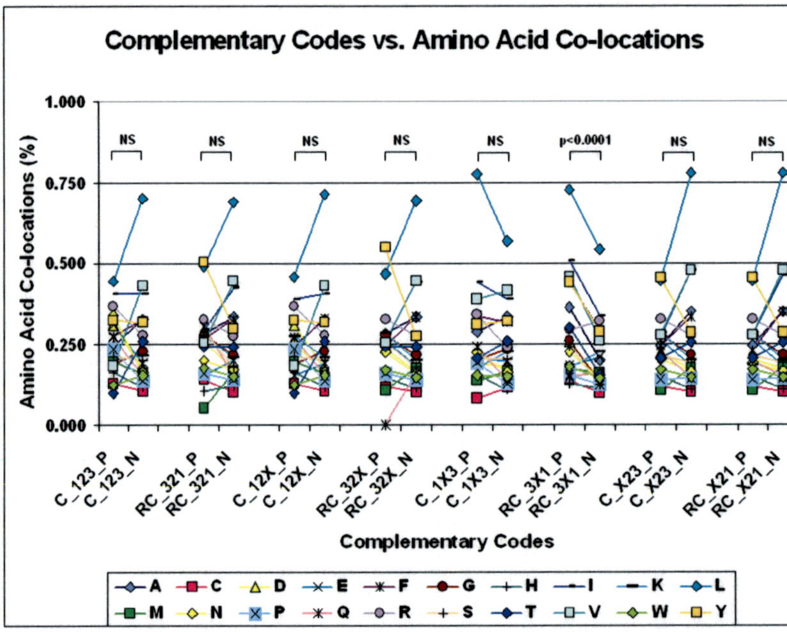

Figure 24. Complementary codes vs amino acid co-locations (modified from [59]). The propensity of the 400 possible amino acid pairs was monitored in 81 different protein structures with the SeqX tool. The tool detected co-locations when two amino acids were within 6 A of each other (neighbors on the same strand were excluded). The total number of co-locations was 34,630. Eight different complementary codes were constructed for the codons (2 optimal and 6 suboptimal). In the two optimal codes, all three codon residues (123) were complementary (C) or reverse complementary (RC) to each other. In the suboptimal codes, only two of three codon residues were C or RC to each other (12, 13, 23), while the third was not necessarily complementary (X). (For example, Complementary Code RC_1X3 means that the first and third codon letters are always complementary, but the not the second and the possible codons are read in reverse orientation. The 400 co-locations were divided into 20 subgroups corresponding to 20 amino acids (one of the co-locating pairs), each group containing the 20 amino acids (corresponding to the other amino acid in the co-locating pair). If the codons of the amino acid pairs followed the predefined complementary code the co-location was regarded as positive (P); if not, the co-location was regarded as negative (N). Each symbol represents the mean frequency of P or N co-locations corresponding to the indicated amino acid. Paired Student's t-test, n=20.

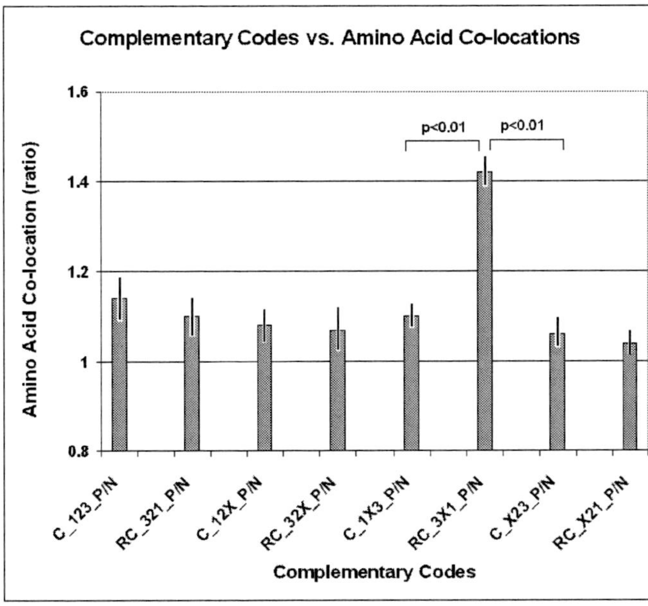

Figure 25. Complementary codes vs amino acid co-locations (ratios) (modified from [59]). The ratio of positive (P) and negative (N) co-locations was calculated on data from Figure 13. Each bar represents the mean±SEM (n=20).

These special amino acid pairs and their frequency are indicated and summarized in a matrix (Figure 26).

Complementary Codes vs. Amino Acid Co-locations

SeqX 80			1st	G	T	G	G	T	G	C	A	A	CT	A	A	C	C	AC	AT	A	G	T	T
			2nd	C	G	A	A	T	G	A	T	A	T	T	A	C	A	G	CG	C	T	G	A
			3rd	X	CT	CT	AG	CT	X	CT	ACT	AG	XAG	G	CT	X	AG	AGX	CTX	X	X	G	CT
1st	2nd	3rd	AA	A	C	D	E	F	G	H	I	K	L	M	N	P	Q	R	S	T	V	W	Y
G	C	X	A	7.5	1.4	2.6	2.9	4.1	3.9	1.5	6.3	3.3	9.9	2.8	2.6	2.0	2.5	3.3	3.7	4.5	7.5	1.6	4.7
T	G	CT	C	1.4	3.1	0.7	0.6	1.3	1.4	0.4	1.2	0.7	1.7	0.5	0.9	0.8	0.4	0.7	1.1	1.3	1.9	0.5	1.3
G	A	CT	D	2.6	0.7	1.6	1.3	1.6	2.7	1.4	2.3	3.1	3.6	0.8	1.5	1.3	1.3	3.5	2.3	2.7	3.0	1.0	2.2
G	A	AG	E	2.9	0.6	1.3	2.2	2.1	2.1	1.2	2.9	3.5	4.1	1.0	1.4	1.3	1.5	3.6	2.3	2.2	3.4	1.0	2.2
T	T	CT	F	4.1	1.3	1.6	2.1	3.3	2.8	1.3	4.8	2.4	6.5	1.9	1.8	2.2	1.7	3.3	2.4	2.4	4.9	1.6	3.0
G	G	X	G	3.9	1.4	2.7	2.1	2.8	4.2	1.3	4.2	2.3	5.2	1.4	2.3	2.1	1.7	3.0	3.0	3.3	4.4	1.3	2.7
C	A	CT	H	1.5	0.4	1.4	1.2	1.3	1.3	0.6	1.5	0.8	2.4	0.5	0.7	0.9	0.7	1.2	1.3	1.2	1.5	0.6	1.3
A	T	ACT	I	6.3	1.2	2.3	2.9	4.8	4.2	1.5	7.0	2.7	12.2	2.5	2.1	1.8	2.2	3.1	3.2	3.7	7.7	1.8	4.4
A	A	AG	K	3.3	0.7	3.1	3.5	2.4	2.3	0.8	2.7	1.5	4.2	1.3	1.6	1.3	1.7	1.7	1.9	2.3	3.4	1.0	2.5
CT	T	XAG	L	9.9	1.7	3.6	4.1	6.5	5.2	2.4	12.2	4.2	19.7	3.9	3.7	3.2	3.5	5.8	4.4	5.0	11.2	3.3	6.4
A	T	G	M	2.8	0.5	0.8	1.0	1.9	1.4	0.5	2.5	1.3	3.9	1.5	0.8	0.9	0.9	1.5	1.2	1.2	2.6	0.6	1.8
A	A	CT	N	2.6	0.9	1.5	1.4	1.8	2.3	0.7	2.1	1.6	3.7	0.8	1.6	1.5	1.4	1.8	2.2	2.1	2.1	1.0	1.9
C	C	X	P	2.0	0.8	1.3	1.3	2.2	2.1	0.9	1.8	1.3	3.2	0.9	1.5	1.5	1.0	2.2	1.4	1.6	2.8	1.2	2.1
C	A	AG	Q	2.5	0.4	1.3	1.5	1.7	1.7	0.7	2.2	1.7	3.5	0.9	1.4	1.0	0.9	1.6	1.3	1.7	2.5	0.7	1.6
AC	G	AGX	R	3.3	0.7	3.5	3.6	3.3	3.0	1.2	3.1	1.7	5.8	1.5	1.8	2.2	1.6	2.4	2.5	2.6	3.7	0.8	2.5
AT	CG	CTX	S	3.7	1.1	2.3	2.3	2.4	3.0	1.3	3.2	1.9	4.4	1.2	2.2	1.4	1.3	2.5	2.5	2.6	3.3	1.2	2.3
A	C	X	T	4.5	1.3	2.7	2.2	2.4	3.3	1.2	3.7	2.3	5.0	1.2	2.1	1.6	1.7	2.6	2.6	3.8	4.4	0.9	2.5
G	T	X	V	7.5	1.9	3.0	3.4	4.9	4.4	1.5	7.7	3.4	11.2	2.6	2.1	2.8	2.5	3.7	3.3	4.4	8.7	2.0	4.6
T	G	G	W	1.6	0.5	1.0	1.0	1.6	1.3	0.6	1.8	1.0	3.3	0.6	1.0	1.2	0.7	0.8	1.2	0.9	2.0	0.9	1.5
T	A	CT	Y	4.7	1.3	2.2	2.2	3.0	2.7	1.3	4.4	2.5	6.4	1.8	1.9	2.1	1.6	2.5	2.3	2.5	4.6	1.5	3.6

RC_3X1 code: 1st and 3rd codon letters are complementary in reverse order, indicated by complementary colors (red, blue); unfavored colocations are gray-colored.
X: any residue; AA: amino acids, one-letter code; Co-location frequency is multiplied by 10, Sum=1000

Figure 26. Complementary codes vs amino acid co-locations. See text for explanation.

It is well known that coding and non-coding DNA sequences (exon/intron) are different and this difference is somehow related to the asymmetry of the codons, i.e., that the third codon letter (wobble) is poorly defined. Many Markov models have been formulated to find this asymmetry and de novo predict coding sequences (genes). These in silico methods work rather well but not perfectly and some scientists remain unconvinced that the codon asymmetry explains the exon–intron differences satisfactorily.

Another codon-related problem is that the well-known, non-overlapping, triplet codon translation is extremely phase-dependent and there is theoretically no tolerance for any phase shift. There are famous examples of how single nucleotide deletion might destroy the meaningful translation of a sequence and which are incompatible with life. However, considering the magnitude and complexity of the eukaryotic proteome, the precision of translation is astonishingly good. Such physical precision is not possible without massive and consistent physico-chemical fundaments. Therefore, discovery of the existence of secondary structure bias (folding energy differences) in coding regions of many organisms [78] was a very welcome observation because it differentiates exons from introns on a physico-chemical basis.

Our experiments with free folding energy (FFE) confirmed that this bias exists. In addition, there is a very consistent and very significant pattern of FFE distribution along the nucleotide sequence. Comparing the FFE of phase-selected subsequences, subsequences comprised of only the 1st or only the 3rd codon letters showed significantly higher FFE than those consisting only of the 2nd letters. This FFE difference was not present in intronic sequences preceding and following the exons, but it was present in exons from different species including viruses. This is an interesting observation because this phenomenon might not only distinguish between exons and introns on a physico-chemical basis, but it might even clearly define the tri-nucleotide codons and thus the phase of the translation. This codon-related phase-specific variation in FFE may explain why mRNAs have greater negative free folding energies than shuffled or codon choice randomized sequences [85].

Free folding energy in nucleic acids is always associated with W-C base pair formation. Higher FFE indicates more W-C pairs (presence of complementarity) and lower FFE indicates fewer W-C pairs (less complementarity). The FFE in the 1st and 3rd codon positions was additive, while the 2nd letter did not contribute to the total FFE; the total FFE of the entire (intact) nucleic acid was the same as subsequences containing only the 1st and 3rd codon letters (2nd deleted). This is an indication that the local RNA secondary structure bias is caused by complementarity of the 1st and 3rd codon residues in local sequences. This partial, local complementarity is more optimal in reverse orientation of the local sequences as expected with loop formations.

It is known that single stranded RNA molecules can form local secondary structures through the interactions of complementary segments. The novel observation here is that these interactions preferentially involve the 1st and 3rd codon residues. This connection between the RNA secondary structure and codons immediately directed attention towards the question of protein folding and its long suspected connection to RNA folding [86,87].

Only about one-third (20/64) of the genetic code is used for protein coding, i.e., there is a great excess of information in the mRNA. At the same time, the information carried by amino acids seems to be insufficient (as stated by some scientists) to complete unambiguous protein folding. Therefore, it is believed that the third codon residue (wobble base) carries some additional information to that already present in the genetic code. A specialized wobble base oriented database, the ISSD [84] was established in an effort to connect different features of protein structure to wobble bases [88] with more or less success.

We found a significant negative correlation between FFE of the 2nd codon residue and the helix content of protein structures, which was not expected even though this possibility is mentioned in the literature [74]. Our previous work on a Common Periodic Table of Codons and Nucleic Acids [52] indicated that the second codon residue is intimately coupled with the known physico-chemical properties of the amino acids. Almost all amino acids show significant positive or negative correlation to the helix content of proteins. Therefore, the real biological meaning and significance of any connection between FFE of the 2nd codon residue and the propensity of a protein structural element is highly questionable.

It was possible to make direct visual comparison of mRNA structure (as statistically predicted by the mfold energy dot-plot) and protein structures (as 2D residue contact maps). This method suggests similarity between nucleic acid and protein structures. It is known that some complex protein structures are very similar even if there is less than 30% sequence similarity. It was interesting to see that the same principle might apply for nucleic acids, and structural similarity might exist even when the sequence similarity is low. Furthermore, significant similarity between nucleic acid and protein structures might exist even without translational connection.

Structure seems to be more preserved, even in nucleic acids, than sequence.

However, even if the matrix comparisons are suggestive, they remain semi-quantitative methods. Better support was necessary.

A working hypotheses grew out of these observations, namely that (a) partial, local reverse complementarity exists in nucleic acids that form the nucleic acid structure; (b) there is some degree of similarity between the folding of nucleic acids and proteins; (c) protein structure determines the amino acid co-locations; (d) as a consequence, amino acids coded by the interacting (partially reverse complementary) codons might show preferential co-locations in the protein structures.

This seems to be the case: codons which contain complementary bases at the 1st and 3rd positions and are translated in reverse orientation result in amino acids which are preferentially co-located (interacting) in the 3D protein structure. Other complementary residue combinations or translation in the same (not reverse) direction (as much as seven combinations in total) did not result in any preferentially co-locating subset of amino acid pairs.

Construction of residue contact maps for protein structures and statistical evaluation of residue co-locations is a frequently used method for visualization and analyses of spatial connections between amino acids [89,90,91. The amino acid co-locations in real protein structures is clearly not random [92,93] and therefore residue co-location matrices are often used to assist in the prediction of novel protein structures [94,95]. We have carefully examined the physico-chemical properties of specifically interacting amino acids in and between protein structures, and we concluded that these interactions follow the well known physico-chemical rules of size, charge and hydrophobe compatibility (unpublished data) well in line with Anfinsen's prediction. A recent study supports the fact that there is a previously unknown connection between the codons of specifically interacting amino acids; those codons are complementary at the 1st and 3rd (but not the 2nd) codon positions.

The idea that sequence complementarity might explain the nature of specific protein–protein interactions is not new and was suggested already in 1981 [9,10,11].

I was never able to experimentally confirm my own original theory, which suggested a perfect complementarity between codons of interacting amino [9,10,11,50], in contrast to others [37]. The explanation is that this codon complementarity is suboptimal and does not involve the 2nd codon residue. Experimental in vitro confirmation is required to validate this recent theoretical and in silico prediction.

Availability

http://www.janbiro.com/downloads: SeqX, SeqForm, SeqPlot, Dotlet.

RNA Assisted Protein Loop Formation

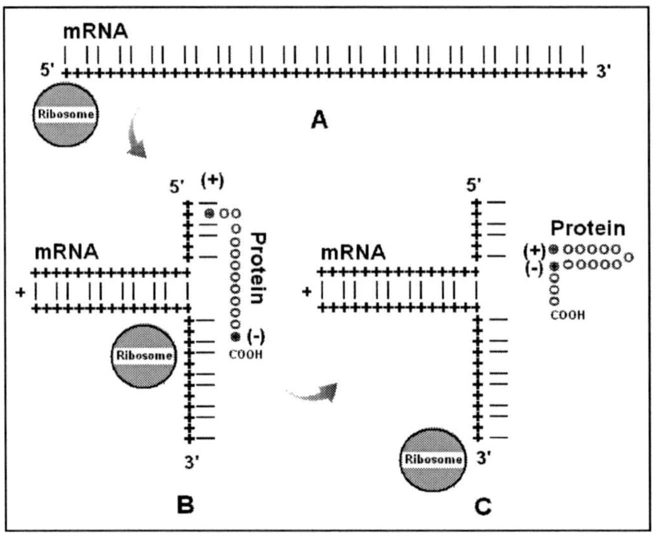

Figure 27. RNA assisted protein loop formation (from [96]. Translation begins with the attachment of the 5′ end of a mRNA to the ribosome (A). Ribonucleotides are indicated by blue "+" and the 1st and 3rd bases in the codons by blue lines; the 2nd base positions are left empty. A positively charged amino acid ((+) and red dots), for example, arginine, remains attached to its codon. The mRNA forms a loop because the 1st and 3rd bases are locally complementary to each other in reverse orientation (B). The growing protein is indicated by red circles. When translation proceeds to an amino acid with especially high affinity to the mRNA-attached arginine, for example a negatively charged Glu or Asp ((–) and blue dot), the charge attraction removes the Arg from its mRNA binding site and the entire protein is released from the mRNA and completes a protein loop (C). The protein continues to grow towards the direction of its carboxy terminal (COOH).

Theory of Nucleic Acid (Chaperons) Assisted Protein Folding.

A series of novel arguments has been presented above to support a deeper connection between nucleic acids and expressed proteins, than the traditional codon translation. The physico-chemical properties of amino acids are clearly associated with the 2nd codon letter as was shown in the Common Periodic Table of Codons and Amino Acids. The co-locating amino acids are preferentially coded by codons that are complementary at the 1st and 3rd codon position. The structure of proteins and their coding nucleic acid are rather similar to each other in many cases. All these observations suggest the co-evolution of codons and amino acids and that protein folding (structural) information is present in the nucleic acids in addition to the canonical genetic code. This immediately raises the possibility of nucleic acid-assisted protein folding, i.e. the possibility of nucleic acid chaperons [96]. This is an exciting possibility, because the protein primary sequence seems not to carry the necessary information for unambiguous protein folding (in contrast to Anfinsen's theorem), while there is a two fold excess of information in the redundant genetic code. A theoretical example of how such nucleic acid-assisted protein folding may look is presented in Figures 27 and 28.

RNA-Assisted (translational) Protein Folding

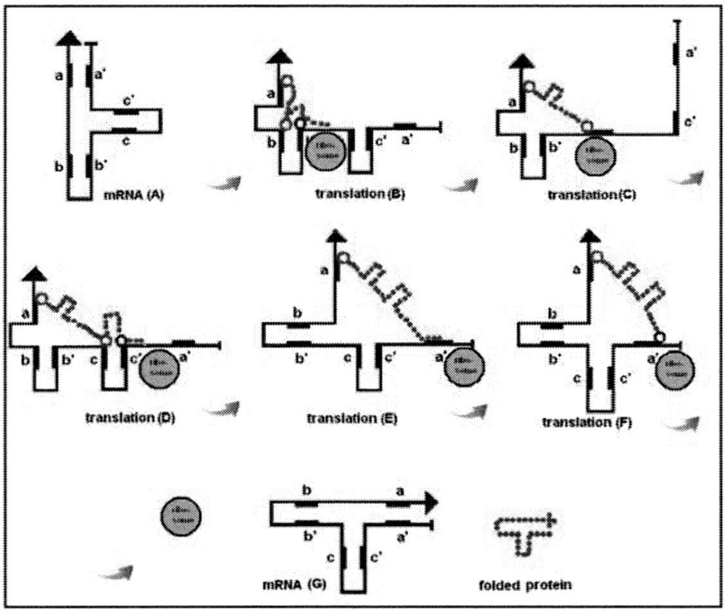

Figure 28. RNA-assisted (translational) protein folding (from [96]. There are three reverse and complementary regions in a mRNA (blue line, A): a–a′, b–b′, c–c′, which fold the mRNA into a T-like shape. During the translation process the mRNA unfolds on the surface of the ribosome, but subsequently refolds, accompanied by its translated and lengthening peptide (red dotted line, B–F). The result of translation is a temporary ribonucleotide complex, which dissociates into two T-shape-like structures: the original mRNA and the properly folded protein product (G). The red circles indicate the specific, temporary attachment points between the RNA and protein (for example a basic amino acid); the blue circles indicate amino acids with exceptionally high affinity for the attachment points (e.g., acidic amino acids); these capture the amino acids at the attachment point and dissociate the ribonucleoprotein complex. Transfer RNAs are of course important participants in translation, but they are not included in this scenario.

Definition of the 2nd Generation Proteomic Code

The Proteomic Code is a set of comprehensive rules by which information in genetic material is transferred into the physico-chemical properties of amino acids and determines how individual amino acids interact with each other during folding and in specific protein–protein interactions. The Proteomic Code is part of the redundant genetic code. The theory of Proteomic Code contains the following observations:

- Co-locating (interacting) amino acids in native proteins are coded by partially (imperfect) complementary codons in reverse (5′→3′/5′→3′) orientation.
- Partial complementarity means that the 1st and 3rd codon bases are complementary (Watson–Crick) bases to each other, while the 2nd bases may or may not be complementary to each other.
- The physico-chemical characteristics of the coded and interacting amino acids are determined mainly by the 2nd (central) codon residues.
- The physico-chemical properties of the interacting amino acids (size, charge, and hydropathy) are compatible with each other at the individual amino acid level.
- Nucleic acids (exons) contain protein folding information in (or in addition to) the redundant genetic code.
- Nucleic acids may directly assist protein folding as chaperons.

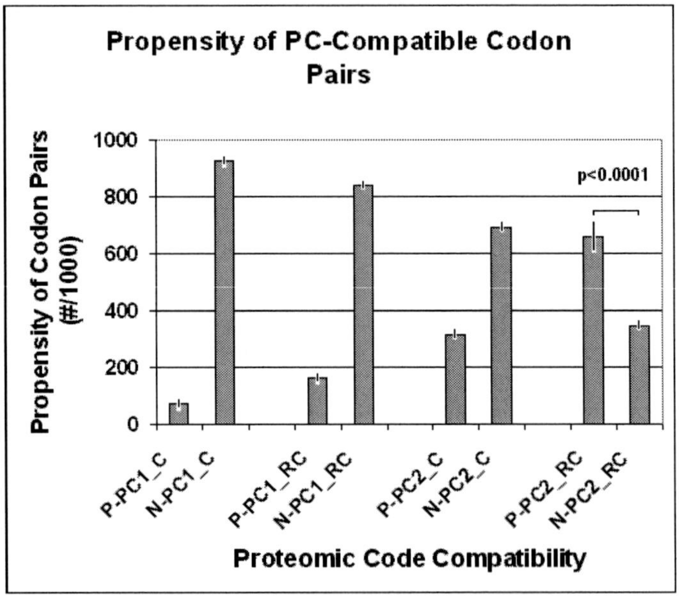

Figure 29. Propensity of PC-compatible codon pairs. There are 4096 possible codon pairs altogether (64×64, including those formed with the 3 stop codons). Codons, which are (P) or are not (N) coded by a specific complementary rule (PC) were counted. The bars represent the mean±SD of 9 independent determinations (*n*=9). PC1 and PC2 indicate complementarity of all 3 or only the first and third codon bases in parallel (C) or anti-parallel (RC) readings.

There are four different Proteomic Codes at this moment. PC1_C and PC1_RC are the original codes based on the perfect complementarity of all three codon bases in complementary (C, 5′→3′/3′→5′) readings; PC2_C and PC2_RC are the recent extended codes requiring base pair complementarity only at the 1st and 3rd codon position, but not necessarily at the 2nd. PC1 is part of PC2. The PC_C variants require 3′→5′ translations, which do not exist, therefore I regard this variant as an artifact, caused by the symmetry of many codons. Only a small fraction of total codon and amino acid pairs belongs to PC1; ~50% of all amino acid pairs and >60% of all codon pairs can be classified into PC2_RC (Figures 29 and 30).

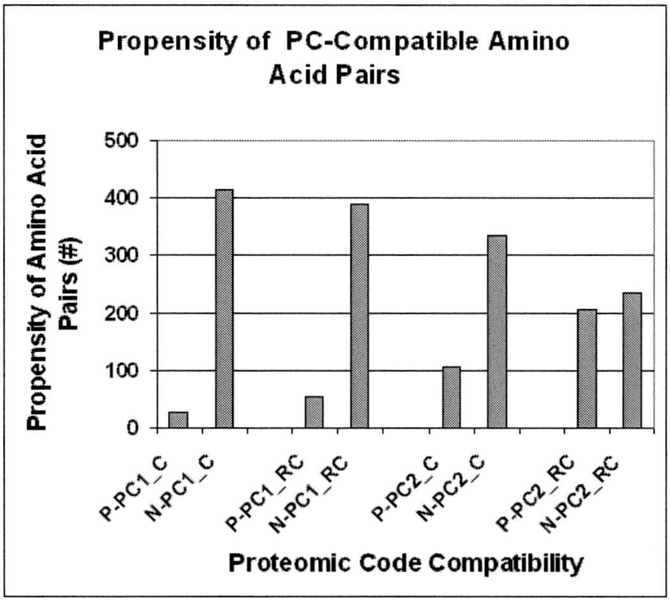

Figure 30. Propensity of PC-compatible amino acid pairs. There are 441 possible amino acid pairs altogether (21×21, including the stop place as the 21st variable). Amino acids, which are (P) or are not (N) coded by a specific complementary rule (Proteomic Code, PC) were counted. PC1 and PC2 indicate complementarity of the respective codon pairs at all 3 or only the first and third codon bases in parallel (C) or anti-parallel (RC) readings.

A list of all possible amino acid pairs (21×21=441, including the virtual pairs formed with the Stop/End signal) are listed in Table 4. The most important physico-chemical parameters (molecular weight (MW), isoelectric point (p*I*), hydropathy (HP)) and the derived values of three compatibility indexes (charge (CCI), size (SCI) and hydropathy (HCI) compatibility indexes [58]) as well as the expected frequency of the amino acid pairs (natural frequency (NF), calculated from codon table 1) are included in Table 4. P (positive) and N (negative) indicate whether an amino acid pair may or may not be coded by a given Proteomic Code.

Table IV.

First	Second	1st	2nd	PC1_C	PC2_C	PC1_RC	PC2_RC	mw-57_1	HP_1	pI_1	mw-57_2	HP_2	pI_2	NF*/1k	SCI	CCI	HCI
Ala	Ala	A	A	N	N	N	P	14.0	1.0	6.0	14.0	1.0	6.0	4.3	6.6	10.4	20.0
Ala	Arg	A	R	P	P	P	P	14.0	1.0	6.0	99.0	-7.5	10.8	6.4	18.6	13.1	4.8
Ala	Asn	A	N	N	N	N	P	14.0	1.0	6.0	57.0	-2.7	5.4	2.1	12.7	10.1	13.4
Ala	Asp	A	D	N	N	N	P	14.0	1.0	6.0	58.0	-3.0	2.8	2.1	12.8	8.6	12.8
Ala	Cys	A	C	N	N	P	P	14.0	1.0	6.0	46.0	0.2	5.1	2.1	11.1	9.9	18.5
Ala	Gln	A	Q	N	P	N	N	14.0	1.0	6.0	71.0	-2.9	5.7	2.1	14.7	10.2	13.0
Ala	Glu	A	E	N	N	N	N	14.0	1.0	6.0	72.0	-2.6	3.2	2.1	14.8	8.9	13.5
Ala	Gly	A	G	N	N	P	P	14.0	1.0	6.0	0.0	0.7	6.0	4.3	4.7	10.4	19.4
Ala	His	A	H	N	P	N	P	14.0	1.0	6.0	80.0	-1.7	7.6	2.1	15.9	11.3	15.2
Ala	Ile	A	I	N	N	N	P	14.0	1.0	6.0	56.0	3.1	5.9	3.2	12.5	10.4	16.2
Ala	Leu	A	L	N	P	N	P	14.0	1.0	6.0	56.0	2.2	6.0	6.4	12.5	10.4	17.8
Ala	Lys	A	K	N	N	N	N	14.0	1.0	6.0	71.0	-4.6	9.7	2.1	14.7	12.5	10.0
Ala	Met	A	M	N	N	N	N	14.0	1.0	6.0	74.0	1.1	5.7	1.1	15.1	10.3	19.8
Ala	Phe	A	F	N	N	N	P	14.0	1.0	6.0	90.0	2.5	5.5	2.1	17.3	10.2	17.3
Ala	Pro	A	P	N	P	N	P	14.0	1.0	6.0	40.0	-0.3	6.5	4.3	10.3	10.7	17.7
Ala	Ser	A	S	N	N	P	P	14.0	1.0	6.0	30.0	-1.1	5.7	6.4	8.9	10.3	16.2
Ala	Thr	A	T	N	N	N	P	14.0	1.0	6.0	44.0	-0.8	5.9	4.3	10.9	10.4	16.9
Ala	Trp	A	W	N	N	N	N	14.0	1.0	6.0	129.0	1.5	6.0	1.1	17.2	10.4	19.1
Ala	Tyr	A	Y	N	N	N	P	14.0	1.0	6.0	106.0	0.1	5.7	2.1	19.6	10.3	18.4
Ala	Val	A	V	N	N	N	P	14.0	1.0	6.0	42.0	2.3	6.0	4.3	10.6	10.4	17.7
Arg	Ala	R	A	P	P	P	P	99.0	-7.5	10.8	14.0	1.0	6.0	6.4	18.6	13.1	4.8
Arg	Arg	R	R	N	N	N	P	99.0	-7.5	10.8	99.0	-7.5	10.8	9.7	9.4	3.1	20.0
Arg	Asn	R	N	N	P	N	N	99.0	-7.5	10.8	57.0	-2.7	5.4	3.2	15.4	14.4	11.4
Arg	Asp	R	D	N	P	N	N	99.0	-7.5	10.8	58.0	-3.0	2.8	3.2	15.2	19.9	11.9
Arg	Cys	R	C	N	P	N	P	99.0	-7.5	10.8	46.0	0.2	5.1	3.2	16.9	15.1	6.3
Arg	Gln	R	Q	N	N	N	P	99.0	-7.5	10.8	71.0	-2.9	5.7	3.2	13.4	13.9	11.8
Arg	Glu	R	E	N	P	N	P	99.0	-7.5	10.8	72.0	-2.6	3.2	3.2	13.2	19.0	11.2
Arg	Gly	R	G	N	P	N	P	99.0	-7.5	10.8	0.0	0.7	6.0	6.4	16.6	13.2	5.4
Arg	His	R	H	N	N	N	P	99.0	-7.5	10.8	80.0	-1.7	7.6	3.2	12.1	9.8	9.6
Arg	Ile	R	I	N	N	N	N	99.0	-7.5	10.8	56.0	3.1	5.9	4.8	15.5	13.3	1.0
Arg	Leu	R	L	N	N	N	P	99.0	-7.5	10.8	56.0	2.2	6.0	9.7	15.5	13.2	2.6
Arg	Lys	R	K	N	N	N	P	99.0	-7.5	10.8	71.0	-4.6	9.7	3.2	13.4	5.2	14.8
Arg	Met	R	M	N	N	N	P	99.0	-7.5	10.8	74.0	1.1	5.7	1.6	13.0	13.7	4.6
Arg	Phe	R	F	N	P	N	P	99.0	-7.5	10.8	90.0	2.5	5.5	3.2	10.7	14.2	2.1
Arg	Pro	R	P	N	N	P	P	99.0	-7.5	10.8	40.0	-0.3	6.5	6.4	17.7	12.1	7.1
Arg	Ser	R	S	P	P	P	P	99.0	-7.5	10.8	30.0	-1.1	5.7	9.7	19.2	13.8	8.5
Arg	Thr	R	T	N	N	P	P	99.0	-7.5	10.8	44.0	-0.8	5.9	6.4	17.2	13.4	7.9
Arg	Trp	R	W	N	N	N	P	99.0	-7.5	10.8	129.0	1.5	6.0	1.6	5.2	13.2	3.9
Arg	Tyr	R	Y	N	P	N	P	99.0	-7.5	10.8	106.0	0.1	5.7	3.2	8.5	13.8	6.4
Arg	Val	R	V	N	P	N	P	99.0	-7.5	10.8	42.0	2.3	6.0	6.4	17.5	13.2	2.4
Asn	Ala	N	A	N	N	N	P	57.0	-2.7	5.4	14.0	1.0	6.0	2.1	12.7	10.1	13.4
Asn	Arg	N	R	N	N	N	N	57.0	-2.7	5.4	99.0	-7.5	10.8	3.2	15.4	14.4	11.4
Asn	Asn	N	N	N	N	N	P	57.0	-2.7	5.4	57.0	-2.7	5.4	1.1	18.7	9.8	20.0
Asn	Asp	N	D	N	N	N	P	57.0	-2.7	5.4	58.0	-3.0	2.8	1.1	18.9	7.2	19.5
Asn	Cys	N	C	N	N	N	N	57.0	-2.7	5.4	46.0	0.2	5.1	1.1	17.2	9.3	14.9
Asn	Gln	N	Q	N	N	N	N	57.0	-2.7	5.4	71.0	-2.9	5.7	1.1	19.3	9.8	19.6
Asn	Glu	N	E	N	N	N	N	57.0	-2.7	5.4	72.0	-2.6	3.2	1.1	19.2	7.6	19.8
Asn	Gly	N	G	N	N	N	P	57.0	-2.7	5.4	0.0	0.7	6.0	2.1	10.7	10.1	14.0
Asn	His	N	H	N	N	N	N	57.0	-2.7	5.4	80.0	-1.7	7.6	1.1	18.0	11.5	18.2
Asn	Ile	N	I	N	N	P	P	57.0	-2.7	5.4	56.0	3.1	5.9	1.6	18.6	10.0	9.6
Asn	Leu	N	L	P	P	N	N	57.0	-2.7	5.4	56.0	2.2	6.0	3.2	18.6	10.1	11.2
Asn	Lys	N	K	N	N	N	N	57.0	-2.7	5.4	71.0	-4.6	9.7	1.1	19.3	13.4	16.6
Asn	Met	N	M	N	N	N	N	57.0	-2.7	5.4	74.0	1.1	5.7	0.5	18.9	9.9	13.2
Asn	Phe	N	F	N	N	N	N	57.0	-2.7	5.4	90.0	2.5	5.5	1.1	16.6	9.6	10.7
Asn	Pro	N	P	N	N	N	N	57.0	-2.7	5.4	40.0	-0.3	6.5	2.1	16.3	10.5	15.7
Asn	Ser	N	S	N	P	N	P	57.0	-2.7	5.4	30.0	-1.1	5.7	3.2	14.9	9.8	17.1
Asn	Thr	N	T	N	N	N	P	57.0	-2.7	5.4	44.0	-0.8	5.9	2.1	16.9	10.0	16.5
Asn	Trp	N	W	N	P	N	N	57.0	-2.7	5.4	129.0	1.5	6.0	0.5	11.1	10.1	12.5
Asn	Tyr	N	Y	N	N	N	N	57.0	-2.7	5.4	106.0	0.1	5.7	1.1	14.4	9.8	15.0
Asn	Val	N	V	N	N	P	P	57.0	-2.7	5.4	42.0	2.3	6.0	2.1	16.6	10.1	11.0
Asp	Ala	D	A	N	N	N	P	58.0	-3.0	2.8	14.0	1.0	6.0	2.1	12.8	8.6	12.8
Asp	Arg	D	R	N	P	N	N	58.0	-3.0	2.8	99.0	-7.5	10.8	3.2	15.2	19.9	11.9
Asp	Asn	D	N	N	N	N	P	58.0	-3.0	2.8	57.0	-2.7	5.4	1.1	18.9	7.2	19.5
Asp	Asp	D	D	N	N	N	P	58.0	-3.0	2.8	58.0	-3.0	2.8	1.1	19.0	0.9	20.0
Asp	Cys	D	C	N	N	N	N	58.0	-3.0	2.8	46.0	0.2	5.1	1.1	17.3	6.4	14.3
Asp	Gln	D	Q	N	P	N	N	58.0	-3.0	2.8	71.0	-2.9	5.7	1.1	19.2	7.8	19.8
Asp	Glu	D	E	N	N	N	N	58.0	-3.0	2.8	72.0	-2.8	3.2	1.1	19.0	2.0	19.3
Asp	Gly	D	G	N	N	N	P	58.0	-3.0	2.8	0.0	0.7	6.0	2.1	10.9	8.6	13.4
Asp	His	D	H	N	N	N	N	58.0	-3.0	2.8	80.0	-1.7	7.6	1.1	17.9	12.4	17.7
Asp	Ile	D	I	N	N	P	P	58.0	-3.0	2.8	56.0	3.1	5.9	1.6	18.7	8.4	9.1
Asp	Leu	D	L	P	P	N	N	58.0	-3.0	2.8	56.0	2.2	6.0	3.2	18.7	8.6	10.7
Asp	Lys	D	K	N	N	N	N	58.0	-3.0	2.8	71.0	-4.6	9.7	1.1	19.2	17.5	17.1
Asp	Met	D	M	N	N	N	N	58.0	-3.0	2.8	74.0	1.1	5.7	0.5	18.7	8.0	12.7
Asp	Phe	D	F	N	N	N	N	58.0	-3.0	2.8	90.0	2.5	5.5	1.1	16.5	7.4	10.1
Asp	Pro	D	P	N	P	N	N	58.0	-3.0	2.8	40.0	-0.3	6.5	2.1	16.5	9.8	15.1
Asp	Ser	D	S	N	N	N	P	58.0	-3.0	2.8	30.0	-1.1	5.7	3.2	15.1	7.9	16.6
Asp	Thr	D	T	N	N	N	P	58.0	-3.0	2.8	44.0	-0.8	5.9	2.1	17.0	8.3	16.0
Asp	Trp	D	W	N	N	N	N	58.0	-3.0	2.8	129.0	1.5	6.0	0.5	11.0	8.8	11.9
Asp	Tyr	D	Y	N	N	N	N	58.0	-3.0	2.8	106.0	0.1	5.7	1.1	14.2	7.8	14.5
Asp	Val	D	V	N	N	P	P	58.0	-3.0	2.8	42.0	2.3	6.0	2.1	16.8	8.6	10.5

Table IV. Continued

First	Second	1st	2nd	PC1_C	PC2_C	PC1_RC	PC2_RC	mw-57_1	HP_1	pI_1	mw-57_2	HP_2	pI_2	NF*/1k	SCI	CCI	HCI
Cys	Ala	C	A	N	N	P	P	46.0	0.2	5.1	14.0	1.0	6.0	2.1	11.1	9.9	18.5
Cys	Arg	C	R	N	P	N	P	46.0	0.2	5.1	99.0	-7.5	10.8	3.2	16.9	15.1	6.3
Cys	Asn	C	N	N	N	N	N	46.0	0.2	5.1	57.0	-2.7	5.4	1.1	17.2	9.3	14.9
Cys	Asp	C	D	N	N	N	N	46.0	0.2	5.1	58.0	-3.0	2.8	1.1	17.3	6.4	14.3
Cys	Cys	C	C	N	N	N	N	46.0	0.2	5.1	46.0	0.2	5.1	1.1	15.6	8.9	20.0
Cys	Gln	C	Q	N	N	N	N	46.0	0.2	5.1	71.0	-2.9	5.7	1.1	19.2	9.5	14.5
Cys	Glu	C	E	N	N	N	P	46.0	0.2	5.1	72.0	-2.6	3.2	1.1	19.3	6.9	15.0
Cys	Gly	C	G	N	N	N	P	46.0	0.2	5.1	0.0	0.7	6.0	2.1	9.2	9.9	19.1
Cys	His	C	H	N	N	N	N	46.0	0.2	5.1	80.0	-1.7	7.6	1.1	19.6	11.6	16.6
Cys	Ile	C	I	N	P	N	P	46.0	0.2	5.1	56.0	3.1	5.9	1.6	17.0	9.8	14.7
Cys	Leu	C	L	N	N	N	N	46.0	0.2	5.1	56.0	2.2	6.0	3.2	17.0	9.9	16.4
Cys	Lys	C	K	N	P	N	P	46.0	0.2	5.1	71.0	-4.6	9.7	1.1	19.2	14.0	11.5
Cys	Met	C	M	N	P	N	N	46.0	0.2	5.1	74.0	1.1	5.7	0.5	19.6	9.6	18.3
Cys	Phe	C	F	N	N	N	N	46.0	0.2	5.1	90.0	2.5	5.5	1.1	18.2	9.4	15.8
Cys	Pro	C	P	N	N	N	N	46.0	0.2	5.1	40.0	-0.3	6.5	2.1	14.8	10.4	19.2
Cys	Ser	C	S	N	N	N	N	46.0	0.2	5.1	30.0	-1.1	5.7	3.2	13.4	9.6	17.7
Cys	Thr	C	T	P	P	P	P	46.0	0.2	5.1	44.0	-0.8	5.9	2.1	15.4	9.8	18.4
Cys	Trp	C	W	N	N	N	N	46.0	0.2	5.1	129.0	1.5	6.0	0.5	12.7	9.9	17.6
Cys	Tyr	C	Y	N	N	N	N	46.0	0.2	5.1	106.0	0.1	5.7	1.1	15.9	9.5	19.8
Cys	Val	C	V	N	N	N	P	46.0	0.2	5.1	42.0	2.3	6.0	2.1	15.1	9.9	16.2
Gln	Ala	Q	A	N	P	N	N	71.0	-2.9	5.7	14.0	1.0	6.0	2.1	14.7	10.2	13.0
Gln	Arg	Q	R	N	N	N	P	71.0	-2.9	5.7	99.0	-7.5	10.8	3.2	13.4	13.9	11.8
Gln	Asn	Q	N	N	N	N	N	71.0	-2.9	5.7	57.0	-2.7	5.4	1.1	19.3	9.8	19.6
Gln	Asp	Q	D	N	P	N	N	71.0	-2.9	5.7	58.0	-3.0	2.8	1.1	19.2	7.8	19.8
Gln	Cys	Q	C	N	N	N	N	71.0	-2.9	5.7	46.0	0.2	5.1	1.1	19.2	9.5	14.5
Gln	Gln	Q	Q	N	N	N	P	71.0	-2.9	5.7	71.0	-2.9	5.7	1.1	17.3	10.0	20.0
Gln	Glu	Q	E	N	N	N	N	71.0	-2.9	5.7	72.0	-2.6	3.2	1.1	17.2	8.1	19.5
Gln	Gly	Q	G	N	P	N	N	71.0	-2.9	5.7	0.0	0.7	6.0	2.1	12.7	10.2	13.6
Gln	His	Q	H	N	N	N	N	71.0	-2.9	5.7	80.0	-1.7	7.6	1.1	16.1	11.4	17.8
Gln	Ile	Q	I	N	N	N	N	71.0	-2.9	5.7	56.0	3.1	5.9	1.6	19.4	10.2	9.2
Gln	Leu	Q	L	N	N	P	P	71.0	-2.9	5.7	56.0	2.2	6.0	3.2	19.4	10.2	10.9
Gln	Lys	Q	K	N	N	N	N	71.0	-2.9	5.7	71.0	-4.6	9.7	1.1	17.3	13.1	17.0
Gln	Met	Q	M	N	N	N	N	71.0	-2.9	5.7	74.0	1.1	5.7	0.5	16.9	10.0	12.8
Gln	Phe	Q	F	N	N	N	N	71.0	-2.9	5.7	90.0	2.5	5.5	1.1	14.7	9.8	10.3
Gln	Pro	Q	P	N	N	N	P	71.0	-2.9	5.7	40.0	-0.3	6.5	2.1	18.3	10.6	15.3
Gln	Ser	Q	S	N	N	N	P	71.0	-2.9	5.7	30.0	-1.1	5.7	3.2	16.9	10.0	16.8
Gln	Thr	Q	T	N	N	N	N	71.0	-2.9	5.7	44.0	-0.8	5.9	2.1	18.9	10.1	16.1
Gln	Trp	Q	W	N	N	N	P	71.0	-2.9	5.7	129.0	1.5	6.0	0.5	9.2	10.2	12.1
Gln	Tyr	Q	Y	N	N	N	N	71.0	-2.9	5.7	106.0	0.1	5.7	1.1	12.4	10.0	14.7
Gln	Val	Q	V	P	P	N	N	71.0	-2.9	5.7	42.0	2.3	6.0	2.1	18.6	10.2	10.7
Glu	Ala	E	A	N	N	N	N	72.0	-2.6	3.2	14.0	1.0	6.0	2.1	14.8	8.9	13.5
Glu	Arg	E	R	N	P	N	P	72.0	-2.6	3.2	99.0	-7.5	10.8	3.2	13.2	19.0	11.2
Glu	Asn	E	N	N	N	N	N	72.0	-2.6	3.2	57.0	-2.7	5.4	1.1	19.2	7.6	19.8
Glu	Asp	E	D	N	N	N	N	72.0	-2.6	3.2	58.0	-3.0	2.8	1.1	19.0	2.0	19.3
Glu	Cys	E	C	N	N	N	P	72.0	-2.6	3.2	46.0	0.2	5.1	1.1	19.3	6.9	15.0
Glu	Gln	E	Q	N	N	N	N	72.0	-2.6	3.2	71.0	-2.9	5.7	1.1	17.2	8.1	19.5
Glu	Glu	E	E	N	N	N	N	72.0	-2.6	3.2	72.0	-2.6	3.2	1.1	17.0	3.0	20.0
Glu	Gly	E	G	N	N	N	N	72.0	-2.6	3.2	0.0	0.7	6.0	2.1	12.8	8.8	14.1
Glu	His	E	H	N	P	N	P	72.0	-2.6	3.2	80.0	-1.7	7.6	1.1	15.9	12.3	18.4
Glu	Ile	E	I	N	N	N	N	72.0	-2.6	3.2	56.0	3.1	5.9	1.6	19.3	8.7	9.8
Glu	Leu	E	L	P	P	P	P	72.0	-2.6	3.2	56.0	2.2	6.0	3.2	19.3	8.8	11.4
Glu	Lys	E	K	N	N	N	N	72.0	-2.6	3.2	71.0	-4.6	9.7	1.1	17.2	16.8	16.4
Glu	Met	E	M	N	N	N	N	72.0	-2.6	3.2	74.0	1.1	5.7	0.5	16.8	8.3	13.4
Glu	Phe	E	F	N	N	P	P	72.0	-2.6	3.2	90.0	2.5	5.5	1.1	14.5	7.8	10.9
Glu	Pro	E	P	N	P	N	P	72.0	-2.6	3.2	40.0	-0.3	6.5	2.1	18.5	9.9	15.9
Glu	Ser	E	S	N	N	N	P	72.0	-2.6	3.2	30.0	-1.1	5.7	3.2	17.0	8.2	17.3
Glu	Thr	E	T	N	N	N	N	72.0	-2.6	3.2	44.0	-0.8	5.9	2.1	19.0	8.6	16.7
Glu	Trp	E	W	N	N	N	N	72.0	-2.6	3.2	129.0	1.5	6.0	0.5	9.0	8.8	12.7
Glu	Tyr	E	Y	N	N	N	P	72.0	-2.6	3.2	106.0	0.1	5.7	1.1	12.3	8.2	15.2
Glu	Val	E	V	N	N	N	N	72.0	-2.6	3.2	42.0	2.3	6.0	2.1	18.7	8.8	11.2
Gly	Ala	G	A	N	N	P	P	0.0	0.7	6.0	14.0	1.0	6.0	4.3	4.7	10.4	19.4
Gly	Arg	G	R	N	P	N	P	0.0	0.7	6.0	99.0	-7.5	10.8	6.4	16.6	13.2	5.4
Gly	Asn	G	N	N	N	N	P	0.0	0.7	6.0	57.0	-2.7	5.4	2.1	10.7	10.1	14.0
Gly	Asp	G	D	N	N	N	P	0.0	0.7	6.0	58.0	-3.0	2.8	2.1	10.9	8.6	13.4
Gly	Cys	G	C	N	N	N	P	0.0	0.7	6.0	46.0	0.2	5.1	2.1	9.2	9.9	19.1
Gly	Gln	G	Q	N	P	N	N	0.0	0.7	6.0	71.0	-2.9	5.7	2.1	12.7	10.2	13.6
Gly	Glu	G	E	N	N	N	N	0.0	0.7	6.0	72.0	-2.6	3.2	2.1	12.8	8.8	14.1
Gly	Gly	G	G	N	N	N	P	0.0	0.7	6.0	0.0	0.7	6.0	4.3	2.7	10.4	20.0
Gly	His	G	H	N	P	N	P	0.0	0.7	6.0	80.0	-1.7	7.6	2.1	13.9	11.3	15.8
Gly	Ile	G	I	N	N	N	P	0.0	0.7	6.0	56.0	3.1	5.9	3.2	10.6	10.4	15.6
Gly	Leu	G	L	N	P	N	P	0.0	0.7	6.0	56.0	2.2	6.0	6.4	10.6	10.4	17.3
Gly	Lys	G	K	N	N	N	N	0.0	0.7	6.0	71.0	-4.6	9.7	2.1	12.7	12.6	10.6
Gly	Met	G	M	N	N	N	N	0.0	0.7	6.0	74.0	1.1	5.7	1.1	13.1	10.3	19.2
Gly	Phe	G	F	N	N	N	P	0.0	0.7	6.0	90.0	2.5	5.5	2.1	15.4	10.1	16.7
Gly	Pro	G	P	P	P	P	P	0.0	0.7	6.0	40.0	-0.3	6.5	4.3	8.3	10.7	18.3
Gly	Ser	G	S	N	N	P	P	0.0	0.7	6.0	30.0	-1.1	5.7	6.4	6.9	10.2	16.8
Gly	Thr	G	T	N	N	P	P	0.0	0.7	6.0	44.0	-0.8	5.9	4.3	8.9	10.3	17.5
Gly	Trp	G	W	N	N	N	N	0.0	0.7	6.0	129.0	1.5	6.0	1.1	19.2	10.4	18.5
Gly	Tyr	G	Y	N	N	N	P	0.0	0.7	6.0	106.0	0.1	5.7	2.1	17.6	10.2	18.9
Gly	Val	G	V	N	N	N	P	0.0	0.7	6.0	42.0	2.3	6.0	4.3	8.6	10.4	17.1

Table IV. Continued

First	Second	1st	2nd	PC1_C	PC2_C	PC1_RC	PC2_RC	mw-57_1	HP_1	pI_1	mw-57_2	HP_2	pI_2	NF'/1k	SCI	CCI	HCI
His	Ala	H	A	N	P	N	P	80.0	-1.7	7.6	14.0	1.0	6.0	2.1	15.9	11.3	15.2
His	Arg	H	R	N	N	N	P	80.0	-1.7	7.6	99.0	-7.5	10.8	3.2	12.1	9.8	9.6
His	Asn	H	N	N	N	N	N	80.0	-1.7	7.6	57.0	-2.7	5.4	1.1	18.0	11.5	18.2
His	Asp	H	D	N	N	N	N	80.0	-1.7	7.6	58.0	-3.0	2.8	1.1	17.9	12.4	17.7
His	Cys	H	C	N	N	N	N	80.0	-1.7	7.6	46.0	0.2	5.1	1.1	19.6	11.6	16.6
His	Gln	H	Q	N	N	N	N	80.0	-1.7	7.6	71.0	-2.9	5.7	1.1	16.1	11.4	17.8
His	Glu	H	E	N	P	N	P	80.0	-1.7	7.6	72.0	-2.6	3.2	1.1	15.9	12.3	18.4
His	Gly	H	G	N	P	N	P	80.0	-1.7	7.6	0.0	0.7	6.0	2.1	13.9	11.3	15.8
His	His	H	H	N	N	N	N	80.0	-1.7	7.6	80.0	-1.7	7.6	1.1	14.8	10.8	20.0
His	Ile	H	I	N	N	N	N	80.0	-1.7	7.6	56.0	3.1	5.9	1.6	18.2	11.4	11.4
His	Leu	H	L	N	N	N	N	80.0	-1.7	7.6	56.0	2.2	6.0	3.2	18.2	11.3	13.0
His	Lys	H	K	N	N	N	P	80.0	-1.7	7.6	71.0	-4.6	9.7	1.1	16.1	10.1	14.8
His	Met	H	M	N	N	P	P	80.0	-1.7	7.6	74.0	1.1	5.7	0.5	15.6	11.4	15.0
His	Phe	H	F	N	N	N	N	80.0	-1.7	7.6	90.0	2.5	5.5	1.1	13.4	11.5	12.5
His	Pro	H	P	N	N	N	N	80.0	-1.7	7.6	40.0	-0.3	6.5	2.1	19.6	11.2	17.5
His	Ser	H	S	N	N	N	N	80.0	-1.7	7.6	30.0	-1.1	5.7	3.2	18.2	11.4	18.9
His	Thr	H	T	N	N	N	P	80.0	-1.7	7.6	44.0	-0.8	5.9	2.1	19.9	11.4	18.3
His	Trp	H	W	N	N	N	N	80.0	-1.7	7.6	129.0	1.5	6.0	0.5	7.9	11.3	14.3
His	Tyr	H	Y	N	N	N	N	80.0	-1.7	7.6	106.0	0.1	5.7	1.1	11.1	11.4	16.8
His	Val	H	V	P	P	P	P	80.0	-1.7	7.6	42.0	2.3	6.0	2.1	19.9	11.3	12.8
Ile	Ala	I	A	N	N	N	P	56.0	3.1	5.9	14.0	1.0	6.0	3.2	12.5	10.4	16.2
Ile	Arg	I	R	N	N	N	N	56.0	3.1	5.9	99.0	-7.5	10.8	4.8	15.5	13.3	1.0
Ile	Asn	I	N	N	N	P	P	56.0	3.1	5.9	57.0	-2.7	5.4	1.6	18.6	10.0	9.6
Ile	Asp	I	D	N	N	P	P	56.0	3.1	5.9	58.0	-3.0	2.8	1.6	18.7	8.4	9.1
Ile	Cys	I	C	N	P	N	P	56.0	3.1	5.9	46.0	0.2	5.1	1.6	17.0	9.8	14.7
Ile	Gln	I	Q	N	N	N	N	56.0	3.1	5.9	71.0	-2.9	5.7	1.6	19.4	10.2	9.2
Ile	Glu	I	E	N	N	N	N	56.0	3.1	5.9	72.0	-2.6	3.2	1.6	19.3	8.7	9.8
Ile	Gly	I	G	N	N	N	P	56.0	3.1	5.9	0.0	0.7	6.0	3.2	10.6	10.4	15.6
Ile	His	I	H	N	N	N	N	56.0	3.1	5.9	80.0	-1.7	7.6	1.6	18.2	11.4	11.4
Ile	Ile	I	I	N	N	N	P	56.0	3.1	5.9	56.0	3.1	5.9	2.4	18.5	10.3	20.0
Ile	Leu	I	L	N	P	N	N	56.0	3.1	5.9	56.0	2.2	6.0	4.8	18.5	10.4	18.4
Ile	Lys	I	K	N	N	N	N	56.0	3.1	5.9	71.0	-4.6	9.7	1.6	19.4	12.7	6.2
Ile	Met	I	M	N	N	N	N	56.0	3.1	5.9	74.0	1.1	5.7	0.8	19.0	10.2	16.4
Ile	Phe	I	F	N	P	N	P	56.0	3.1	5.9	90.0	2.5	5.5	1.6	16.8	10.1	18.9
Ile	Pro	I	P	N	N	N	N	56.0	3.1	5.9	40.0	-0.3	6.5	3.2	16.2	10.7	13.9
Ile	Ser	I	S	N	P	N	P	56.0	3.1	5.9	30.0	-1.1	5.7	4.8	14.8	10.2	12.5
Ile	Thr	I	T	N	N	N	P	56.0	3.1	5.9	44.0	-0.8	5.9	3.2	16.8	10.3	13.1
Ile	Trp	I	W	N	P	N	N	56.0	3.1	5.9	129.0	1.5	6.0	0.8	11.3	10.4	17.1
Ile	Tyr	I	Y	P	P	P	P	56.0	3.1	5.9	106.0	0.1	5.7	1.6	14.5	10.2	14.6
Ile	Val	I	V	N	N	N	P	56.0	3.1	5.9	42.0	2.3	6.0	3.2	16.5	10.4	18.6
Leu	Ala	L	A	N	P	N	P	56.0	2.2	6.0	14.0	1.0	6.0	6.4	12.5	10.4	17.8
Leu	Arg	L	R	N	N	N	P	56.0	2.2	6.0	99.0	-7.5	10.8	9.7	15.5	13.2	2.6
Leu	Asn	L	N	P	P	N	N	56.0	2.2	6.0	57.0	-2.7	5.4	3.2	18.6	10.1	11.2
Leu	Asp	L	D	P	P	N	N	56.0	2.2	6.0	58.0	-3.0	2.8	3.2	18.7	8.6	10.7
Leu	Cys	L	C	N	N	N	N	56.0	2.2	6.0	46.0	0.2	5.1	3.2	17.0	9.9	18.4
Leu	Gln	L	Q	N	N	P	P	56.0	2.2	6.0	71.0	-2.9	5.7	3.2	19.4	10.2	10.9
Leu	Glu	L	E	P	P	P	P	56.0	2.2	6.0	72.0	-2.6	3.2	3.2	19.3	8.8	11.4
Leu	Gly	L	G	N	P	N	P	56.0	2.2	6.0	0.0	0.7	6.0	6.4	10.6	10.4	17.3
Leu	His	L	H	N	N	N	N	56.0	2.2	6.0	80.0	-1.7	7.6	3.2	18.2	11.3	13.0
Leu	Ile	L	I	N	P	N	N	56.0	2.2	6.0	56.0	3.1	5.9	4.8	18.5	10.4	18.4
Leu	Leu	L	L	N	N	N	P	56.0	2.2	6.0	56.0	2.2	6.0	9.7	18.5	10.4	20.0
Leu	Lys	L	K	N	N	P	P	56.0	2.2	6.0	71.0	-4.6	9.7	3.2	19.4	12.6	7.8
Leu	Met	L	M	N	N	N	P	56.0	2.2	6.0	74.0	1.1	5.7	1.6	19.0	10.3	18.0
Leu	Phe	L	F	N	N	N	N	56.0	2.2	6.0	90.0	2.5	5.5	3.2	16.8	10.1	19.5
Leu	Pro	L	P	N	N	N	P	56.0	2.2	6.0	40.0	-0.3	6.5	6.4	16.2	10.7	15.5
Leu	Ser	L	S	N	P	N	P	56.0	2.2	6.0	30.0	-1.1	5.7	9.7	14.8	10.2	14.1
Leu	Thr	L	T	N	P	N	P	56.0	2.2	6.0	44.0	-0.8	5.9	6.4	16.8	10.4	14.7
Leu	Trp	L	W	N	N	N	P	56.0	2.2	6.0	129.0	1.5	6.0	1.6	11.3	10.4	18.7
Leu	Tyr	L	Y	N	N	N	N	56.0	2.2	6.0	106.0	0.1	5.7	3.2	14.5	10.2	16.2
Leu	Val	L	V	N	P	N	P	56.0	2.2	6.0	42.0	2.3	6.0	6.4	16.5	10.4	19.8
Lys	Ala	K	A	N	N	N	N	71.0	-4.6	9.7	14.0	1.0	6.0	2.1	14.7	12.5	10.0
Lys	Arg	K	R	N	N	N	P	71.0	-4.6	9.7	99.0	-7.5	10.8	3.2	13.4	5.2	14.8
Lys	Asn	K	N	N	N	N	N	71.0	-4.6	9.7	57.0	-2.7	5.4	1.1	19.3	13.4	18.6
Lys	Asp	K	D	N	N	N	N	71.0	-4.6	9.7	58.0	-3.0	2.8	1.1	19.2	17.5	17.1
Lys	Cys	K	C	N	P	N	P	71.0	-4.6	9.7	46.0	0.2	5.1	1.1	19.2	14.0	11.5
Lys	Gln	K	Q	N	N	N	N	71.0	-4.6	9.7	71.0	-2.9	5.7	1.1	17.3	13.1	17.0
Lys	Glu	K	E	N	N	N	N	71.0	-4.6	9.7	72.0	-2.6	3.2	1.1	17.2	16.8	16.4
Lys	Gly	K	G	N	N	N	N	71.0	-4.6	9.7	0.0	0.7	6.0	2.1	12.7	12.6	10.6
Lys	His	K	H	N	N	N	P	71.0	-4.6	9.7	80.0	-1.7	7.6	1.1	16.1	10.1	14.8
Lys	Ile	K	I	N	N	N	N	71.0	-4.6	9.7	56.0	3.1	5.9	1.6	19.4	12.7	6.2
Lys	Leu	K	L	N	N	P	P	71.0	-4.6	9.7	56.0	2.2	6.0	3.2	19.4	12.6	7.8
Lys	Lys	K	K	N	N	N	N	71.0	-4.6	9.7	71.0	-4.6	9.7	1.1	17.3	6.9	20.0
Lys	Met	K	M	N	N	N	N	71.0	-4.6	9.7	74.0	1.1	5.7	0.5	16.9	12.9	9.8
Lys	Phe	K	F	P	P	P	P	71.0	-4.6	9.7	90.0	2.5	5.5	1.1	14.7	13.3	7.3
Lys	Pro	K	P	N	N	N	P	71.0	-4.6	9.7	40.0	-0.3	6.5	2.1	18.3	11.8	12.3
Lys	Ser	K	S	N	P	N	P	71.0	-4.6	9.7	30.0	-1.1	5.7	3.2	16.9	13.0	13.7
Lys	Thr	K	T	N	N	N	N	71.0	-4.6	9.7	44.0	-0.8	5.9	2.1	18.9	12.7	13.1
Lys	Trp	K	W	N	N	N	N	71.0	-4.6	9.7	129.0	1.5	6.0	0.5	9.2	12.6	9.1
Lys	Tyr	K	Y	N	P	N	P	71.0	-4.6	9.7	106.0	0.1	5.7	1.1	12.4	13.1	11.6
Lys	Val	K	V	N	N	N	N	71.0	-4.6	9.7	42.0	2.3	6.0	2.1	18.6	12.6	7.6

Table IV. Continued

First	Second	1st	2nd	PC1_C	PC2_C	PC1_RC	PC2_RC	mw-57_1	HP_1	pI_1	mw-57_2	HP_2	pI_2	NF*/1k	SCI	CCI	HCI
Met	Ala	M	A	N	N	N	N	74.0	1.1	5.7	14.0	1.0	6.0	1.1	15.1	10.3	19.8
Met	Arg	M	R	N	N	N	P	74.0	1.1	5.7	99.0	-7.5	10.8	1.6	13.0	13.7	4.8
Met	Asn	M	N	N	N	N	N	74.0	1.1	5.7	57.0	-2.7	5.4	0.5	18.9	9.9	13.2
Met	Asp	M	D	N	N	N	N	74.0	1.1	5.7	58.0	-3.0	2.8	0.5	18.7	8.0	12.7
Met	Cys	M	C	N	P	N	N	74.0	1.1	5.7	46.0	0.2	5.1	0.5	19.6	9.6	18.3
Met	Gln	M	Q	N	N	N	N	74.0	1.1	5.7	71.0	-2.9	5.7	0.5	16.9	10.0	12.8
Met	Glu	M	E	N	N	N	N	74.0	1.1	5.7	72.0	-2.6	3.2	0.5	16.8	8.3	13.4
Met	Gly	M	G	N	N	N	N	74.0	1.1	5.7	0.0	0.7	6.0	1.1	13.1	10.3	19.2
Met	His	M	H	N	N	P	P	74.0	1.1	5.7	80.0	-1.7	7.6	0.5	15.6	11.4	15.0
Met	Ile	M	I	N	N	N	N	74.0	1.1	5.7	56.0	3.1	5.9	0.8	19.0	10.2	16.4
Met	Leu	M	L	N	N	N	P	74.0	1.1	5.7	56.0	2.2	6.0	1.6	19.0	10.3	18.0
Met	Lys	M	K	N	N	N	N	74.0	1.1	5.7	71.0	-4.6	9.7	0.5	16.9	12.9	9.8
Met	Met	M	M	N	N	N	N	74.0	1.1	5.7	74.0	1.1	5.7	0.3	16.5	10.1	20.0
Met	Phe	M	F	N	P	N	N	74.0	1.1	5.7	90.0	2.5	5.5	0.5	14.2	9.9	17.5
Met	Pro	M	P	N	N	N	P	74.0	1.1	5.7	40.0	-0.3	6.5	1.1	18.7	10.6	17.5
Met	Ser	M	S	N	P	N	N	74.0	1.1	5.7	30.0	-1.1	5.7	1.6	17.3	10.1	18.1
Met	Thr	M	T	N	N	N	N	74.0	1.1	5.7	44.0	-0.8	5.9	1.1	19.3	10.2	16.7
Met	Trp	M	W	N	N	N	N	74.0	1.1	5.7	129.0	1.5	6.0	0.3	8.7	10.3	19.3
Met	Tyr	M	Y	P	P	N	N	74.0	1.1	5.7	106.0	0.1	5.7	0.5	12.0	10.1	18.2
Met	Val	M	V	N	N	N	N	74.0	1.1	5.7	42.0	2.3	6.0	1.1	19.0	10.3	17.8
Phe	Ala	F	A	N	N	N	P	90.0	2.5	5.5	14.0	1.0	6.0	2.1	17.3	10.2	17.3
Phe	Arg	F	R	N	P	N	P	90.0	2.5	5.5	99.0	-7.5	10.8	3.2	10.7	14.2	2.1
Phe	Asn	F	N	N	N	N	N	90.0	2.5	5.5	57.0	-2.7	5.4	1.1	16.6	9.6	10.7
Phe	Asp	F	D	N	N	N	N	90.0	2.5	5.5	58.0	-3.0	2.8	1.1	16.5	7.4	10.1
Phe	Cys	F	C	N	N	N	N	90.0	2.5	5.5	46.0	0.2	5.1	1.1	18.2	9.4	15.8
Phe	Gln	F	Q	N	N	N	N	90.0	2.5	5.5	71.0	-2.9	5.7	1.1	14.7	9.8	10.3
Phe	Glu	F	E	N	N	P	P	90.0	2.5	5.5	72.0	-2.6	3.2	1.1	14.5	7.8	10.9
Phe	Gly	F	G	N	N	N	P	90.0	2.5	5.5	0.0	0.7	6.0	2.1	15.4	10.1	16.7
Phe	His	F	H	N	N	N	N	90.0	2.5	5.5	80.0	-1.7	7.6	1.1	13.4	11.5	12.5
Phe	Ile	F	I	N	P	N	P	90.0	2.5	5.5	56.0	3.1	5.9	1.6	16.8	10.1	18.9
Phe	Leu	F	L	N	N	N	N	90.0	2.5	5.5	56.0	2.2	6.0	3.2	16.8	10.1	19.5
Phe	Lys	F	K	P	P	P	P	90.0	2.5	5.5	71.0	-4.6	9.7	1.1	14.7	13.3	7.3
Phe	Met	F	M	N	P	N	N	90.0	2.5	5.5	74.0	1.1	5.7	0.5	14.2	9.9	17.5
Phe	Phe	F	F	N	N	N	N	90.0	2.5	5.5	90.0	2.5	5.5	1.1	12.0	9.7	20.0
Phe	Pro	F	P	N	N	N	N	90.0	2.5	5.5	40.0	-0.3	6.5	2.1	19.0	10.6	15.0
Phe	Ser	F	S	N	N	N	N	90.0	2.5	5.5	30.0	-1.1	5.7	3.2	19.6	9.9	13.5
Phe	Thr	F	T	N	P	N	P	90.0	2.5	5.5	44.0	-0.8	5.9	2.1	18.5	10.0	14.2
Phe	Trp	F	W	N	N	N	N	90.0	2.5	5.5	129.0	1.5	6.0	0.5	6.5	10.1	18.2
Phe	Tyr	F	Y	N	N	N	N	90.0	2.5	5.5	106.0	0.1	5.7	1.1	9.7	9.9	15.7
Phe	Val	F	V	N	N	N	P	90.0	2.5	5.5	42.0	2.3	6.0	2.1	18.7	10.1	19.6
Pro	Ala	P	A	N	P	N	P	40.0	-0.3	6.5	14.0	1.0	6.0	4.3	10.3	10.7	17.7
Pro	Arg	P	R	N	N	P	P	40.0	-0.3	6.5	99.0	-7.5	10.8	6.4	17.7	12.1	7.1
Pro	Asn	P	N	N	N	N	N	40.0	-0.3	6.5	57.0	-2.7	5.4	2.1	16.3	10.5	15.7
Pro	Asp	P	D	N	P	N	N	40.0	-0.3	6.5	58.0	-3.0	2.8	2.1	16.5	9.8	15.1
Pro	Cys	P	C	N	N	N	N	40.0	-0.3	6.5	46.0	0.2	5.1	2.1	14.8	10.4	19.2
Pro	Gln	P	Q	N	N	N	P	40.0	-0.3	6.5	71.0	-2.9	5.7	2.1	18.3	10.6	15.3
Pro	Glu	P	E	N	P	N	P	40.0	-0.3	6.5	72.0	-2.6	3.2	2.1	18.5	9.9	15.9
Pro	Gly	P	G	P	P	P	P	40.0	-0.3	6.5	0.0	0.7	6.0	4.3	8.3	10.7	18.3
Pro	His	P	H	N	N	N	N	40.0	-0.3	6.5	80.0	-1.7	7.6	2.1	19.6	11.2	17.5
Pro	Ile	P	I	N	N	N	N	40.0	-0.3	6.5	56.0	3.1	5.9	3.2	16.2	10.7	13.9
Pro	Leu	P	L	N	N	N	P	40.0	-0.3	6.5	56.0	2.2	6.0	6.4	16.2	10.7	15.5
Pro	Lys	P	K	N	N	N	P	40.0	-0.3	6.5	71.0	-4.6	9.7	2.1	18.3	11.8	12.3
Pro	Met	P	M	N	N	N	P	40.0	-0.3	6.5	74.0	1.1	5.7	1.1	18.7	10.6	17.5
Pro	Phe	P	F	N	N	N	N	40.0	-0.3	6.5	90.0	2.5	5.5	2.1	19.0	10.6	15.0
Pro	Pro	P	P	N	N	N	P	40.0	-0.3	6.5	40.0	-0.3	6.5	4.3	13.9	10.8	20.0
Pro	Ser	P	S	N	N	N	P	40.0	-0.3	6.5	30.0	-1.1	5.7	6.4	12.5	10.6	18.5
Pro	Thr	P	T	N	N	N	P	40.0	-0.3	6.5	44.0	-0.8	5.9	4.3	14.5	10.7	19.2
Pro	Trp	P	W	N	N	P	P	40.0	-0.3	6.5	129.0	1.5	6.0	1.1	13.5	10.7	16.8
Pro	Tyr	P	Y	N	N	N	N	40.0	-0.3	6.5	106.0	0.1	5.7	2.1	16.8	10.6	19.3
Pro	Val	P	V	N	P	N	P	40.0	-0.3	6.5	42.0	2.3	6.0	4.3	14.2	10.7	15.4
Ser	Ala	S	A	N	N	P	P	30.0	-1.1	5.7	14.0	1.0	6.0	6.4	8.9	10.3	16.2
Ser	Arg	S	R	P	P	P	P	30.0	-1.1	5.7	99.0	-7.5	10.8	9.7	19.2	13.8	8.5
Ser	Asn	S	N	N	P	N	P	30.0	-1.1	5.7	57.0	-2.7	5.4	3.2	14.9	9.8	17.1
Ser	Asp	S	D	N	N	N	P	30.0	-1.1	5.7	58.0	-3.0	2.8	3.2	15.1	7.9	16.6
Ser	Cys	S	C	N	N	N	N	30.0	-1.1	5.7	46.0	0.2	5.1	3.2	13.4	9.6	17.7
Ser	Gln	S	Q	N	N	N	P	30.0	-1.1	5.7	71.0	-2.9	5.7	3.2	16.9	10.0	16.8
Ser	Glu	S	E	N	N	N	P	30.0	-1.1	5.7	72.0	-2.6	3.2	3.2	17.0	8.2	17.3
Ser	Gly	S	G	N	N	P	P	30.0	-1.1	5.7	0.0	0.7	6.0	6.4	6.9	10.2	16.8
Ser	His	S	H	N	N	N	N	30.0	-1.1	5.7	80.0	-1.7	7.6	3.2	18.2	11.4	18.9
Ser	Ile	S	I	N	P	N	P	30.0	-1.1	5.7	56.0	3.1	5.9	4.8	14.8	10.2	12.5
Ser	Leu	S	L	N	P	N	P	30.0	-1.1	5.7	56.0	2.2	6.0	9.7	14.8	10.2	14.1
Ser	Lys	S	K	N	P	N	P	30.0	-1.1	5.7	71.0	-4.6	9.7	3.2	16.9	13.0	13.7
Ser	Met	S	M	N	P	N	N	30.0	-1.1	5.7	74.0	1.1	5.7	1.6	17.3	10.1	16.1
Ser	Phe	S	F	N	N	N	N	30.0	-1.1	5.7	90.0	2.5	5.5	3.2	19.6	9.9	13.5
Ser	Pro	S	P	N	N	N	P	30.0	-1.1	5.7	40.0	-0.3	6.5	6.4	12.5	10.6	18.5
Ser	Ser	S	S	P	P	N	P	30.0	-1.1	5.7	30.0	-1.1	5.7	9.7	11.1	10.0	20.0
Ser	Thr	S	T	N	P	P	P	30.0	-1.1	5.7	44.0	-0.8	5.9	6.4	13.1	10.2	19.4
Ser	Trp	S	W	N	P	N	N	30.0	-1.1	5.7	129.0	1.5	6.0	1.6	14.9	10.2	15.3
Ser	Tyr	S	Y	N	N	N	N	30.0	-1.1	5.7	106.0	0.1	5.7	3.2	18.2	10.0	17.9
Ser	Val	S	V	N	N	N	P	30.0	-1.1	5.7	42.0	2.3	6.0	6.4	12.8	10.2	13.9

Table IV. Continued

First	Second	1st	2nd	PC1_C	PC2_C	PC1_RC	PC2_RC	mw-57_1	HP_1	pI_1	mw-57_2	HP_2	pI_2	NF'/1k	SCI	CCI	HCI
Thr	Ala	T	A	N	N	N	P	44.0	-0.8	5.9	14.0	1.0	6.0	4.3	10.9	10.4	16.9
Thr	Arg	T	R	N	N	P	P	44.0	-0.8	5.9	99.0	-7.5	10.8	6.4	17.2	13.4	7.9
Thr	Asn	T	N	N	N	N	P	44.0	-0.8	5.9	57.0	-2.7	5.4	2.1	16.9	10.0	16.5
Thr	Asp	T	D	N	N	N	P	44.0	-0.8	5.9	58.0	-3.0	2.8	2.1	17.0	8.3	16.0
Thr	Cys	T	C	P	P	P	P	44.0	-0.8	5.9	46.0	0.2	5.1	2.1	15.4	9.8	18.4
Thr	Gln	T	Q	N	N	N	N	44.0	-0.8	5.9	71.0	-2.9	5.7	2.1	18.9	10.1	16.1
Thr	Glu	T	E	N	N	N	N	44.0	-0.8	5.9	72.0	-2.6	3.2	2.1	19.0	8.6	16.7
Thr	Gly	T	G	N	N	P	P	44.0	-0.8	5.9	0.0	0.7	6.0	4.3	8.9	10.3	17.5
Thr	His	T	H	N	N	N	P	44.0	-0.8	5.9	80.0	-1.7	7.6	2.1	19.9	11.4	18.3
Thr	Ile	T	I	N	N	N	P	44.0	-0.8	5.9	56.0	3.1	5.9	3.2	16.8	10.3	13.1
Thr	Leu	T	L	N	P	N	P	44.0	-0.8	5.9	56.0	2.2	6.0	6.4	16.8	10.4	14.7
Thr	Lys	T	K	N	N	N	N	44.0	-0.8	5.9	71.0	-4.6	9.7	2.1	18.9	12.7	13.1
Thr	Met	T	M	N	N	N	N	44.0	-0.8	5.9	74.0	1.1	5.7	1.1	19.3	10.2	16.7
Thr	Phe	T	F	N	P	N	P	44.0	-0.8	5.9	90.0	2.5	5.5	2.1	18.5	10.0	14.2
Thr	Pro	T	P	N	N	N	P	44.0	-0.8	5.9	40.0	-0.3	6.5	4.3	14.5	10.7	19.2
Thr	Ser	T	S	N	P	P	P	44.0	-0.8	5.9	30.0	-1.1	5.7	6.4	13.1	10.2	19.4
Thr	Thr	T	T	N	N	N	P	44.0	-0.8	5.9	44.0	-0.8	5.9	4.3	15.1	10.3	20.0
Thr	Trp	T	W	P	P	N	N	44.0	-0.8	5.9	129.0	1.5	6.0	1.1	13.0	10.4	16.0
Thr	Tyr	T	Y	N	P	N	P	44.0	-0.8	5.9	106.0	0.1	5.7	2.1	16.2	10.1	18.5
Thr	Val	T	V	N	N	N	P	44.0	-0.8	5.9	42.0	2.3	6.0	4.3	14.8	10.3	14.5
Trp	Ala	W	A	N	N	N	N	44.0	-0.8	5.9	14.0	1.0	6.0	1.1	17.2	10.4	19.1
Trp	Arg	W	R	N	N	N	P	129.0	1.5	6.0	99.0	-7.5	10.8	1.6	5.2	13.2	3.9
Trp	Asn	W	N	N	P	N	N	129.0	1.5	6.0	57.0	-2.7	5.4	0.5	11.1	10.1	12.5
Trp	Asp	W	D	N	N	N	N	44.0	-0.8	5.9	58.0	-3.0	2.8	0.5	11.0	8.6	11.9
Trp	Cys	W	C	N	N	N	N	44.0	-0.8	5.9	46.0	0.2	5.1	0.5	12.7	9.9	17.6
Trp	Gln	W	Q	N	N	N	P	129.0	1.5	6.0	71.0	-2.9	5.7	0.5	9.2	10.2	12.1
Trp	Glu	W	E	N	N	N	N	44.0	-0.8	5.9	72.0	-2.6	3.2	0.5	9.0	8.8	12.7
Trp	Gly	W	G	N	N	N	N	44.0	-0.8	5.9	0.0	0.7	6.0	1.1	19.2	10.4	18.5
Trp	His	W	H	N	N	N	N	44.0	-0.8	5.9	80.0	-1.7	7.6	0.5	7.9	11.3	14.3
Trp	Ile	W	I	N	P	N	N	129.0	1.5	6.0	56.0	3.1	5.9	0.8	11.3	10.4	17.1
Trp	Leu	W	L	N	N	N	P	129.0	1.5	6.0	56.0	2.2	6.0	1.6	11.3	10.4	18.7
Trp	Lys	W	K	N	N	N	N	44.0	-0.8	5.9	71.0	-4.6	9.7	0.5	9.2	12.6	9.1
Trp	Met	W	M	N	N	N	N	44.0	-0.8	5.9	74.0	1.1	5.7	0.3	8.7	10.3	19.3
Trp	Phe	W	F	N	N	N	N	44.0	-0.8	5.9	90.0	2.5	5.5	0.5	6.5	10.1	18.2
Trp	Pro	W	P	N	N	P	P	129.0	1.5	6.0	40.0	-0.3	6.5	1.1	13.5	10.7	16.8
Trp	Ser	W	S	N	P	N	N	129.0	1.5	6.0	30.0	-1.1	5.7	1.6	14.9	10.2	15.3
Trp	Thr	W	T	P	P	N	N	129.0	1.5	6.0	44.0	-0.8	5.9	1.1	13.0	10.4	16.0
Trp	Trp	W	W	N	N	N	N	44.0	-0.8	5.9	129.0	1.5	6.0	0.3	1.0	10.4	20.0
Trp	Tyr	W	Y	N	N	N	N	44.0	-0.8	5.9	106.0	0.1	5.7	0.5	4.2	10.2	17.5
Trp	Val	W	V	N	N	N	N	44.0	-0.8	5.9	42.0	2.3	6.0	1.1	13.2	10.4	18.6
Tyr	Ala	Y	A	N	N	N	P	106.0	0.1	5.7	14.0	1.0	6.0	2.1	19.6	10.3	18.4
Tyr	Arg	Y	R	N	P	N	P	106.0	0.1	5.7	99.0	-7.5	10.8	3.2	8.5	13.8	6.4
Tyr	Asn	Y	N	N	N	N	N	106.0	0.1	5.7	57.0	-2.7	5.4	1.1	14.4	9.8	15.0
Tyr	Asp	Y	D	N	N	N	N	106.0	0.1	5.7	58.0	-3.0	2.8	1.1	14.2	7.8	14.5
Tyr	Cys	Y	C	N	N	N	N	106.0	0.1	5.7	46.0	0.2	5.1	1.1	15.9	9.5	19.8
Tyr	Gln	Y	Q	N	N	N	N	106.0	0.1	5.7	71.0	-2.9	5.7	1.1	12.4	10.0	14.7
Tyr	Glu	Y	E	N	N	N	P	106.0	0.1	5.7	72.0	-2.6	3.2	1.1	12.3	8.2	15.2
Tyr	Gly	Y	G	N	N	N	P	106.0	0.1	5.7	0.0	0.7	6.0	2.1	17.6	10.2	18.9
Tyr	His	Y	H	N	N	N	N	106.0	0.1	5.7	80.0	-1.7	7.6	1.1	11.1	11.4	16.8
Tyr	Ile	Y	I	P	P	P	P	106.0	0.1	5.7	56.0	3.1	5.9	1.6	14.5	10.2	14.6
Tyr	Leu	Y	L	N	N	N	N	106.0	0.1	5.7	56.0	2.2	6.0	3.2	14.5	10.2	16.2
Tyr	Lys	Y	K	N	P	N	P	106.0	0.1	5.7	71.0	-4.6	9.7	1.1	12.4	13.1	11.6
Tyr	Met	Y	M	P	P	N	N	106.0	0.1	5.7	74.0	1.1	5.7	0.5	12.0	10.1	18.2
Tyr	Phe	Y	F	N	N	N	N	106.0	0.1	5.7	90.0	2.5	5.5	1.1	9.7	9.9	15.7
Tyr	Pro	Y	P	N	N	N	N	106.0	0.1	5.7	40.0	-0.3	6.5	2.1	16.8	10.6	19.3
Tyr	Ser	Y	S	N	N	N	N	106.0	0.1	5.7	30.0	-1.1	5.7	3.2	18.2	10.0	17.9
Tyr	Thr	Y	T	N	P	N	P	106.0	0.1	5.7	44.0	-0.8	5.9	2.1	16.2	10.1	18.5
Tyr	Trp	Y	W	N	N	N	N	106.0	0.1	5.7	129.0	1.5	6.0	0.5	4.2	10.2	17.5
Tyr	Tyr	Y	Y	N	N	N	N	106.0	0.1	5.7	106.0	0.1	5.7	1.1	7.5	10.0	20.0
Tyr	Val	Y	V	N	N	P	P	106.0	0.1	5.7	42.0	2.3	6.0	2.1	16.5	10.2	16.0
Val	Ala	V	A	N	N	N	P	42.0	2.3	6.0	14.0	1.0	6.0	4.3	10.6	10.4	17.7
Val	Arg	V	R	N	P	N	P	42.0	2.3	6.0	99.0	-7.5	10.8	6.4	17.5	13.2	2.4
Val	Asn	V	N	N	N	P	P	42.0	2.3	6.0	57.0	-2.7	5.4	2.1	16.6	10.1	11.0
Val	Asp	V	D	N	N	P	P	42.0	2.3	6.0	58.0	-3.0	2.8	2.1	16.8	8.6	10.5
Val	Cys	V	C	N	N	N	P	42.0	2.3	6.0	46.0	0.2	5.1	2.1	15.1	9.9	16.2
Val	Gln	V	Q	P	P	N	N	42.0	2.3	6.0	71.0	-2.9	5.7	2.1	18.6	10.2	10.7
Val	Glu	V	E	N	N	N	N	42.0	2.3	6.0	72.0	-2.6	3.2	2.1	18.7	8.8	11.2
Val	Gly	V	G	N	N	N	P	42.0	2.3	6.0	0.0	0.7	6.0	4.3	8.6	10.4	17.1
Val	His	V	H	P	P	P	P	42.0	2.3	6.0	80.0	-1.7	7.6	2.1	19.9	11.3	12.8
Val	Ile	V	I	N	N	N	P	42.0	2.3	6.0	56.0	3.1	5.9	3.2	16.5	10.4	18.6
Val	Leu	V	L	N	P	N	P	42.0	2.3	6.0	56.0	2.2	6.0	6.4	16.5	10.4	19.8
Val	Lys	V	K	N	N	N	N	42.0	2.3	6.0	71.0	-4.6	9.7	2.1	18.6	12.6	7.6
Val	Met	V	M	N	N	N	N	42.0	2.3	6.0	74.0	1.1	5.7	1.1	19.0	10.3	17.8
Val	Phe	V	F	N	N	N	P	42.0	2.3	6.0	90.0	2.5	5.5	2.1	18.7	10.1	19.6
Val	Pro	V	P	N	P	N	P	42.0	2.3	6.0	40.0	-0.3	6.5	4.3	14.2	10.7	15.4
Val	Ser	V	S	N	N	N	P	42.0	2.3	6.0	30.0	-1.1	5.7	6.4	12.8	10.2	13.9
Val	Thr	V	T	N	N	N	P	42.0	2.3	6.0	44.0	-0.8	5.9	4.3	14.8	10.3	14.5
Val	Trp	V	W	N	N	N	N	42.0	2.3	6.0	129.0	1.5	6.0	1.1	13.2	10.4	18.6
Val	Tyr	V	Y	N	N	P	P	42.0	2.3	6.0	106.0	0.1	5.7	2.1	16.5	10.2	16.0
Val	Val	V	V	N	N	N	P	42.0	2.3	6.0	42.0	2.3	6.0	4.3	14.5	10.4	20.0

System and Method to Obtain Oligo-Peptides with Specific High Affinity to Query Proteins

I have developed a system and method to obtain oligo-peptides with specific high affinity to query proteins [97]. The method is built on the second generation Proteomic Code which is based on the partially complementary coding of interacting amino acids.

Figures 31 and 32 show the steps for producing the target proteins of this system with a high affinity for query proteins, wherein the primary structure is known for the query proteins.

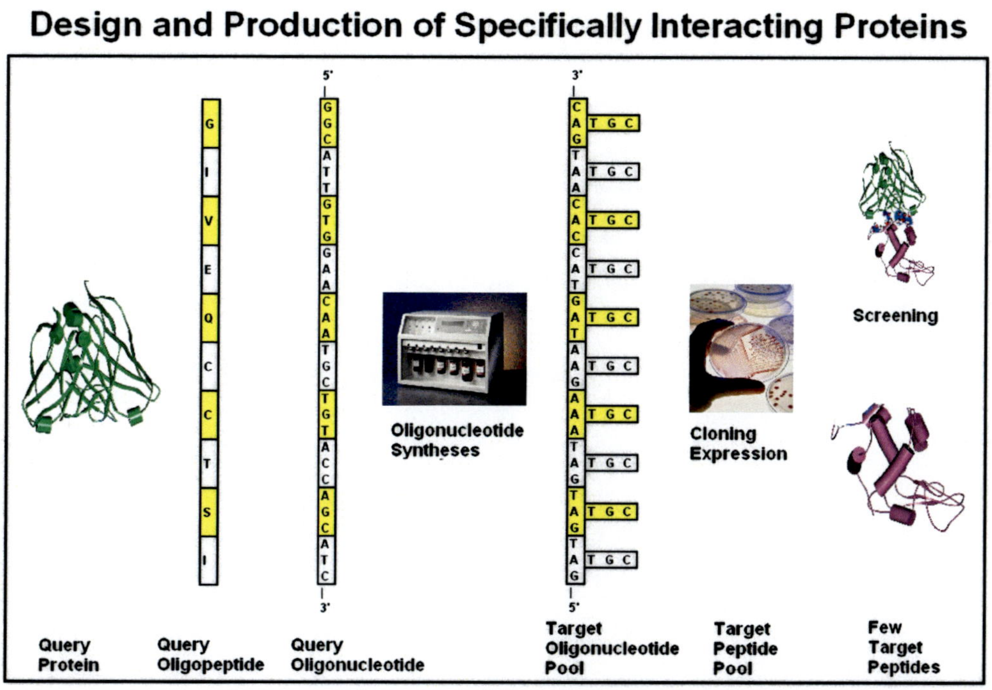

Figure 31. Design and production of specifically interacting proteins (see text for details).

There is no limitation to the size of the query; however, preferably the sequence is from about 5 amino acid residues to about 40 amino acid residues, and better from about 7 to 15 amino acid residues. Preferably, the real and natural coding sequence is known for the query protein. However, there might be some special cases when the sequence is not exactly known, for example in case of designed or artificially modified proteins. Thus, it is possible to fabricate a virtual coding sequence with back translation, using Codon Usage Frequency Tables. The present method relies on the entire information carried by the naturally occurring DNA/mRNA and not only that used for coding of the protein primary sequence.

A Novel Method to Obtain Oligo-peptides with Specific High Affinityto Query Proteins

1. Query Selection (Q1)
2. Target (T) RNA Pool Prediction
   (using partial codon-complementarity rule)
3. Syntheses of RNAs in the Target RNA Pool
   (modified method of RNA syntheses)
4. Cloning of the Target (T) RNA Pool
5. Target (T) Protein Pool Expression Library
6. Screening for Q-T interactions
7. Clone Selection
   (Few, unique, mono-clones) >>> Large scale protein production
8. Sequencing the best monoclonal RNAs and Proteins
9. Refine Protein - Protein Interaction Code >>> Gain knowledge
10. Iterate 1-9 on another query (Q2)

Figure 32. A novel method to obtain oligo-peptides with specific high affinity to query proteins. The steps required to practice the method for obtaining oligo-peptides with specific high affinity to query proteins.

The query sequence should be a "promising" domain of the query protein and specific domains are more important, including domains that: (a) are known to be antigenic; (b) are located on the surface of the query protein; (c) are not simple (repetitive) sequences; d) contain less frequent amino acids; and e) contain charged amino acid residues.

Once the promising area of the known amino acid sequence is chosen and the nucleotide sequence is determined, then construction of nucleic acid sequences encoding for the target proteins is initiated. As used here, the term "nucleotide sequence" means a sequence of nucleotides connected by phosphodiester linkages. Nucleotide sequences are presented here in the direction from the 5′ to the 3′ direction and can be a deoxyribonucleic acid (DNA) molecule or ribonucleic acid (RNA) molecule. Relevant nucleotide bases are indicated here by a single letter code: adenine (A), guanine (G), thymine (T), cytosine (C), inosine (I) and uracil (U). The target nucleotide (RNA or DNA) prediction should follow a simple rule, namely that the 1st and 3rd codon letters of the target nucleotide sequences should have reverse complementary to the 1st and 3rd codon nucleotide residues of the query nucleotide sequence, but the middle, 2nd residue should be any of the four possible nucleotides. The expected number of predicted target RNAs will be 4^n, where n is the number of amino acids (=number of codons, =number of 2nd codon letters).

Synthesis of the nucleotide sequences can be readily prepared by, for example, directly synthesizing the fragment by chemical means, by application of nucleic acid reproduction technology, such as the PCR or by excising selected DNA fragments from recombinant plasmids containing appropriate inserts and suitable restriction enzyme sites. However, synthesis of predicted (max. 4^n) sequences on a one by one basis does not seem practical.

Thus, a simple mass-production is needed which will result in a mixture, containing all possible sequences in the predicted RNA/DNA pool. Fortunately, the regular nature of the nucleotides in the pool makes it possible to synthesize the entire pool of nucleotide sequences as if were only one single nucleotide sequence. For example, the usual step-by-step (base by base) protocol can be followed except at the positions for the synthesis of the 2nd codon residue. At those points in the synthesis process, an equal mixture of the four nucleotides should be provided instead of a single nucleotide. The result of this modified oligo-nucleotide synthesis should be a mixture of the desired potential target RNAs.

The step to clone the predicted and synthesized RNAs in the pool is the regular cloning procedure which involves insertion of RNA into vector (plasmid or other carrier) and multiplying the sequences in bacteria or yeast as described in the literature. Expression vectors of the system may comprise polynucleotides operatively linked to an enhancer-promoter, such as a prokaryotic or eukaryotic promoter. Further, an enhancer may be included in the vector. A major function of an enhancer is to increase the level of transcription of a coding sequence in a cell that contains one or more transcription factors that bind to that enhancer. Unlike a promoter, an enhancer can function when located at variable distances from transcription start sites so long as a promoter is present.

Expression vectors of the present system comprise polynucleotides that encode the target peptides of the pool. Where expression of recombinant polypeptide of the present system is desired and a eukaryotic host is contemplated, it is most desirable to employ a vector, such as a plasmid, that incorporates a eukaryotic origin of replication. In addition, for the purposes of expression in eukaryotic systems, it is desired to position the peptide encoding sequence adjacent to and under the control of an effective eukaryotic promoter such as those used in combination with Chinese hamster ovary cells. To bring a coding sequence under the control of a promoter, whether it is eukaryotic or prokaryotic, what is generally needed is to position the 5′ end of the translation initiation side of the proper translational reading frame of the polypeptide between about 1 and 50 nucleotides 3′ of or downstream with respect to the promoter chosen. Furthermore, where eukaryotic expression is anticipated, one would typically desire to incorporate an appropriate polyadenylation site into the transcriptional unit which includes the different target peptides.

The pRc/CMV vector (available from Invitrogen) is an exemplary vector for expressing a peptide in mammalian cells, particularly COS and CHO cells. Target polypeptides of the present invention under the control of a CMV promoter can be efficiently expressed in mammalian cells. The pCMV plasmids are a series of mammalian expression vectors of particular utility in the present system. The vectors are designed for use in essentially all cultured cells and work extremely well in SV40-transformed simian COS cell lines. The pCMV1, 2, 3, and 5 vectors differ from each other in certain unique restriction sites in the polylinker region of each plasmid. The pCMV4 vector differs from these 4 plasmids in containing a translation enhancer in the sequence prior to the polylinker. While they are not directly derived from the pCMV1-5 series of vectors, the functionally similar pCMV6b and c vectors, available from the Chiron Corp. (Emeryville, CA), are identical except for the orientation of the polylinker region which is reversed in one relative to the other. The pCMV vectors have been successfully expressed in simian COS cells, mouse L cells, CHO cells, and HeLa cells.

Means of transforming or transfecting cells with exogenous polynucleotide such as the nucleotide molecules of the present system are well known and include techniques such as calcium-phosphate- or DEAE-dextran-mediated transfection, protoplast fusion, electroporation, liposome-mediated transfection, direct microinjection and adenovirus infection.

The most widely used method is transfection mediated by either calcium phosphate or DEAE-dextran. Although the mechanism remains obscure, it is believed that the transfected DNA enters the cytoplasm of the cell by endocytosis and is transported to the nucleus. Depending on the cell type, up to 90% of a population of cultured cells can be transfected at any one time. Because of its high efficiency, transfection mediated by calcium phosphate or DEAE-dextran is the method of choice for experiments that require transient expression of the foreign DNA in large numbers of cells. Calcium phosphate-mediated transfection is also used to establish cell lines that integrate copies of the foreign DNA, which are usually arranged in head-to-tail tandem arrays into the host cell genome.

The application of brief, high-voltage electric pulses to a variety of mammalian and plant cells leads to the formation of nanometer-sized pores in the plasma membrane. DNA is taken directly into the cell cytoplasm either through these pores or as a consequence of the redistribution of membrane components that accompanies closure of the pores. Electroporation can be extremely efficient and can be used both for transient expression of cloned genes and for establishment of cell lines that carry integrated copies of the gene of interest. Electroporation, in contrast to calcium phosphate-mediated transfection and protoplast fusion, frequently gives rise to cell lines that carry one, or at most a few, integrated copies of the foreign DNA.

Liposome transfection involves encapsulation of DNA or RNA within liposomes, followed by fusion of the liposomes with the cell membrane. The mechanism of how DNA or RNA is delivered into the cell is unclear but transfection efficiencies can be as high as 90%.

Direct microinjection of a DNA molecule into nuclei has the advantage of not exposing DNA to cellular compartments such as low pH endosomes. Microinjection is therefore used primarily as a method to establish lines of cells that carry integrated copies of the DNA of interest. A transfected cell can be prokaryotic or eukaryotic.

In addition to prokaryotes, eukaryotic microbes, such as yeast can also be used. *Saccharomyces cerevisiae* or baker's yeast is the most commonly used among eukaryotic microorganisms, although a number of other strains are also available. For expression in *Saccharomyces*, the plasmid YRp7, for example, is commonly used. This plasmid already contains the trpl gene which provides a selection marker for a mutant strain of yeast lacking the ability to grow in tryptophan, for example ATCC No. 44076 or PEP4-1. The presence of the trpl lesion as a characteristic of the yeast host cell genome then provides an effective environment for detecting transformation by growth in the absence of tryptophan. Suitable promoter sequences in yeast vectors include the promoters for 3-phosphoglycerate kinase or other glycolytic enzymes such as enolase, glyceraldehyde-3-phosphate dehydrogenase, hexokinase, pyruvate decarboxylase, phosphofructokinase, glucose-6-phosphate isomerase, 3-phosphoglycerate mutase, pyruvate kinase, triosephosphate isomerase, phosphoglucose isomerase, and glucokinase. In constructing suitable expression plasmids, the termination sequences associated with these genes are also introduced into the expression vector downstream from the sequences to be expressed to provide polyadenylation of the mRNA and termination. Other promoters, which have the additional advantage of transcription controlled by growth conditions, are the promoter region for alcohol dehydrogenase 2, isocytochrome *c*, acid phosphatase, degradative enzymes associated with nitrogen metabolism, and the aforementioned glyceraldehyde-3-phosphate dehydrogenase, and enzymes responsible for maltose and galactose utilization. Any plasmid vector containing a yeast-compatible promoter, origin or replication and termination sequences is suitable.

In addition to microorganisms, cultures of cells derived from multicellular organisms can also be used as hosts. In principle, any such cell culture is workable, whether from vertebrate or invertebrate culture. However, interest has been greatest in vertebrate cells, and propagation of vertebrate cells in culture (tissue culture) has become a routine procedure in recent years. Examples of such useful host cell lines are AtT-20, VERO and HeLa cells, Chinese hamster ovary (CHO) cell lines, and W138, BHK, COSM6, COS-1, COS-7, 293 and MDCK cell lines. Expression vectors for such cells ordinarily include (if necessary) an origin of replication, a promoter located upstream of the gene to be expressed, along with any necessary ribosome binding sites, RNA splice sites, polyadenylation site, and transcriptional terminator sequences.

For use in mammalian cells, the control functions on the expression vectors are often derived from viral material. For example, commonly used promoters are derived from polyoma, Adenovirus 2, Cytomegalovirus and most frequently Simian Virus 40 (SV40). The early and late promoters of SV40 virus are particularly useful because both are obtained easily from the virus as a fragment that also contains the SV40 viral origin of replication. Smaller or larger SV40 fragments can also be used, provided they include the approximately 250 bp sequence extending from the HindIII site towards the BglI site located in the viral origin of replication. It is also possible, and often desirable, to utilize promoter or control sequences normally associated with the desired gene sequence, provided such control sequences are compatible with the host cell systems.

Following transfection, the cell is maintained under culture conditions for a period of time sufficient for expression of the target proteins of the pool. Culture conditions are well known and include ionic composition and concentration, temperature, pH, etc. Typically, transfected cells are maintained under culture conditions in a culture medium. Suitable medium for various cell types are well known. Temperature is preferably from about 20°C to about 50°C. pH is preferably from about a value of 6.0 to a value of about 8.0, better at about 6.8–7.8 and, better again at about 7.4. Other biological conditions needed for transfection and expression of an encoded protein are well known.

Transfected cells are maintained for a period of time sufficient for expression of the target proteins. A suitable time depends inter alia upon the cell type used and is readily determinable by a skilled artisan. Typically, maintenance time is from about 2 to 14 days. Recovery of the target proteins comprises isolating and purifying the recombinant polypeptides. Isolation and purification techniques for polypeptides are well known and include such procedures as precipitation, filtration, chromatography, electrophoresis, etc.

The target proteins are preferably arranged in a library assay system for screening with samples of the query protein. Any method that detects specific, high affinity protein–protein interactions is theoretically useful to perform the screening.

Selecting the best clones with the most specific and highest affinity interacting proteins can be followed by repeated screenings, thus leading to the most desired target proteins having the highest binding affinity for the query protein. The target proteins with the highest affinity are suitable for large-scale target protein production.

These aspects and embodiments of the present system are further described in the following examples. However, the present system is not limited by the following example, and variations will be apparent to those skilled in the art without departing from the scope of the present setup.

Example 1

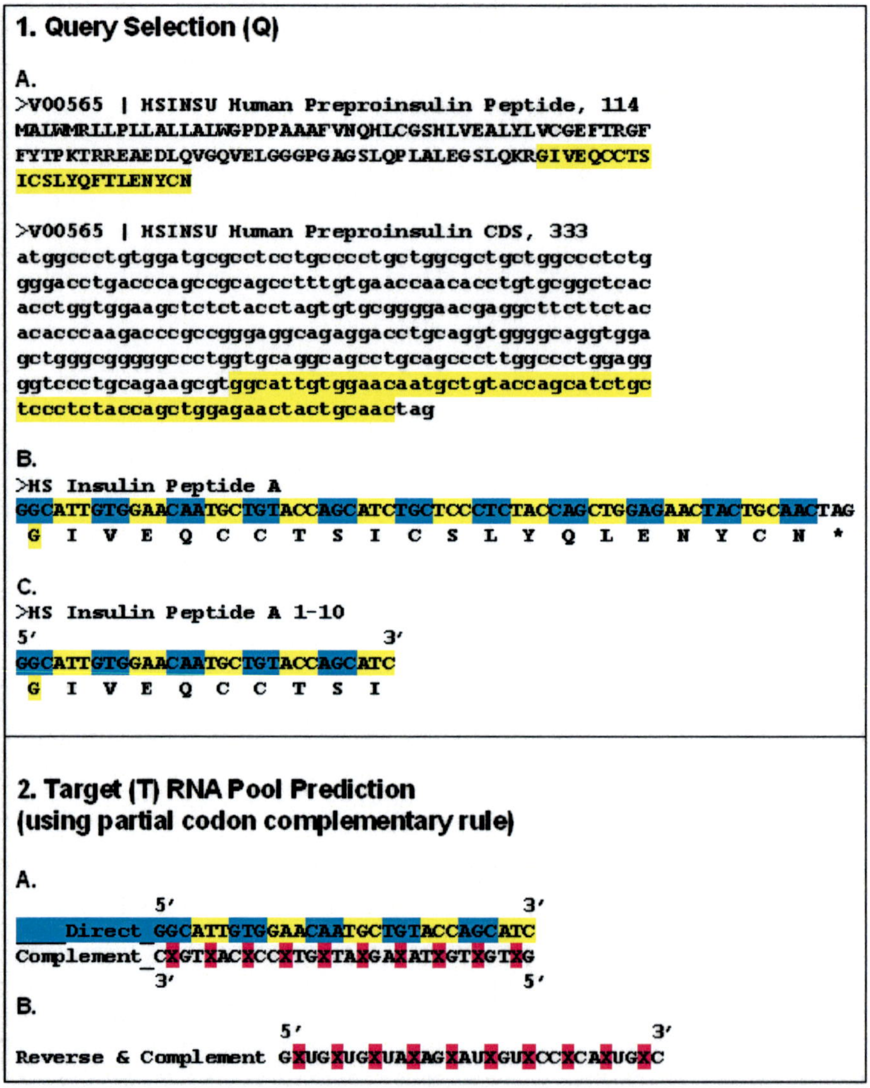

Figure 33. Representative query amino acid sequences and preparation of reverse and complement sequences wherein the second nucleotide of each codon is replaced with a variable X nucleotide.

Figure 33 shows the use of the present system to obtain a specific high affinity protein with binding affinity for a section of the A-peptide in human insulin.

Starting with the known protein and nucleic acid sequence of the entire Pre-pro-insulin, 1–10 residues of the A peptide and the corresponding nucleic acid sequence are selected. The selected part of the peptide, called the query, is used for screening of the target protein expression library. Therefore, this sequence should be available in pure peptide form.

Next, a sequence is created which is complementary to the query nucleotide sequence at the 1st and 3rd codon positions but leaving the 2nd position undefined (X). The complementary sequence is reversed and in this particular example, the bases T are changed to U. The second (central) codon position remains undefined and this undefined X position can be any one of the possible nucleotides (A, U, G, C). Therefore, this prediction method defines many different target RNA sequences. In the case of a sequence including 30 nucleotide bases, the expected number of possible target sequences will be about $4^{10}=10^{6}$.

The predicted pool of target RNAs is synthesized by following the usual step-by-step (base by base) protocol, known to those skilled in the art, except the syntheses of the X positions. At the X position, a mixture of nucleotide bases is provided (which contain equal amounts of A and U and G and C). The result of this modified oligo-nucleotide synthesis is a mixture of the desired potential target RNAs as shown in Figure 34. The target RNAs are cloned and transfected, via an expression vector, into a cell for expression therein of the encoded protein. An expression library of the expressed target protein is created for screening for query protein/target protein affinity binding. When binding complexes are found to meet the affinity binding levels, the target protein may be cloned for large-scale production.

Syntheses of RNAs in Target RNA Pool (modified method of RNA syntheses)

1. Step	2. Step	5. Step	6. Step	7. Step
5′	GA	GAUGA	GAUGAU	GAUGAUG
G	GU	GUUGA	GUUGAU	GUUGAUG
	GG	GGUGA	GGUGAU	GGUGAUG
	GC	GCUGA	GCUGAU	GCUGAUG
		GAUGU	GAUGUU	GAUGUUG
	3. Step	GUUGU	GUUGUU	GUUGUUG
	GAU	GGUGU	GGUGUU	GGUGUUG
	GUU	GCUGU	GCUGUU	GCUGUUG
	GGU	GAUGG	GAUGGU	GAUGGUG
	GCU	GUUGG	GUUGGU	GUUGGUG
		GGUGG	GGUGGU	GGUGGUG
	4. Step	GCUGG	GCUGGU	GCUGGUG
	GAUG	GAUGC	GAUGCU	GAUGCUG
	GUUG	GUUGC	GUUGCU	GUUGCUG
	GGUG	GGUGC	GGUGCU	GGUGCUG
	GCUG	GCUGC	GCUGCU	GCUGCUG

Colored background: addition of nucleotide mixture

Figure 34. Synthesis pattern for construction of target sequences and the progression of the permutations depending on the number of amino acid residues.

These steps may be repeated numerous times by modifying the length of the query sequence and/or using another domain area of the query protein that may be of interest.

Example 2. Example for designing and characterization of a specific protein–protein interaction

The BacterioMatch™ two-hybrid system (Stratagene, 11011 N. Torrey Pines Road, La Jolla, CA 92037) was used to quickly detect protein–protein interactions designed by the recent method. It is a simple alternative or complement to yeast two-hybrid systems for in vivo detection of protein–protein interactions. Because the two-hybrid assay is performed in bacteria, the results are obtained more easily and quickly than in yeast. The system is based on transcriptional activation of a primary ampicillin-resistant reporter and a secondary β-galactosidase reporter for validation. The BacterioMatch two-hybrid system is based on a methodology developed by Dove, Joung, and Hochschild of Harvard Medical School.

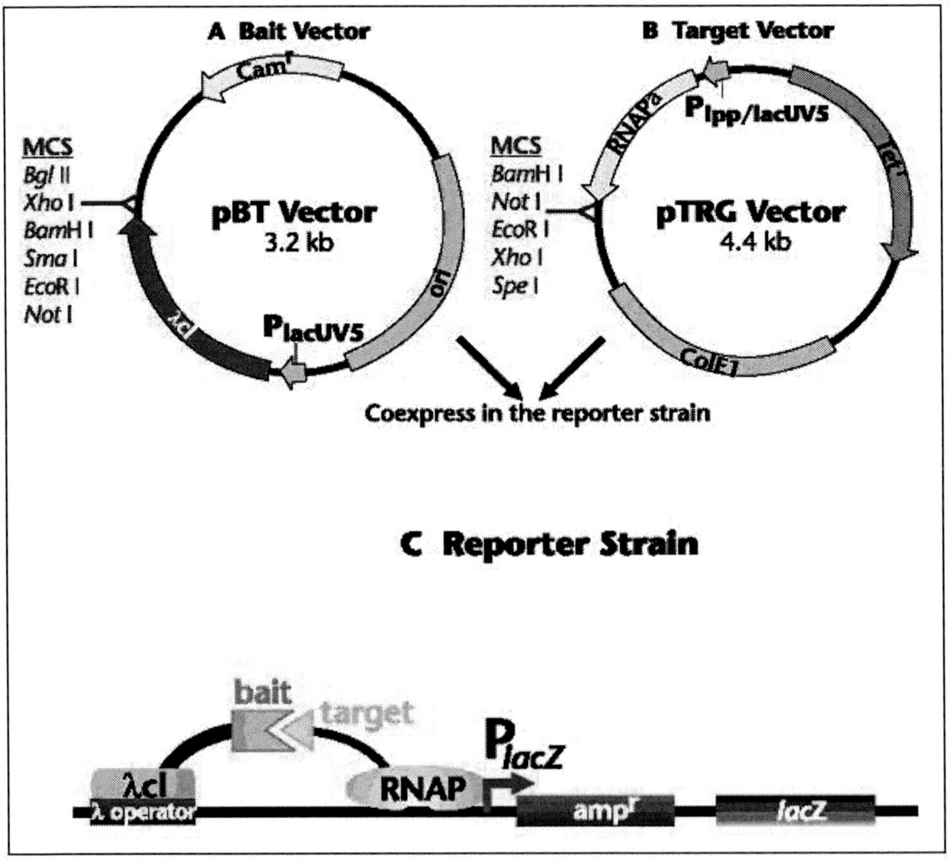

Figure 35. The BacterioMatch™. Two-hybrid system (reproduced from http://www.strategene.com).

The BacterioMatch two-hybrid system is based on transcriptional activation (Figure 35). A protein of interest – the bait – is fused to the full-length bacteriophage repressor protein (λcI). The corresponding target protein is fused to the amino-terminal domain of the α-subunit of RNA polymerase (RNAPα). The bait is tethered to the x operator sequence upstream of the reporter promoter through the DNA-binding domain of λcI. If the bait and target interact, they recruit and stabilize the binding of RNA polymerase close to the promoter and activate the transcription of the ampicillin-resistant reporter gene in the BacterioMatch two-hybrid reporter strain. The β-galactosidase reporter gene provides an additional mechanism to validate putative protein–protein interactions.

(a) *Bait vector*. The bait vector, pBT encodes the full-length bacterial phage cI protein under the control of the strong lacUV5 promoter. A protein of interest is fused to the bacterial phage λcI protein by inserting its gene into the multiple cloning site at the 3′ end of the λcI gene. The presence of a multiple cloning site present makes it convenient to subclone a bait gene that is already present in many yeast two-hybrid bait plasmids.
(b) *Target vector*. The target plasmid, pTRG, is compatible with Stratagene's cDNA library construction kit. The target plasmid directs transcription of the amino-terminal domain of RNA polymerase α-subunit and linker region under the control of tandem promoters, lpp and lacUV5. The target gene is fused in-frame to the α-subunit NTD through a multiple cloning site at the 3′ end of the α-subunit gene.
(c) *Reporter strain*. The reporter strain is derived from XL1-Blue MRF′. The strain lacks all restriction systems in order to be compatible with current cDNA library construction methods. The lac I^q gene located on the F′ episome represses synthesis of the bait and target until induction. The reporter cassette is also located on the F′ episome in the cell. The lacZ gene serves as a secondary reporter to provide a visible phenotype for identifying positive protein–protein interactions.

Definitions

Query (or bait) is one protein sequence with which the target protein, designed and produced with the method, will specifically interact. *Target protein* is one or more protein sequence(s) designed by the method to specifically interact with the query protein sequence. The target is expected to be present in a pool of protein sequences, called the *target pool*. The target pool is designed using a *target template*, which is a nucleic acid sequence containing 2/3 defined and 1/3 undefined (any) nucleotides (X) (a target template, which contains 15 undefined nucleic acid residues, will result in $4^{15}=10^9$ different oligonucleotides which will be translated into the corresponding number of proteins). The target pool is synthesized using a *target oligo template* (TOT) which has a *constant* (C) and *variable* (V) part. The TOT-C is necessary to synthesize dsDNA of the target pool sequences and it is ~20 nucleotides long. The TOT-V (target template) is about 30–45 nucleotide long, 2/3rd of nucleotides are unambiguously defined, while 1/3rd are not (X). The X residues should be synthesized by adding a mixture of nucleotides (equal amounts of A+T+G+C) to the reaction during oligo synthesis.

The results (number of highly, moderately, slightly positive clones) are evaluated by visual inspection. The positive clones are saved for further experiments. If there are no positive clones, it is necessary to validate the orientation and translation frame in the target mRNAs. This is possible by sequencing some target mRNAs. The sequence should show the residue pattern.

Both TARGET TEMPLATE to ESRLERLEQLFLLIF (GAL4 09–23AA) and TARGET TEMPLATE to QLFLLIFPREDLDMI (GAL4 17–31AA) contained numerous positive bacterial clones growing on double selective medium. Sequencing of DNA from the vectors in randomly selected positive clones confirmed that:

- they contained the characteristic TOT pattern, i.e. defined 1st and 3rd codon residues;
- the nucleic acids differed only in the 2nd codon positions, while they were the same regarding the 1st and 3rd codon positions;
- The restriction endonuclease recognition sequences were present;
- the start and stop codons were present;
- the sequences were inserted into the correct, sense DNA strands;
- the codon frames were correct in relation to the start codon and were read in the correct frames.

Some positive TARGET TEMPLATE to ESRLERLEQLFLLIF (GAL4 09–23AA) clones were further processed to monoclonal colonies and proteins were extracted. Characterization of the binding properties of fluorescent labelled GAL4 peptide to the protein extract indicated the presence of saturable binding sites in the protein extracts from positive clones and the absence of saturable binding sites in the negative clones.

The Experiment

The experiment below is specifically designed for the BacterioMatch (Stratagene) two-hybrid system. This system uses:

- a bait vector (pBT) and the manufacturer's standard are used as insert, the dimerization domain of 1HBW REGULATORY PROTEIN GAL4;
- a target vector (pTRG) and the manufacturer's standard are used as insert, and ~90 aa long mutant form of Gal11.

SHARP Sequence Design

```
>TARGET TEMPLATE to ESRLERLEQLFLLIF (GAL4 09-23AA)-ssDNA-sense
5'-GGATCC-ATG-[AxAAxTCxGTxGAxATxGCxGTxCCxGTxTTxCTxGCxTTxATxC]-
ATT-CGGCCGC-3'
|----BamHI--Met-[------------VARIABLE SEQUENCE, TOT-V ~30-45 NA-----------]-
Stop-----NotI------|

>TARGET TEMPLATE to QLFLLIFPREDLDMI (GAL4 17-31AA)-ssDNA-sense
5'-GGATCC-ATG-[AxTCxTGxCAxGGxCTxCTxGAxGAxAAxTCxGTxGAxATxGCxG]-
ATT-CGGCCGC-3'
|----BamHI--Met-[------------VARIABLE SEQUENCE, TOT-V ~30-45 NA-------------]-
Stop-----NotI------|
```

Figure 36. Sequences, designed by this method, were expected to produce proteins (when transcribed and translated) with the potential to specifically interact with the indicated domains of the Gal4 protein. The 1st and 3rd codon letters in these target templates are complementary to the 3rd and 1st codon letters in the Gal4 coding sequences (reverse reading direction) while the 2nd codon letter is undefined (A or T or G or C).

In the experiment below the target oligo pool is used instead of Gall 1 in the pTRG vector.

The query in this experiment is the dimerization domain of 1HBW REGULATORY PROTEIN GAL4 inserted into pBT (as provided and described by Stratagene). The target oligo templates (TOT-V) were designed to specifically interact with K01486_SCGAL4_DIMDOM-171/9-23 and K01486_SCGAL4_DIMDOM-171/17/31 sequences.

The sequences below are sense, ssDNA sequences which means that the TOT-V in this sequence is the same as the sequence in the expected mRNAs (except T/U conversion). The TOT-C is not indicated here, BPD can decide which TOT-C to use for this purpose (Figure 36).

The experiment consists of the following steps:

1) Sequence the Gal 4 DNA (provided by Stratagene) to make sure that the query sequence is as expected.
2) Synthesize the target pool using the target oligo templates. This is a single run routine oligo synthesis. Residue X is equal amounts of A+T+G+C.
3) Make dsDNAs. This is a single run PCR.
4) Make restriction enzyme cuts on the target oligo pool sequences. This is a single run RE reaction.
5) Insert the oligo pool sequences into the pTRG vector. $\sim 10^9$ different vectors are expected. Make sure that the orientation of the target oligo-s is correct and the transcription will result in the following mRNA. The target oligo pool insertion is a single run ligase reaction (Figure 37). Transcription of TOT dsDNA will result in TOT mRNA. A 45 nucleic acid long TOT will be translated into 4^{15} different oligopeptides, each 15 amino acids long. Some of these oligopeptides are expected to specifically interact with the respective GAL4 targets.
6) Insert the vectors into bacteria.
7) Perform the BacterioMatch two-hybrid assay accordingly to the Stratagene manual.

The K_d of the binding sites varied between 1 and 100 nM indicating the presence of a limited number of high affinity binding sites. Unlabelled GAL4 inhibited the binding of labeled GAL4 to the proteins from positive clones while other randomly chosen proteins (insulin, growth hormone, prolactin) were ineffective competitors even in much higher concentrations.

This experiment indicates that it is possible to design specifically interacting oligo-peptides (target) to any oligo-peptide (query) and detect the interaction in a bacterial two-hybrid system (like BacterioMatch™. The method is quick; it takes only a few days to obtain interacting monoclonal proteins. The designed protein–protein interaction is highly specific and has high affinity (K_d ~1–100 nM).

Further details can be found in the following references [58, 59, 96].

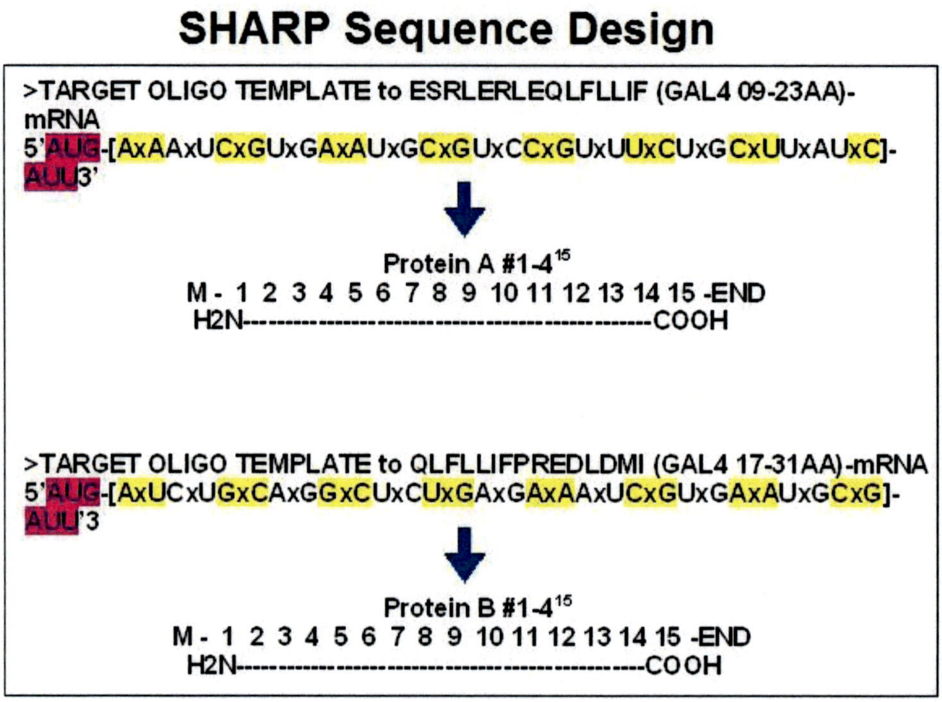

Figure 37. The transcription of TOT dsDNA will result in TOT mRNA. A 45 nucleic acid long TOT will be translated into 4^{15} different oligopeptides, each 15 amino acids long. Some of these oligopeptides are expected to specifically interact with the respective GAL4 targets.

Conclusion

The Promises of the Second Generation Proteomic Code

Industrial Applications

The second generation Proteomic Code and the method to develop SHARP have some potential advantages that are not obvious in the recent antibody developing methods:

- provide quick access to interacting peptides;
- provide direct and permanent access to monoclonal sources for large-scale production;
- SHARP is small (MW <2000 Da) compared to antibodies (155 kDa) or affibodies (which gives therapeutic and manufacturing advantage), no need for humanization;
- might be the key to the mass production of interactive oligopeptides (similar to the on-demand synthesis of nucleotide oligo-s;
- a self-learning method; every single successful SHARP can contribute to a more and more exact amino acid interaction table.

Scientific Potential

The present system is a unique in silico method of identifying the most effective binding proteins to interact with reactive epitopes on a respective protein antigen. Epitopes of a protein antigen represent the sites that are recognized as binding sites by certain immunoglobulin molecules, as antibodies.

The benefits of the system are widespread and beneficial to biotechnology, and are useful, for example, in developing drugs for treatment of viral diseases such as AIDS and influenza, as well as diseases such as Alzheimer's disease and bovine spongiform encephalopathy disease. In addition to medical research and drug development, this system has applications related to environmental health and public safety, including for example the detection of bacteria, viruses, toxins, etc. in air, water, and food supplies.

By way of further specific examples, the present system has applications in the following areas:

1) improving health care, by providing a new and easily implemented approach to the development of diagnostic kits and therapeutic drugs;
2) improving the environment, by providing new and economic approaches for detecting environmental pathogens;
3) improving working conditions by providing economic and effective ways to detect environmental pathogens; and
4) improving homeland security, by providing rapid detection of known as well as new pathogens in air, water, food, etc.

The Vision of a Proteome-Sensor Chip

Detection and measurement of proteins is a fundamental procedure in life sciences. Many diagnostic procedures are already based on this technique and many more will follow:

- detection of hormones and enzymes for diagnosis of organ failure;
- detection of pathogen-derived antigens or antibody responses for diagnosis of infections;
- detection of allergens for allergy diagnoses;
- tumor markers to detect and evaluate neoplasias.

Considering that the number of hormones, enzymes, pathogens, and markers is very large, it is easy to recognize that the demand for specifically interacting diagnostic proteins is large. However, there are limitations to satisfy the demand. It takes several weeks to develop one antibody with the traditional methods. Therefore it is not cheap. The traditional one protein–one kit method is simply no longer feasible. Some kind of integration is necessary. Development points to protein chips which permit the simultaneous, parallel detection of hundreds or thousands of protein signals and the computerized integrated evaluation of the results. Chip-based protein detection requires a large number of easy to produce, cheap interacting proteins.

The Proteomic Code-based method described here could significantly contribute to the industrial production of these interacting peptides.

The SHARP chip-based technology opens the way to the real possibility of monitoring the proteome, i.e. obtain detailed information on the qualitative and quantitative state of a large number of different proteins simultaneously, including even some information about the splicing and configuration changes of individual proteins (Figure 38).

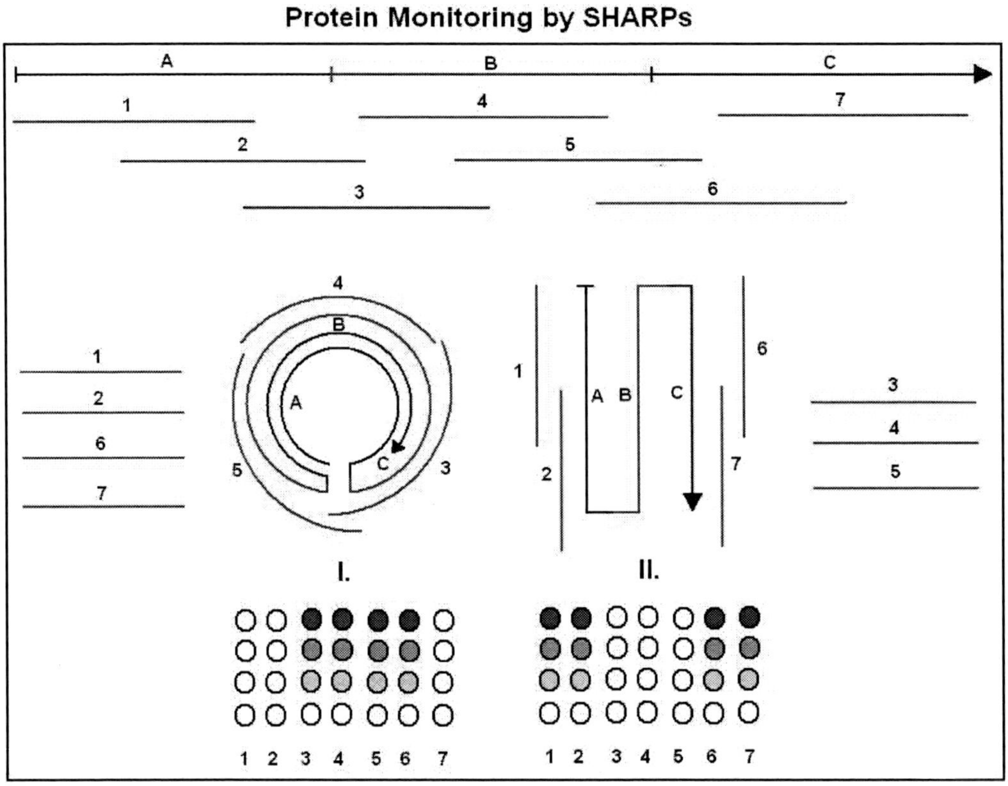

Figure 38. Protein monitoring by SHARPs. Specific and high affinity reacting proteins (SHARPs, 1–7) are designed and produced against a protein. The protein has 3 domains (A, B, C) and two main conformations (I, II). Domain B is exposed on the surface by conformation I and that form is detected by SHARPs 3–6. Domain B is sandwiched by domains A and B in configuration II and therefore this form is detected by SHARPs 1–2 and 6–7. A concentration and configuration dependent response is detected by a slot corresponding to the protein by the SHARPs CHIP (dots).

The human genome contains 3×10^9 base pairs. About 2% of this is located in about 30,000 genes which are expressed, spliced into ~100,000 different proteins containing ~6 million amino acids. The 2× coverage of this size proteome with 15 amino acid long complementary sequences requires 800,000 different oligopeptides. (This is probably the upper estimate, because the proteome consists of many similar or identical domains.) Proteome monitoring with proteomic chips is a huge challenge which is only possible if SHARP proteins are accessible in much larger number and for a much lower price than is possible today.

The Vision of a New Physiology

The receptor – ligand and antigen – antibody type or interaction has a fundamental role in the physiology of humans and animals. Peptides which are able to specifically interact with regulatory pathway have of course significant potential in manipulating these pathways. Most drugs are effective because of their interaction with regulatory pathways (e.g., GPCR or 7TM receptor system). Therefore, easy and inexpensive access to designed interacting proteins will have an impact on further development of physiological and pharmacological research and drug discovery. Some kind of industrial-scale, standardized, semi-automated physiological research is also desirable. The traditional one-by-one approach is too slow and to expensive for the complexity of life.

The Vision of New Protein-Based Therapeutic Approaches

Proteins/peptides are underutilized in medical therapy. Only insulin is used to treat a common disease (Type I diabetes); a few more proteins are used to treat relatively rare disease, such as growth hormone (GH) deficiency and hemophilia. Veterinary use of GH (lean meat production) or misuse in sports and cosmetology far exceeds its medical indication. Protein therapy is expensive and requires daily injections which is not attractive to most patients. Most pharmaceutical companies have large accumulated knowledge (and patent base) regarding traditional, simple molecular drug design and treatments while their knowledge in proteomics is still undergoing development. Simple molecule-based drugs are "simple"; not much specificity is provided by these structures. Protein-based, highly specific treatments are not too far off in the future and easy access to biologically active proteinaceous substances will facilitate physiological evaluation and medical application of these more complex peptide molecules.

SHARPs, complementary or not, are obvious candidates for receptor agonist or antagonist functions. However, there is an even more exciting and less expected therapeutic application of complementary peptides. Some experiments suggest that immunization with complementary peptides to receptors induces production of ligand-like antibodies, with ligand-like biological effects. Just imagine the possibility that inducing insulin-like antibodies by immunization with designed complementary peptides to insulin-receptors and obtaining insulin-like effects (regulation will of course be a problem to solve). Or treat GH deficiency by raising GH-like antibodies using GH-receptor complementary peptides for immunization. This seems like science fiction today, but for example Blalocks' experiments are already pointing in that direction [27].

Is protein therapy an alternative to gene therapy? Yes. Gene therapy is technically still difficult to perform and the effect is irreversible. The effects of protein therapy are short-lived and reversible (if immunization does not occur).

Software

SeqForm, http://janbiro.com/Downloads.html
SeqPlot http://janbiro.com/Downloads.html

Dotlet http://janbiro.com/Downloads.html
mfold http://frontend.bioinfo.rpi.edu/applications/mfold/cgi-bin/rna-form1.cgi
SeqX: http://janbiro.com/Downloads.html

Patents

Root-Bernstein

Blalock JE (Birmingham, AL), Bost KL (Birmingham, AL), Smith EM (Galveston, TX): Polypeptides complementary to peptides or proteins having an amino acid sequence or nucleotide coding sequence at least partially known and methods of design therefor. United States Patent 5,212,072, May 18, 1993.

Omichinski JG (Potomac, MD), Fassina G (Padova, IT), Olson AD (Silver Spring, MD), Thorgeirsson SS (Bethesda, MD): Computer-assisted design of anti-peptides based on the amino acid sequence of a target peptide. United States Patent 5,081,584, January 14, 1992.

Biro J: System and method to obtain oligo-peptides with specific high affinity to query proteins. US Patent Application Publication US 2007/0015189 A1, January 18, 2007.

Abbreviations

amino acid complementarity:the physico-chemical complementarity (size, charge, hydropathy) of amino acid pairs as well as their origin from complementary codons

complementary amino acids:amino acid pairs, coded by complementary codons

NA:nucleic acid

P:protein

PC:Proteomic Code, the comprehensive rules describing the origin and nature of amino acid complementarity

SHARP:Specific High Affinity Reacting Peptides

W-C base pairs:Watson–Crick's complementary base pairs.

Competing Interests

The author of this article has pending US and PCT patent applications for the system and the method of obtaining oligo-peptides with specific high affinity to query proteins.

Authors' Contributions

Acknowledgement

In memory of Dr David F. Horrobin, Founder and editor of Medical Hypotheses, who helped and encouraged my first publication of a non-sense idea of anti-sense translation (1981) that later developed into the concept of the Proteomic Code. The support and editorial help provided by Dr Paul S. Agutter (editor of TBioMed) in the publication of this and many related articles is greatly appreciated and acknowledged. Marilyn Juengling provided me with very valuable personal support for the completion of this review. This review is a preprint publication of a chapter in a textbook Biro JC: Principia Biroinformatica, 2008.

References (ex Table I)

[1] Pawson T, Gish GD: Protein-protein interactions: a common theme in cell biology. In *Protein-Protein Interactions*, Edited by Golemis EA, Adams PD. New York: CSHL Press; 2005.

[2] Watson JD, Baker TA, Bell SP, Gann A, Levine M, Losick R: The structure of DNA and RNA. Chapter 6. In *Molecular Biology of the Gene*, 5th edition. Cold Spring Harbor, New York: Cold Spring Harbor Laboratory Press; 2004:1-33.

[3] Pauling L: *The Nature of the Chemical Bond and the Structure of Molecules and Crystals: An Introduction to Modern Structural Chemistry*, 2nd edition. Ithaca, NY: Cornell University Press; London: Humphrey Milford, Oxford University Press: 1940: 450 pp.

[4] Pauling L, Corey RB, Yakel Jr HL, Marsh RE: Calculated form factors for the 18-residue 5-turn _-helix. *Acta Crystallogr* 1955, 8:853-855.

[5] Pauling L, Corey RB: Specific hydrogen-bond formation between pyrimidines and purines in deoxyribonucleic acids. (In Linderstrom-Lang Festschrift). *Arch Biochem Biophys* 1956, 56:164-181.

[6] Biro JC: Speculations about alternative DNA structures. *Med Hypotheses* 2003, 61:86-97.

[7] Woese CR: Translation: in retrospect and prospect. *RNA* 2001, 7:1055-1067.

[8] Dayhoff MO: *Atlas of Protein Sequence and Structure*. National Biomedical Research Foundation, 1966.

[9] Biro J: Comparative analysis of specificity in protein-protein interactions. Part I: A theoretical and mathematical approach to specificity in protein-protein interactions. *Med Hypotheses* 1981, 7:969-979.

[10]Biro J: Comparative analysis of specificity in protein-protein interactions. Part II: The complementary coding of some proteins as the possible source of specificity in protein-protein interactions. . *Med Hypotheses* 1981, 7:981-993.

[11]Biro J: Comparative analysis of specificity in protein-protein interactions. Part III: Models of the gene expression based on the sequential complementary coding of some pituitary proteins. *Med Hypotheses* 1981, 7:995-1007.

[12]Gamow G: Possible mathematical relation between deoxyribonucleic acid and proteins. *Kgl Danske Videnskab Selskab Biol Medd* 1954, 22:1-13.

[13]Gamow G: Possible relation between deoxyribonucleic acid and protein structures. *Nature* 1954, 173:318.

[14]Gamow G: Information transfer in the living cell. *Sci Am* 1955, 193:70-78.

[15]Gamow G, Ycas M: Statistical correlation of protein and ribonucleic acid composition. *Proc Natl Acad Sci USA* 1955, 41:1011-1019.

[16]Gamow G, Rich A, Ycas M: The problem of information transfer from nucleic acids to proteins. In *Advances in Biological and Medical Physics, Volume 4.* New York: Academic Press; 1956:23-68.

[17]Gamow G, Ycas M: *Mr. Tompkins Inside Himself: Adventures in the New Biology.* New York: Viking Press; 1967.

[18]Biro JC, Biro JM: Frequent occurrence of recognition site-like sequences in the restriction endonucleases. *BMC Bioinformatics* 2004, 16:30.

[19]Biro JC, Biro JM: The BlastNP: a novel, sensitive sequence similarity searching method using overlappingly translated sequences. *Conf Proc IEEE Eng Med Biol Soc* 2004, 4:2777-2780.

[20]Mekler LB: Specific selective interaction between amino acid groups of polypeptide chains. *Biofizika* 1969, 14:581-584.

[21]Mekler LB, Idlis RG: *VINITI Deposited Doc* 1981, 1476-1481.

[22]Blalock JE, Smith EM: Hydropathic anti-complementarity of amino acids based on the genetic code. *Biochem Biophys Res Commun* 1984, 121:203-207.

[23]Blalock JE, Bost KL: Binding of peptides that are specified by complementary RNAs. *Biochem J* 1986, 234:679-683.

[24]Bost KL, Smith EM, Blalock JE: Similarity between the corticotrophin (ACTH) receptor and a peptide encoded by an RNA that is complementary to ACTH mRNA. *Proc Natl Acad Sci USA* 1985, 82:1372-1375.

[25]Blalock JE, Smith EM: A complete regulatory loop between the immune and neuroendocrine systems. *Fed Proc* 1985, 44(1 Pt 1):108-111.

[26]Blalock JE, Harbour-McMenamin D, Smith EM: Peptide hormones shared by the neuroendocrine and immunologic systems. *J Immunol* 1985, 135(Suppl 2):858s-861s.

[27]Bost KL, Blalock JE: Production of anti-idiotypic antibodies by immunization with a pair of complementary peptides. *J Mol Recognit* 1989, 1:179-183.

[28]Blalock JE: Department of Physiology and Biophysics, The University of Alabama at Birmingham. http://www.physiology.uab.edu/Blalock.htm#Research

[29]Root-Bernstein RS: Amino acid pairing. *J Theor Biol* 1982, 94:885-894.

[30]Root-Bernstein RS: Protein replication by amino acid pairing. *J Theor Biol* 1983, 100:99-106.

[31]Siemion IZ, Stefanowicz P: Periodical changes of amino acid reactivity within the genetic code. *Biosystems* 1992, 27:77-84.

[32]Siemion IZ, Cebrat M, Kluczyk A: The problem of amino acid complementarity and antisense peptides. *Curr Protein Peptide Sci* 2004, 5:507-527.

[33]Siemion IZ, Zbozień-Pacamaj R, Stefanowicz P: New hypothesis on amino acid complementarity and its evaluation on TGF-β_2-related peptides. *J Mol Recognit* 2001, 14:1-12.

[34]Heal JR, Roberts GW, Christie G, Miller AD: Inhibition of beta-amyloid aggregation and neurotoxicity by complementary (antisense) peptides. *Chembiochemistry* 2002, 3:86-92.

[35]Heal JR, Bino S, Roberts GW, Raynes JG, Miller AD: Mechanistic investigation into complementary (antisense) peptide mini-receptor inhibitors of cytokine interleukin-1. *Chembiochemistry* 2002, 3:76-85.

[36]Bhakoo A, Raynes JG, Heal JR, Keller M, Miller AD: De-novo design of complementary (antisense) peptide mini-receptor inhibitor of interleukin 18 (IL-18). *Mol Immunol* 2004, 41:1217-1224.

[37]Heal JR, Roberts GW, Raynes JG, Bhakoo A, Miller AD: Specific interactions between sense and complementary peptides: the basis for the proteomic code. *Chembiochemistry* 2002, 3:136-151 (Review. Erratum in: *Chembiochemistry* 2002, 3:271).

[38]Heal JR, Bino S, Ray KP, Christie G, Miller AD, Raynes JG: A search within the IL-1 type I receptor reveals a peptide with hydropathic complementarity to the IL-1beta trigger loop which binds to IL-1 and inhibits in vitro responses. *Mol Immunol* 1999, 36:1141-1148.

[39]Zull JE, Smith SK: Is genetic code redundancy related to retention of structural information in both DNA strands? *Trends Biochem Sci* 1990, 15:257-261.

[40]Zull JE, Taylor RC, Michaels GS, Rushforth NB: Nucleic acid sequences coding for internal antisense peptides: are there implications for protein folding and evolution? *Nucleic Acids Res* 1994, 22:3373-3380.

[41]Rother KI, Clay OK, Bourquin JP, Silke J, Schaffner W: Long non-stop reading frames on the antisense strand of heat shock protein 70 genes and prion protein (PrP) genes are conserved between species. *Biol Chem* 1997, 378:1521-1530.

[42]Duax WL, Huether R, Pletnev VZ, Langs D, Addlagatta A, Connare S, Habegger L, Gill J: Rational genomics I: antisense open reading frames and codon bias in short-chain oxido reductase enzymes and the evolution of the genetic code. *Proteins* 2005, 61:900-906.

[43]Forsdyke DR: Sense in antisense? *J Mol Evol* 1995, 41:582-586.

[44]Okada H: The possible role of sense and antisense peptide interactions in the generation and maintenance of the tertiary structure of a protein. *Anticancer Res* 1998, 18:3927-3930.

[45]McGuigan JE: Antibodies to complementary peptides as probes for receptors. *Immunomethods* 1994, 5:158-166.

[46]Baranyi L: Antisense homology boxes in proteins. http://citeseer.ist.psu.edu/16596.html (year unknown).

[47]Baranyi L, Campbell W, Ohshima K, Fujimoto S, Boros M, Okada H: The antisense homology box: a new motif within proteins that encodes biologically active peptides. *Nat Med* 1995, 1:894-901.

[48]Baranyi L, Campbell W, Ohshima K, Fujimoto S, Boros M, Kaszaki J, Okada H: Antisense homology box-derived peptides represent a new class of endothelin receptor inhibitors. *Peptides* 1998, 19:211-223.

[49]Baranyi L, Campbell W, Okada H: Antisense homology boxes in C5a receptor and C5a anaphylatoxin: a new method for identification of potentially active peptides. *J Immunol* 1996, 157:4591-4601.

[50]Segerstéen U, Nordgren H, Biro JC: Frequent occurrence of short complementary sequences in nucleic acids. *Biochem Biophys Res Commun* 1986, 139:94-101.

[51]Weigent DA, Clarke BL, Blalock JE: Peptide design using a genetically patterned binary code: growth hormone-releasing hormone as a model. *Immunomethods* 1994, 5:91-97.

[52]Biro JC, Benyo B, Sansom C, Szlavecz A, Fordos G, Micsik T, Benyo Z: A common periodic table of codons and amino acids. *Biochem Biophys Res Commun* 2003, 306:408-415.

[53]Benyo B, Biro JC, Benyo Z: Codes in the codons: construction of a codon/amino acid periodic table and a study of the nature of specific nucleic acid-protein interactions. *Conf Proc IEEE Eng Med Biol Soc* 2004, 4:2860-2863.

[54]Woese CR: Chapters 6-7. In: *The Genetic Code. The Molecular Basis for Genetic Expression.* New York: Harper & Row: 1967:156-160.

[55]Crick FHC: The origin of the genetic code. *J Mol Biol* 1968, 38:367-379.

[56]Wong JT: Evolution of the genetic code. *Microbiology* 1988, 5:174-181.
[57]Biro JC, Fordos G: SeqX: a tool to detect, analyze and visualize residue co-locations in protein and nucleic acid structures. *BMC Bioinformatics* 2005, 6:170.
[58]Biro JC: Amino acid size, charge, hydropathy indices and matrices for protein structure analysis. *Theor Biol Med Model* 2006, 3:15.
[59]Biro JC: Indications that "codon boundaries" are physico-chemically defined and that protein-folding information is contained in the redundant exon bases. *Theor Biol Med Model* 2006, 3:28.
[60]Anfinsen CB, Redfield RR, Choate WI, Page J, Carroll WR: Studies on the gross structure, cross-linkages, and terminal sequences in ribonuclease. *J Biol Chem* 1954, 207:201-210.
[61]Biro JC: A novel intra-molecular protein-protein interaction code based on partial complementary coding of co-locating amino acids, *Med Hypotheses* 2006; 66:137-142.
[62]Neher E: How frequent are correlated changes in families of protein sequences. *Proc Natl Acad Sci USA* 1994, 91:98-102.
[63]Levinthal C: How to fold graciously in Mossbauer spectroscopy in biological systems. In *Proceedings of a Meeting held at Allerton House, Monticello, IL*. Edited by Debrunner P, Tsibris JCM, Munck E. Urbana, IL: University of Illinois Press; 1969:22-24.
[64]Isogai Y, Ota M, Ishii A, Ishida M, Nishikawa K: Identification of amino acids involved in protein structural uniqueness: implication for de novo protein design. *Protein Eng* 2002, 15:555–560 (doi: 10.1093/protein/15.7.555).
[65]Zuker M: Mfold web server for nucleic acid folding and hybridization prediction. *Nucleic Acids Res* 2003, 31:3406-3415.
[66]Brunak S, Engelbrecht J: Protein structure and the sequential structure of mRNA: alpha-helix and beta-sheet signals at the nucleotide level. *Proteins* 1996, 25:237-252.
[67]Gu W, Zhou T, Ma J, Sun X, Lu Z: Folding type specific secondary structure propensities of synonymous codons. *IEEE Trans Nanobiosci* 2003, 2:150-157.
[68]Gupta SK, Majumdar S, Bhattacharya TK, Ghosh TC: Studies on the relationships between the synonymous codon usage and protein secondary structural units. *Biochem Biophys Res Commun* 2000, 269:692-696.
[69]Gokhale KC, Newnam GP, Sherman MY, Chernoff YO: Modulation of prion-dependent polyglutamine aggregation and toxicity by chaperone proteins in the yeast model. *J Biol Chem* 2005, 280:22809-22818.
[70]Fan H, Mark AE: Mimicking the action of folding chaperones in molecular dynamics simulations: application to the refinement of homology-based protein structures. *Protein Sci* 2004, 13:992-999.
[71]Walter S, Buchner J: Molecular chaperones – cellular machines for protein folding. *Angew Chem Int Ed Engl* 2002, 41:1098-1113.
[72]Komar AA, Kommer A, Krasheninnikov IA, Spirin AS: Cotranslational folding of globin. *J Biol Chem* 1997, 272:10646-10651.
[73]Thanaraj TA, Argos P: Protein secondary structural types are differentially coded on messenger RNA. *Protein Sci* 1996, 5:1973-1983.
[74]Chiusano ML, Alvarez-Valin F, Di Giulio M, D'Onofrio G, Ammirato G, Colonna G, Bernardi G: Second codon positions of genes and the secondary structures of proteins. Relationships and implications for the origin of the genetic code. *Gene* 2000;261:63-69.

[75]Gu W, Zhou T, Ma J, Sun X, Lu Z: The relationship between synonymous codon usage and protein structure in *Escherichia coli* and *Homo sapiens*. *Biosystems* 2004, 73:89-97.
[76]Ermolaeva O: Synonymous codon usage in bacteria. *Curr Issues Mol Biol* 2001, 3:91-97.
[77]Biro JC, Biro JM, Biro AM: Hidden massages in hidden sub-sequences: a study on collagens. In *30th FEBS Congress – 9th IUBMB Conference, Budapest, Hungary, 2–7 July 2005*; 2005:abstract.
[78]Katz L, Burge CB: Widespread selection for local RNA secondary structure in coding regions of bacterial genes. *Genome Res* 2003, 13:2042-2051.
[79]Biro JC: SeqForm. 2005 [http://www.janbiro.com/downloads].
[80]Biro JC: SeqPlot. 2005 [http://www.janbiro.com/downloads].
[81]Junier T, Pagni M: Dotlet: diagonal plots in a web browser. *Bioinformatics* 2000, 16:178-179 [http://www.isrec.isb-sib.ch/java/dotlet/Dotlet.html].
[82]Berman HM, Westbrook J, Feng Z, Gilliland G, Bhat TN, Weissig H, Shindyalov IN, Bourne PE: The Protein Data Bank. *Nucleic Acids Res* 2000, 28:235-242.
[83]Berman HM, Olson WK, Beveridge DL, Westbrook J, Gelbin A, Demeny T, Hsieh SH, Srinivasan AR, Schneider B: The Nucleic Acid Database: a comprehensive relational database of three-dimensional structures of nucleic acids. *Biophys J* 1992, 63:751–759.
[84]Adzhubei IA, Adzhubei AA: ISSD Version 2.0: taxonomic range extended. *Nucleic Acids Res* 1999, 27:268-271.
[85]Seffens W, Digby D: mRNA have greater negative folding free energies than shuffled or codon choice randomized sequences. *Nucleic Acids Res* 1999, 27:1578-1584.
[86]Oresic M, Dehn M, Korenblum D, Shalloway D: Tracing specific synonymous codon-secondary structure correlations through evolution. *J Mol Evol* 2003, 56:473-484.
[87]D'Onofrio G, Ghosh TC, Bernardi G: The base composition of the genes is correlated with the secondary structures of the encoded proteins. *Gene* 2002, 300:179-187.
[88]Xie T, Ding D: The relationship between synonymous codon usage and protein structure. *FEBS Lett* 1998, 434:93-96.
[89]Kumarevel TS, Gromiha MM, Ponnuswamy MN: Distribution of amino acid residues and residue-residue contacts in molecular chaperons. *Prep Biochem Biotechnol* 2001, 31:163-183.
[90]Eilers M, Patel AB, Liu W, Smith SO: Comparison of helix interactions in membrane and soluble alpha-bundle proteins. *Biochem J* 2002, 82:2720-2736.
[91]Glaser F, Steinberg DM, Vakser IA, Ben-Tal N: Residue frequencies at protein-protein interfaces. *Proteins Struct Funct Genet* 2001, 43:89-102.
[92]Naor D, Fisher D, Jernigan RL, Wolfson H, Nussinov R: Amino acid pair interchanges at spatially conserved locations. *J Mol Biol* 1996, 256:924-938.
[93]Azarya-Sprinzak E, Naor D, Wolfson HJ, Nussinov R: Interchanges of spatially neighboring residues in structurally conserved environment. *Protein Eng* 1997, 10:1109-1122.
[94]Singer MS, Vriend G, Bywater RP: Prediction of protein residue contacts with a PDB-derived likelihood matrix. *Protein Eng* 2002, 15:721-725.
[95]Shao Y, Bystroff C: Predicting inter-residue contacts using templates and pathways. *Proteins Struct Funct Genet* 2003, 53:497-502.
[96]Biro JC: Nucleic acid chaperons: a theory of an RNA-assisted protein folding. *Theor Biol Med Model* 2005, 2:35.

[97]Biro JC: System and method to obtain oligo-peptides with specific high affinity to query proteins. US Patent Application Publication, US 2007/0015189 A1, Jan 18, 2007.

References to Table I

[1] Araga S, Kishimoto M, Doi S, Nakashima K: A complementary peptide vaccine that induces T cell anergy and prevents experimental allergic neuritis in Lewis rats. *J Immunol* 1999, 163:476-482.

[2] Araga S, LeBoeuf RD, Blalock JE: Prevention of experimental autoimmune myasthenia gravis by manipulation of the immune network with a complementary peptide for the acetylcholine receptor. *Proc Natl Acad Sci USA* 1993, 90:8747-8751.

[3] Bajpai A, Hooper KP, Ebner KE: Interactions of antisense peptides with ovine prolactin. *Biochem Biophys Res Commun* 1991, 180:1312-1317.

[4] Baranyi L, Campbell W, Ohshima K, Fujimoto S, Boros M, Okada H: The antisense homology box: a new motif within proteins that encodes biologically active peptides. *Nat Med* 1995, 1:894-901.

[5] Baranyi L, Campbell W, Okada H: Antisense homology boxes in C5a receptor and C5a anaphylatoxin: a new method for identification of potentially active peptides. *J Immunol* 1996, 157:4591-4601.

[6] Bhakoo A, Raynes JG, Heal JR, Keller M, Miller AD. De-novo design of complementary (antisense) peptide mini-receptor inhibitor of interleukin 18 (IL-18). Mol Immunol. 2004 Nov; 41(12):1217-24.

[7] Blalock JE, Bost KL: Binding of peptides that are specified by complementary RNAs. *Biochem J* 1986, 234:679-683.

[8] Borovsky D, Powell CA, Nayar JK, Blalock JE, Hayes TK: Characterization and localization of mosquito-gut receptors for trypsin modulating oostatic factor using a complementary peptide and immunocytochemistry. *FASEB J* 1994, 8:350-355.

[9] Bost KL, Blalock JE: Production of anti-idiotypic antibodies by immunization with a pair of complementary peptides. *J Mol Recognit* 1989, 1:179-183.

[10]Bost KL, Smith EM, Blalock JE: Similarity between the corticotrophin (ACTH) receptor and a peptide encoded by an RNA that is complementary to ACTH mRNA. *Proc Natl Acad Sci USA* 1985, 82:1372-1375.

[11]Bost KL, Smith EM, Blalock JE: Similarity between the corticotrophin (ACTH) receptor and a peptide encoded by an RNA that is complementary to ACTH mRNA. *Proc Natl Acad Sci USA* 1985, 82:1372-1375.

[12]Brentani RR, Ribeiro SF, Potocnjak P, Pasqualini R, Lopes JD, Nakaie CR: Characterization of the cellular receptor for fibronectin through a hydropathic complementarity approach. *Proc Natl Acad Sci USA* 1988, 85:364-367.

[13]Bret-Dibat JL, Zouaoui D, Déry O, Zerari F, Grassi J, Maillet S, Conrath M, Couraud JY: Antipeptide polyclonal antibodies that recognize a substance P-binding site in mammalian tissues: a biochemical and immunocytochemical study. *J Neurochem* 1994, 63:333-343.

[14]Campbell-Thompson M, McGuigan JE: Canine parietal cell binding by antibodies to the complementary peptide of somatostatin. *Am J Med Sci* 1993, 305:365-373.

[15]Carr DJ, Blalock JE, Bost KL: Monoclonal antibody against a peptide specified by [Met]-enkephalin complementary RNA recognizes the delta-class opioid receptor. *Immunol Lett* 1989, 20:181-186.

[16]Carr DJ, Bost KL, Blalock JE: An antibody to a peptide specified by an RNA that is complementary to gamma-endorphin mRNA recognizes an opiate receptor. *J Neuroimmunol* 1986, 12:329-337.

[17]Castronovo V, Taraboletti G, Sobel ME: Laminin receptor complementary DNA-deduced synthetic peptide inhibits cancer cell attachment to endothelium. *Cancer Res* 1991, 51:5672-5678.

[18]Derrick JM, Taylor DB, Loudon RG, Gartner TK: The peptide LSARLAF causes platelet secretion and aggregation by directly activating the integrin alphaIIbbeta3. *Biochem J* 1997, 325(Pt 2):309-313.

[19]Dillon J, Woods WT, Guarcello V, LeBoeuf RD, Blalock JE: A peptide mimetic of calcium. *Proc Natl Acad Sci USA* 1991, 88:9726-9729.

[20]Elton TS, Oparil S, Blalock JE: The use of complementary peptides in the purification of an angiotensin II binding protein. *J Hypertens* 1988, 6(Suppl):S404-407.

[21]Fassina G, Cassani G, Corti A: Binding of human tumor necrosis factor alpha to multimeric complementary peptides. *Arch Biochem Biophys* 1992, 296:137-143.

[22]Fassina G, Cassani G, Gnocchi P, Fornasiero MC, Isetta AM: Inhibition of interleukin-2/p55 receptor subunit interaction by complementary peptides. *Arch Biochem Biophys* 1995, 318:37-45.

[23]Fassina G, Cassani G, Corti A: Binding of human tumor necrosis factor alpha to multimeric complementary peptides. *Arch Biochem Biophys* 1992, 296:137-143.

[24]Fassina G, Consonni R, Zetta L, Cassani G. Design of hydropathically complementary peptides for Big Endothelin affinity purification. *Int J Peptide Protein Res* 1992, 39:540-548.

[25]Fassina G, Corti A, Cassani G: Affinity enhancement of complementary peptide recognition. *Int J Peptide Protein Res* 1992, 39:549-556.

[26]Fassina G, Roller PP, Olson AD, Thorgeirsson SS, Omichinski JG: Recognition properties of peptides hydropathically complementary to residues 356-375 of the c-raf protein. *J Biol Chem* 1989, 264:11252-11257.

[27]Fassina G, Zamai M, Brigham-Burke M, Chaiken IM: Recognition properties of antisense peptides to Arg8-vasopressin/bovine neurophysin II biosynthetic precursor sequences. *Biochemistry* 1989, 28:8811-8818.

[28]Fujita E, Farkas I, Campbell W, Baranyi L, Okada H, Okada N: Inactivation of C5a anaphylatoxin by a peptide that is complementary to a region of C5a. *J Immunol* 2004, 172:6382-6387.

[29]Gartner TK, Loudon R, Taylor DB: The peptides APLHK, EHIPA and GAPL are hydropathically equivalent peptide mimics of a fibrinogen binding domain of glycoprotein IIb/IIIa. *Biochem Biophys Res Commun* 1991, 180:1446-1452.

[30]Gartner TK, Taylor DB: The peptide Glu-His-Ile-Pro-Ala binds fibrinogen and inhibits platelet aggregation and adhesion to fibrinogen and vitronectin. *Proc Soc Exp Biol Med* 1991, 198:649-655.

[31]Ghiso J, Saball E, Leoni J, Rostagno A, Frangione B: Binding of cystatin C to C4: the importance of sense-antisense peptides in their interaction. *Proc Natl Acad Sci USA* 1990, 87:1288-1291.

[32]Gho YS, Chae CB: Anti-angiogenin activity of the peptides complementary to the receptor-binding site of angiogenin. *J Biol Chem* 1997, 272:24294-24299.

[33]Gho YS, Lee JE, Oh KS, Bae DG, Chae CB: Development of antiangiogenin peptide using a phage-displayed peptide library. *Cancer Res* 1997, 57:3733-3340.

[34]Heal JR, Bino S, Ray KP, Christie G, Miller AD, Raynes JG: A search within the IL-1 type I receptor reveals a peptide with hydropathic complementarity to the IL-1beta trigger loop which binds to IL-1 and inhibits in vitro responses. *Mol Immunol* 1999, 36:1141-1148.

[35]Holsworth DD, Kiely JS, Root-Bernstein RS, Overhiser RW: Antisense-designed peptides: a comparative study focusing on possible complements to angiotensin II. *Peptide Res* 1994, 7:185-193.

[36]Imai M, Okada N, Okada H: Inhibition of HIV-1 infection by an intramolecular antisense peptide to T20 in gp160. *Microbiol Immunol* 2000, 44:205-12.

[37]Johnson HM, Langford MP, Lakhchaura B, Chan TS, Stanton GJ: Neutralization of native human gamma interferon (HuIFN gamma) by antibodies to a synthetic peptide encoded by the 5' end of HuIFN gamma cDNA. *J Immunol* 1982, 129:2357-2359.

[38]Kang CY, Brunck TK, Kieber-Emmons T, Blalock JE, Kohler H: Inhibition of self-binding antibodies (autobodies) by a VH-derived peptide. *Science* 1988, 240:1034-1036.

[39]Kelly JM, Trinder D, Phillips PA, Casley DJ, Kemp B, Mooser V, Johnston CI: Vasopressin antisense peptide interactions with the V1 receptor. *Peptides* 1990, 11:857-862.

[40]Knigge KM, Piekut DT, Berlove D: Immunocytochemistry of a vasopressin (AVP) receptor with anti-idiotype antibody: inhibition of staining with a peptide (PVA) encoded by an RNA that is complementary to AVP mRNA. *Neurosci Lett* 1988, 86:269-271.

[41]Knutson VP: Insulin-binding peptide. Design and characterization. *J Biol Chem* 1988, 263:14146-14151.

[42]Kwak JW, Kim HK, Chae CB: Potential lead for an Alzheimer drug: a peptide that blocks intermolecular interaction and amyloid beta protein-induced cytotoxicity. *J Med Chem* 2006, 49:4813-4817.

[43]Lu FX, Aiyar N, Chaiken I: Affinity capture of [Arg8]vasopressin-receptor complex using immobilized antisense peptide. *Proc Natl Acad Sci USA* 1991, 88:3642-3646.

[44]Ludwig LB, Ambrus JL, Krawczyk KA, Sharma S, Brooks S, Hsiao CB, Schwartz SA: Human Immunodeficiency Virus-Type 1 LTR DNA contains an intrinsic gene producing antisense RNA and protein products. *Retrovirology* 2006, 3:80.

[45]Luo J, Zhang Q, Huang Y, Liu G, Zhao R: Quartz crystal microbalance biosensor for recombinant human interferon-beta detection based on antisense peptide approach. *Anal Chim Acta* 2007, 590:91-97.

[46]Martins VR, Graner E, Garcia-Abreu J, de Souza SJ, Mercadante AF, Veiga SS, Zanata SM, Neto VM, Brentani RR: Complementary hydropathy identifies a cellular prion protein receptor. *Nat Med* 1997, 3:1376-1382.

[47]McGuigan JE, Campbell-Thompson M: Complementary peptide to the carboxyl-terminal tetrapeptide of gastrin. *Gastroenterology* 1992, 103:749-758.

[48]McGuigan JE: Antibodies to complementary peptides as probes for receptors. *Immunomethods* 1994, 5:158-166.

[49]Moore GJ, Ganter RC, Franklin KJ: Angiotensin 'antipeptides': (-)messenger RNA complementary to human angiotensin II (+)messenger RNA encodes an angiotensin receptor antagonist. *Biochem Biophys Res Commun* 1989 160:1387-1391.

[50]Mulchahey JJ, Neill JD, Dion LD, Bost KL, Blalock JE: Antibodies to the binding site of the receptor for luteinizing hormone-releasing hormone (LHRH): generation with a synthetic decapeptide encoded by an RNA complementary to LHRH mRNA. *Proc Natl Acad Sci USA* 1986, 83:9714-718.

[51]Okada N, Asai S, Hotta A, Miura N, Ohno N, Farkas I, Hau L, Okada H: Increased inhibitory capacity of an anti-C5a complementary peptide following acetylation of N-terminal alanine. *Microbiol Immunol* 2007, 1:439-443.

[52]Papamattheou MG, Routsias JG, Karagouni EE, Sakarellos C, Sakarellos-Daitsiotis M, Moutsopoulos HM, Tzioufas AG, Dotsika EN: T cell help is required to induce idiotypic-anti-idiotypic autoantibody network after immunization with complementary epitope 289-308aa of La/SSB autoantigen in non-autoimmune mice. *Clin Exp Immunol* 2004, 135:416-426.

[53]Pascual DW, Blalock JE, Bost KL: Antipeptide antibodies that recognize a lymphocyte substance P receptor. *J Immunol* 1989, 143:3697-3702.

[54]Pasqualini R, Chamone DF, Brentani RR: Determination of the putative binding site for fibronectin on platelet glycoprotein IIb-IIIa complex through a hydropathic complementarity approach. *J Biol Chem* 1989, 264:14566-14570.

[55]Pfister RR, Haddox JL, Blalock JE, Sommers CI, Coplan L, Villain M: *Synthetic complementary peptides inhibit a neutrophil chemoattractant found in the alkali-injured cornea. Cornea* 2000, 19:384-389.

[56]Root-Bernstein RS, Westall FC: Bovine pineal antireproductive tripeptide binds to luteinizing hormone-releasing hormone: a model for peptide modulation by sequence specific peptide interactions? *Brain Res Bull* 1986, 17:519-528.

[57]Routsias JG, Dotsika E, Touloupi E, Papamattheou M, Sakarellos C, Sakarellos-Daitsiotis M, Moutsopoulos HM, Tzioufas AG: Idiotype-anti-idiotype circuit in non-autoimmune mice after immunization with the epitope and complementary epitope 289-308aa of La/SSB: implications for the maintenance and perpetuation of the anti-La/SSB response. *J Autoimmun* 2003, 21:17-26.

[58]Routsias JG, Touloupi E, Dotsika E, Moulia A, Tsikaris V, Sakarellos C, Sakarellos-Daitsiotis M, Moutsopoulos HM, Tzioufas AG: Unmasking the anti-La/SSB response in sera from patients with Sjogren's syndrome by specific blocking of anti-idiotypic antibodies to La/SSB antigenic determinants. *Mol Med* 2002, 8:293-305.

[59]Sakarellos-Daitsiotis M, Cung MT, Sakarellos C, El Hilali Z, Kosmopoulou A, Voitharou C: Complementary peptide epitopes and anti-idiotypic antibodies in autoimmunity. *Protein Peptide Lett* 2004, 11:367-375.

[60]Sautebin L, Rombolà L, Di Rosa M, Caliendo G, Perissutti E, Grieco P, Severino B, Santagada V: Synthesis and structure-activity of antisense peptides corresponding to the region for CaM-binding domain of the inducible nitric oxide synthase. *Eur J Med Chem* 2000, 35:727-732.

[61]Scapol L, Rappuoli P, Viscomi GC: Purification of recombinant human interferon-beta by immobilized antisense peptides. *J Chromatogr* 1992, 600:235-242.

[62]Shahabi NA, Bost KL, Madhok TC, Sharp BM: Characterization of antisera to the naloxone-insensitive receptor for beta-endorphin on U937 cells generated by using the complementary peptide strategy. *J Pharmacol Exp Ther* 1992, 263:876-883.
[63]Shai Y, Brunck TK, Chaiken IM: Antisense peptide recognition of sense peptides: sequence simplification and evaluation of forces underlying the interaction. *Biochemistry* 1989, 28:8804-8811.
[64]Shai Y, Flashner M, Chaiken IM: Anti-sense peptide recognition of sense peptides: direct quantitative characterization with the ribonuclease S-peptide system using analytical high-performance affinity chromatography. *Biochemistry* 1987, 26:669-675.
[65]Wu X, Richards NT, Johns EJ, Kohsaka T, Nakamura A, Okada H: Influence of ETR-p1/fl antisense peptide on endothelin-induced constriction in rat renal arcuate arteries. *Br J Pharmacol* 1997, 122:316-320.

In: Protein Conformation: New Research
Editor: Linda B. Roswell, pp. 125-159

ISBN: 978-1-60456-219-4

Chapter 3

NATURAL CONFORMATION OF HUMAN SERUM ALBUMIN AND ITS CHANGES IN PATHOLOGY

Yuriy A. Gryzunov and Gennady E. Dobretsov
Russian State Medical University
Dept. of Biophysics, Research Institute for Physical Chemical
Medicine Malaya Pirogovskaya, 1-A
Moscow 119998, Russia

Abstract

In the review (including authors' experimental data) the questions of natural serum albumin conformation and its changes due to pathology are considered. Particular attention is given to a method of albumin natural conformational changes detection using a new fluorescent probe, CAPIDAN.

Human serum albumin (HSA) is a protein of human body liquids which possesses many important functions. Its primary and secondary structure are known. At the same time its tertiary structure is uncertain because of flexibility of this molecule. Crystallographic data show a heart-shaped structure with long axis of 8-9 nm but that is a special crystal-induced shape. Hydrodynamic data in water solutions show that HSA molecule is not a compact globular sphere and are in accordance with two alternative (or co-existing?) conformations: prolate ellipsoid of 11-14 nm or/and oblate ellipsoid with long axis of approximately 9 nm. Our neutron scattering data let us suggest conformation of albumin molecule in solution to be between more short prolate ellipsoud of 11 nm or less and an oblate ellipsoid with diameter of 8-9 nm

One of the most important HSA functions is adsorption of many toxic metabolites in tissues and their transport to organs of detoxification. There are special sites in the HSA molecule for those metabolites. These sites are also able to bind many dyes, and that approach is very effective in studies of the site properties. Especially convenient are fluorescent probes proposed at 1952 and developed in further. After appearance of) several fluorescent probes for albumin study, a new probe, N-carboxyphenylimine of dimethylaminonaphthalic acid (CAPIDAN or K-35), has been proposed in 1990-s. CAPIDAN fluorescence in diluted human serum can be attributed (>96%) to HSA-bound probe molecules. A technique has been proposed to measure competition of the probe with specific markers of drug binding sites as well as with many other HSA ligands. The technique is based on fluorescence energy transfer from a covalent fluorescent label (donor) bound to HSA amino groups to CAPIDAN molecule (an acceptor).

A fluorescent test has been proposed to detect the state of HSA main binding sites in natural serum. The advantage of the proposed test is that it can be used in intact serum, at natural HSA conformation. In the majority of serum samples of healthy volunteers, CAPIDAN fluorescence intensity is proportional of HSA concentration in serum. However, significant deviations from this rule have been detected in sera of patients. The correlation was revealed between those fluorescence deviations, i.e. changes of HSA natural conformation, on one hand, and the severity of diseases, on the other hand. Examples and perspectives of detection of albumin conformational changes in clinical practice are discussed.

Key words: human serum albumin, natural conformation, fluorescent probes, changes in pathology

Introduction

Albumin (by mass) is the second protein of human blood after hemoglobin. Albumin is responsible for many functions, including transport ones: it transfers the main mass of nonesterified fatty acids and many different metabolites including drugs. Albumin has special binding sites for these ligands. Albumin is a globular protein but this globule is rather flexible. Therefore the sites influence each other. Moreover, conformation and physical chemical properties of its molecule are changed in pathology. These intramolecular interactions are due to due to a small, a few Å, shifts of nearest domains and groups, and, therefore, detecting these shifts are very problematic. Many physical methods require isolation of albumin from its native environment in blood, and a problem of maintenance of its natural conformation arises.

In clinical laboratories the measurement of albumin mass concentration in serum is obligatory in diagnostic purposes. It is commonly clear that low level of albumin is a negative factor in disease prognosis and is associated with low lifespan [1-3].

On the other hand, recently it has been shown that, besides of (of) quantity, albumin "quality", its capacity for binding and transferring a lot of molecules (different metabolites, fatty acids, hormones, xenobiotics including drugs etc), its ability to clean the body could be very important.

Intensive study of human serum albumin (HSA) has been raised due to development of critical care therapy as well as military medicine. HSA is an object of a huge number of publications which results are reviewed, e.g. in monographs and reviews (e.g. [4-7]). A special Internet site, www.albumin.org, accumulates current results in this field. Therefore some general problems are not included in this Review: we are focused on problem of HSA binding sites conformation in their natural state, mainly in blood plasma or serum. Data on the state and its changes due to pathological processes have been obtained mainly using dyes, especially fluorescent probes.

Some General Words on Human Serum Albumin

Human serum albumin (HSA) relates to a multigene family of proteins which includes also α-fetoprotein, group-specific component and vitamin D-binding (D-binding) protein. A single copy of the HSA gene is placed in proximal end of a long shoulder of fourth chromosome [8]. In medicine, the term "HSA" means only single protein. That is a globular protein with molecular mass of 66400 Da which is soluble in semi-saturated ammonium sulfate at pH 6-7.

Metabolism of the HSA

The main place of HSA synthesis is liver, however few quantity of HSA are synthesized in embryonic kidney, pancreas, lungs, heart and even in retina [9].

A HSA precursor, prealbumin, is synthesized as one polypeptide chain within about 1.5 min, and formation of each of 17 intramolecular disulfide bonds needs 30 s. After it, the N-terminal hexapeptide is splits off, and HSA leaves the cell. All process takes 20-30 min. Daily synthesis is 10-15 g HSA [8].

HSA catabolism takes place in almost all organs and tissues. Mean lifetime of HSA in the body is close to 20 days [4, 6, 8].

Distribution and Circulation

Mean HSA concentration in blood plasma (or serum) is about 45 g/L, that is more than half of total mass of plasma proteins. Other body fluids contain less HSA: 15-36 g/L in lymph, typically 3-10 g/L in interstitial fluid, 1-4 g/L in tears, 0.3 g/L in liquor, 0.1 g/L in aqueous humor. Body vessels contain 120-130 g HSA while about 300 g are outside the vessels [6, 8]

HSA pools of tissues and plasma are permanently exchangeable. In norm, 5% of plasma HSA leave the vessels and permeate capillary wall going into tissues (pervade tissues through capillary wall). Then as a part of interstitial fluid HSA reaches lymph and through thoracic duct returns into blood. All HSA molecules pass this way daily [4, 10]; may be, this exchanges takes much - 2-5 days [11].

Naturally, rate of that exchange is different in various tissues [12]. That HSA circulation and distribution are disease-dependent. Pharmacokinetics of drugs depends on that distribution of HSA between vessels and tissues in more extent than on changes of HSA concentration in plasma [13].

Molecular Structure

The single polypeptide chain of HSA (66400 Da) contains 585 amino acid residues and, free of hydrocarbons, leaves the cell. 50-67 % of residues form α-helixes [14]].

The turns of the chain shapes 9 loops that are fixed by 17 disulfide bonds. The packing of the loops repeats and forms 3 domains. Each of these three similar domains contains about 190 residues. These three domains form a total globule due to hydrophobic and electrostatic interactions [6, 8, 14].

Charge and Electrical Properties

The isoelectric point of. the HSA was reported as 4.8) Hence, total charge is negative at physiological ph 7.3 and is close to -15. At the same time local charge nearby the binding sites can be changed due to binding fatty acids, other ligands or changes of pK of charged groups [15, 16].

Albumin Binding Sites

HSA molecule is able to bind reversibly a lot of different small organic and inorganic molecules (ligands). Some molecules bind chemically (covalently), e.g. hydrocarbons (especially important glucose binding at diabetes), aldehydes, thiols, nitric oxide etc. [6, 17]

The most prevalent, however, is reversible binding to HSA. Special sites for reversible binding of some organic ligands exist. That interaction is commonly characterized by two parameters: mean number, n, of that kind of binding sites per one HSA molecule and binding constant, K [18].

The most important sites are drug-binding sites I and II. They are localized in a region included residues of second and third domains. Other sites bind nonesterified fatty acids, bilirubin, metal ions (Fig.1).

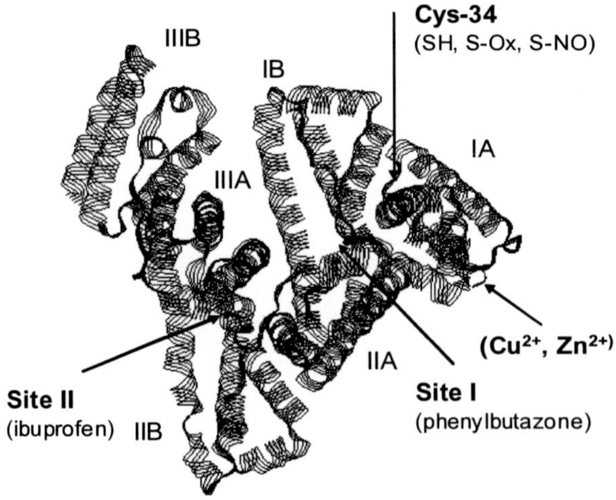

Figure 1. Localization of domains and some binding sites in HSA globule. Ligands bound to specific binding sites are shown in brackets. The picture was drown using RasTop2.1 software, PDB # 1HA2 [19].

Albumin Functions in the Body

HSA performs many functions which are important for homeostasis. Therefore lowering HSA concentration in some diseases may lead to hard complications.

Maintenance of Plasma Oncotic Pressure

In norm, HSA is responsible for 80-85 % of plasma oncotic pressure. That is a consequence of (a) high quantity of water bound with (to) HSA charged groups, on one hand, and low permeability of capillary walls for HSA, on the other hand. Many diseases are accompanied by elevated capillary permeability for HSA, and in some cases (e.g., acute renal failure, hepatocyrrhosis, hemolytic disease) quantity of HSA-bound water changes [20] that influence the oncotic pressure.

Ligand Binding

Testing drug binding to HSA is an obligatory step in any new drug studies. One HSA molecule can bind many (up to tens) ligand molecules with molecular mass about 350 Da, up to 1 g ligand per 1L of blood. That high level of ligands is not typical for actual blood, therefore HSA as a rule is not saturated with them, and sometimes HSA is considered as non-saturable carrier [21].

However cases of high concentrations of metabolites and drugs comparable with HSA concentration are reported. E.g., in newborns high bilirubin level up to 0.4 mM can be found in neonatal holestasis, and that is very close to HSA level (0.6 mM). Moreover, antibiotic cephalosporin replaces bilirubin competitively releasing bilirubin, and this competition can lead to bilirubinic encephalopathy [22, 23].

HSA is important element of detoxification system because it binds many metabolites (e.g., bilirubin, uremic toxin) that are toxic in free form but lose their toxicity in bound state. This HSA property is especially important at renal and hepatic insufficiency [24].

Participation in Red-ox Reactions

HSA plays an important role in preventing toxic effects of free radicals [25-27]: HSA binds transition metals lowering their radical-producing potential by hundreds times. Normal HSA has one thiol group of Cys-34 (Fig 1). About 80 % of total plasma oxidable thiols are the HSA thiols. These thiol groups are easily oxidable in free radical reactions. That oxidation results in a formation of stable products (disulfides, sulfonic acids). Therefore HSA breaks chains of free radical reactions [28, 29]. There are some reasons to believe that a positive effect of HSA infusions at critical states can be attributed to appearing new free thiol groups [30]. Obviously, some other groups (phenolic, amino) also might be able to react with free radicals and therefore to prevent their toxic effects and oxidative stress [31].

Quantity of HSA is rather high (c.a. 499-500 g), lifetime is rather low (20 days), therefore HSA may play a role of an antioxidant-sacrifice of free radicals [26, 32]. In some cases, e.g. at uremia, this "sacrificial" protein meets itself the main attack of oxidative stress [33].

HSA interaction with free radicals can lead to changes of its ability to bind drugs and other ligands) [34-36]]. And, on the other hand, ligands (including fatty acids) bound with HSA can affect and regulate reversibly its antioxidative function by changing HSA conformation [37].

Plastic Function and Interaction with Proteases

HSA catabolism in tissues produces amino acids that are a plastic material and energetic source for cells. HSA is a substrate for proteases. But, on the other hand, HSA influences their activity (as a rule as an inhibitor). Peptides of hydrolyzed HSA are biologically active and, perhaps, might play an important role in pathology) [38, 39] 71].

Regulation of Haemostasis

HSA influences hemocoagulation: it supresses thrombocyte aggregation, and has heparin-like activity in relation to antithrombin III activation [40, 41] Partly its anticoagulant effect may be a result of HSA ability to form S-nitrosothiols and, as a consequence, to suppress nitrogen oxide inactivation while nitrogen oxide depress thrombocyte aggregation [42]. At the same time there is a procoagulant fraction of HSA which after isolation from normal plasma can induce some endothelial factors of coagulation. Probably this fraction is a HSA-phospholipid complex [43]. Glycated HSA is also able to accelerate thrombocyte aggregation [44]; the glycated fraction significantly rises at diabetes.

Participating Maintainance of Ionic Homeostasis

At pH 7, HSA molecule contains about 185 ionic groups. Hence, their concentration in plasma is 0.11 M that is comparable with the level of inorganic ions, 0.3 M. Therefore HSA influence ionic equilibria in blood.

Albumin and Apoptosis

There are data on HSA ability to affect cell apoptosis as a rule to suppress it [45, 46]. May be, this HSA property is due to the state of its thiol group and presence of HSA ligands, particularly fatty acids and other [47].

Endothelium Permeability and Microcirculation

One assumes that HSA is bound in subendothelium and intersticial matrix affecting their permeability for other molecules [48, 49]. HSA influences membrane permeability. In cell monelayers, HSA absence elevates, as a rule, membrane permeability but its recovering takes place after HSA addition of as small as 1/4 of its normal level [50]. On the other hand, tissue permeability for HSA is regulated by inflammation mediators: bradykinin increases it while prostaglandins, vice versa, suppress it [51]. In course of experimental hemorrhagic shock, HSA suppresses the inflammation and improve(s) microcirculation as compared with physiologic solution without HSA [52]

Table 1. Data on size and shape of albumin molecules. *a* and *b* are semiaxes of prolate or oblate ellipsoid.SAXS is small-angle X-ray scattering*, SANS is small-angle neutron scattering

Protein	Conditions	Method	Results	Bibliography
HSA		Dielectric dispersion	Prolate ellipsoid	Oncley, 1942, [53]
HSA	pH 6.4	Viscometry, sedimentation	Prolate ellipsoid 150·38·38 Å	Oncley et al., 1947, [54]
BSA		SAXS	R_g=26.6 Å: prolate ellipsoid a/b=2.7 or oblate one a/b=3.4	Ritland et al., 1950, [55]
HSA;mercapto-albumin		X-ray diffraction	Prism 145·45·32 Å	Low, 1952, [56]
HSA		SAXS	R_g=31 Å, oblate ellipsoid	Anderegg et al.,1952, [57]
HSA		SAXS	R_g=29.8 Å prolate ellipsoid a/b=3.5	Anderegg et al.., 1955, [58]
HSA		Viscometry, sedimentation	Sphere	Loeb, Sheraga, 1956, [59]
BSA	pH 5.6	Viscometry	Prolate ellipsoid a/b=3.3	Tanford, Buzzel, 1955, [60]
BSA		Viscometry	Prolate ellipsoid a/b=3.9	Champagne, 1957,[61]
BSA		SAXS	R_g = 29.8 Åprolate ellipsoid a/b= 120 Å/48 Åor oblate a/b= 30 Å/94 Å	Luzzati, 1961, [62]
Pig albumin		SAXS	Prolate ellipsoid	Laggner 1971, [63]
BSA		Viscometry, sedimentation	Prolate ellipsoid a/b=2.4	Creeth, Knight, 1965, [64]
BSA		Rotational diffusion	Prolate ellipsoid a/b=3.0	Moser et al., 1966, [65]
BSA	pH 4.55-5.15	Sedimentation	Prolate ellipsoid a/b=3.5= 140 Å /40 Å	Squire et al., 1968, [66]

Table 1. Continued

Protein	Conditions	Method	Results	Bibliography
HSA	pH 7.4	Fluorescence energy transfer	Distances (Å): Trp214–Tyr411 = 252; Tyr411–Cys34 =25.6; Cys34–Trp214=31.8.	Hagag et al., 1983, [67]
HSA		Scanning tunnel microscopy	Sphere	Feng et al., 1989, [68]
Recombinant HSA	Crystal	X-ray diffraction	Prolate molecule	Carter, 1989, [69]
Recombinant HSA	Crystal	X-ray diffraction	R_g = 26.7 Å Heart-like shape	Carter, 1990, 1992, 1994 [14, 70, 71]
HSA	pD 7.4, ionic strength I=1 M	SANS	Prolate ellipsoid a/b=3.7= 136 Å /38 Å	Sjoberg, 1994, [72]
Recombinant HSA	Crystal	X-ray diffraction	Heart-like shape 80 Å /30 Å	Sugio, 1999, [73]
Recombinant HSA	Crystal	X-ray diffraction	Heart-like shape 80 Å /30 Å	Curry et al.,1998, 2000, [74, 75]
HSA	pD 7.1, ionic strength I=0.15 M	SANS	R_g = 27.4 ±0,4 Åprolate ellipsoid $2a \leq 108$ Å or oblate $2b \leq 76$ Å	Киселев et al., 2001, [76]
BSA	Solution	Fluorescene polarization of erythrosine label	Triangle prism 84x84x84x31.5 Å	Ferrer, 2001, [77]

- The complete list of PDB structures of HSA can be found at Internet site www.albumin.org. Atomic coordinates for the tertiary structure of human albumin are available at the GenBank DNA database at www.ncbi.nlm.nih.gov . Choose "Structures". To view coordinates, use RasM.1 Chime (Sayle, RA, TIBS 20, 374, 1995.)

As can be seen from the brief review, HSA is polyfunctional participant of many processes in the body. These HSA activities depend on ligands of different kind that bind to HSA. These ligands obviously act indirectly on the processes: a chain ligand⇒HSA⇒activity should exist. A connecting chain link is, probably, HSA conformation - always or at least in many important cases. The role of HSA conformation is studied insufficiently. Perhaps, one of the general causes of this situation is a very limited arsenal of physical methods acceptable for studying natural conformation of HSA in its native medium - in blood plasma. The most data are obtained by different versions of fluorescent probe methods.

HSA Conformation

Problem of describing the size and shape of HSA molecules in solutions has at least 65 years of age. The structure of recombinant HSA crystals is described in detail while the size and shape of natural HSA in solutions and, more over, in natural plasma still is an object of discussions. The history of the problem is shown in brief in Table 1.

In general, a choice between two alternative - or coexisting - models should be made: first one is a prolate ellipsoid with long axis of 11-15 nm; second one is an oblate ellipsoid with diameter about 8-9 nm. If HSA is a compact sphere, its gyration radius should be near Rg = 21 Å. However all methods give greater values (Table 1).

HSA Conformation in Crystals

This conformation with resolution of 6 and then of 3.1 Å has been shown in 1989-1994 by crystallography. According to most results, the HSA molecule is heart-shaped and close to an oblate ellipsoid with diameter of 8 nm [14, 69-71]. The studies have been followed by further research of better (2.2 – 2.6 Å) resolution [19, 74, 75].

HSA Conformation in Solution

Table 1 shows that there are problems with description of HSA shape in solutions. It is clear that the shape is not spherical. However it is not clear, whether the shape is closer to oblate (heart-shaped) or prolate ellipsoid. Some recent measurements by neutron scattering technique have been made in Dubna [76]. Results of these experiments are presented in Fig.2, where $d\Sigma$ is a number of neutrons scattered in a solid angle $d\Omega$, q is an angle parameter where θ is the angle of scattering and λ is wavelength of the de Broglie wave of neutrons (close to 10 Å). These data presented as a Kratky-Porod plot allow to calculate HSA hyration radius: Rg = 27.4 ±0.4 Å This value is the same after increasing pD to 8.9 though the conformational transition N↔B. Binding one NEFA molecule (palmitic acid) slightly increases Rg - by about 1 Å; small conformational changes induced by NEFA have been observed recently by crystallography [71] and molecular dynamics calculations [78].

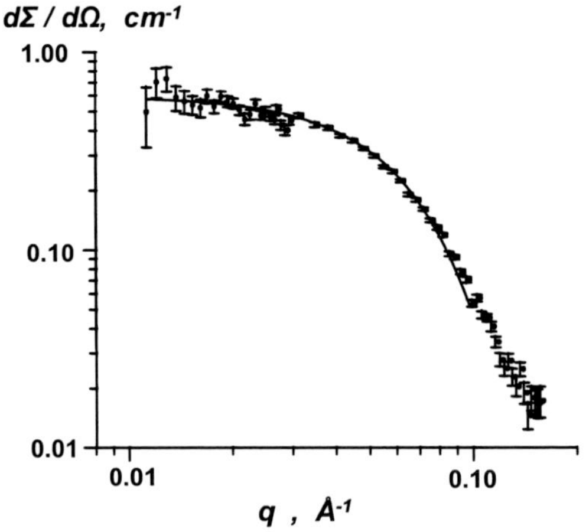

Figure 2. Small-angle neutron scattering by HSA solutions. Fatty acid free HSA (0.19 mM) was dissolved in 0.15 M NaCl in D_2O, pD 7.1.

This value Rg = 27.4 Å is consistent with HSA shapes between prolate ellipsoids of length below 108 Å, as upper limit of HSA size, and oblate ellipsoid of diameter 80 Å as other limit for HSA conformation. The last model is very close to crystallographic sizes of "heart"-like HSA with Rg = 26.7 Å [71].

It seems, a hypothesis cannot be excluded that different HSA molecules in solutions have different conformations because the HSA molecule is flexible and very sensitive to its environment; then the gyration radius R_g = 27.4 is an averaged value for HSA in a variety of conformations.

In water solutions HSA conformation is pH-dependent. This fact can be illustrated by uncharged fluorescent probes that are not sensitive to charges but their fluorescence is influenced by pH-dependent conformational changes of the protein [79-81] (Fig. 3).

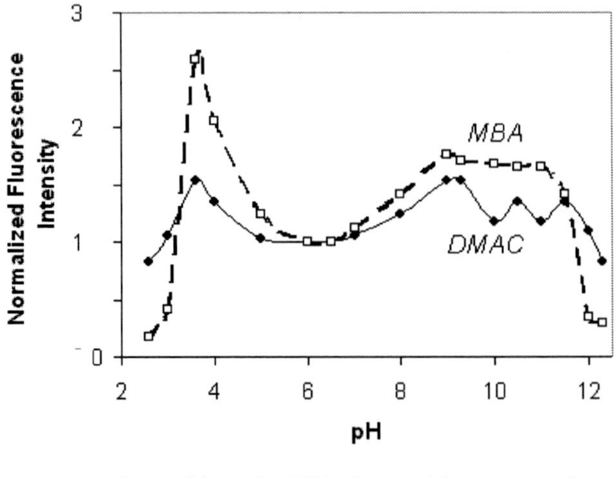

Figure 3. pH-dependent structural transitions in HSA detected by two uncharged fluorescent probes, DMAC and MBA. Fluorescence intensity at any pH was normalized on intensity at pH 6.5.

HSA study by fluorescent probes. Albumin fluorescent test.

Dye and Fluorescent Probe Studies: History

Interaction between dyes and protens has been studied since the end of 19 century (see a review [4]. In 1925, it has been find out that a dye, phenol red, binds to albumin much better than with other plasma proteins [82]. Further studies have shown that albumin has the most ability to bind a lot of dyes as well as drugs than other plasma proteins. As a result, HSA plays an important role in pharmacokinetics [83-85]

Interest to molecular mechanisms of drug-albumin interaction has raised after the Second World War while when has been found out that penicillin is transported in blood by HSA [83, 86]. In these studies it was found that the interaction is not covalent, it is reversible, and that drugs and dyes can occupy the same HSA sites competitively. However dyes are much more convenient for that research because their absorption spectra changes significantly at binding with HSA [87-89]. Moreover, some dyes including a bromocresol group [90] (Fig 4) prefer to bind with HSA in whole (or serum) in spite of the presence of many other proteins. This observation allowed proposing a clinical method of HSA assay which is used today worldwide [91].

However as early as in the 1950-1960 some abnormalities of dye binding have been observed in various diseases. A few researchers started investigation of these abnormalities. One of the dyes used on the early steps of the research was congo red (Fig. 4). Its ability to bind to plasma proteins has been known for a long time (Sorensen 1909: citation from [4]). Csögör has proposed a technique to measure HSA-bound congo red and has shown that its quantity is disease-sensitive [4]. Some other researchers used this technique and confirmed sensitivity of congo red test [92] and several other dyes (e.g., [93]). Unfortunately, these methods were rather laborious, therefore systematic clinical studies were not performed.

In 1952, Laurence has studied interaction of fluorescent dyes with proteins. An aniline dye, N-phenyl-1-amino-8-sulphonaphthalene (ANS, Fig. 4.), had very low fluorescence in water, but its fluorescence intensity raised many times in protein, especially albumin solutions [94, 95]. In clinical samples of sera, ANS fluorescence intensity was roughly proportional to HSA concentration but in some samples, especially of jaundice patients, the intensity was decreased [96]. These results had no a continuation. Ten years later ANS has been used as a fluorescent probe for studying BSA [97-100] and some other proteins [101]. It has been shown that BSA has 15 or more binding sites for ANS, but only 5 of them have rather high binding constant of the order of $1.1 \cdot 10^6$ M^{-1} [102]. The 5 sites of high affinity are really different: occupation of the first one induces a circular dichroism (CD) species of ANS while occupation of other 4 sites induces other CD species [99].

It is interesting that fluorescence of unbound ANS is fully quenched by water, therefore its decay time, τ, is < 1 ns while all 5 molecules of strongly bound ANS have very high τ close to 15 ns [100, 103]. It seems, this high τ allows suggesting that ANS is fully hidden in HSA. However our data on ANS quenching by a water soluble quencher, nitrate ion, shows τ decrease due to collisions of nitrate with HSA-bound ANS. Perhaps, ANS amino group is really hidden while some places of naphthalene ring are acceptable (accessibile) for water (see below). In whole serum, about 73 % ANS are bound with HSA but these molecules are responsible for 87-90% of total ANS fluorescence [104, 105].

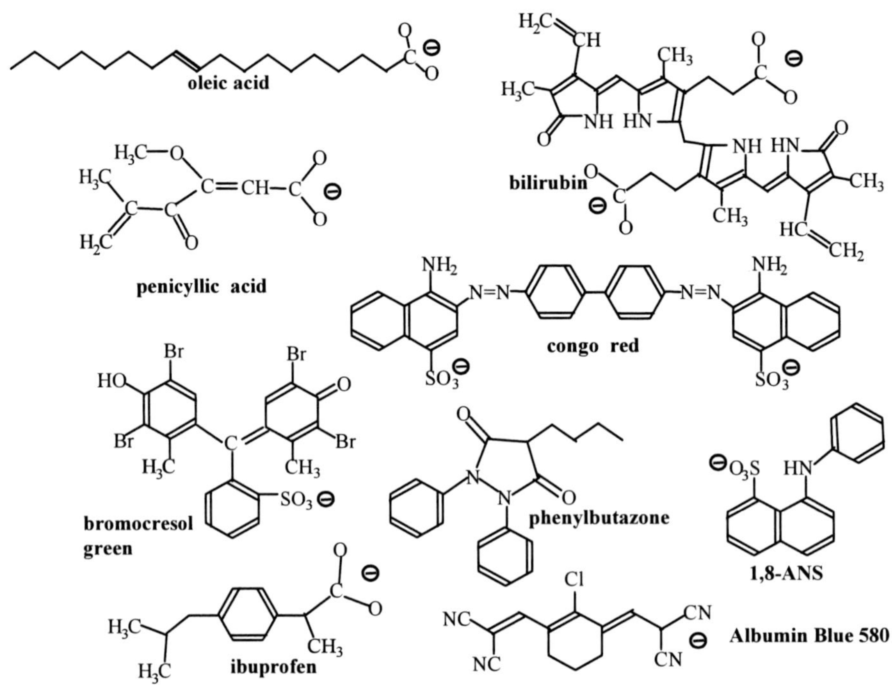

Figure 4. Ligands of human serum albumin: natural substances (oleic acid and bilirubin), dyes (bromocresol red, congo red), fluorescent probes (1,8-ANS, Albumin blue 580) and some drugs (penicillic acid; phenylbutazone as a marker of Site I; ibuprofen as a marker of Site II). Negative charges are shown in their charged forms at pH near 7.4.

The pH decrease from 7.4 to 4 leads to HSA conformational transition N↔F. This transition does not induce changes of ANS accessibility to nitrate but number of binding sites is raised significantly [103, 106].

After first studies of HSA by ANS, other fluorescent probes have been also proposed (e.g., [107-118]. Obviously, fluorescent probes which fluorescence can be attributed to HSA in whole serum, are much more convenient and informative as compared with absorption dyes (congo red etc.), however potential usefulness of that approach has not been studied in detail during 35 years after the first attempt in 1954.

Fluorescent Probe CAPIDAN (K-35)

Abnormal Behavior of Dyes in Serum at Diseases

Some cases of abnormal interaction of dyes with HSA in serum of patients are published perhaps starting in 1951 [119-121]. Decrease of serum ability to bind dye has been detected for azorubin at hepatitis [119], for phenolphthalein at hemolytic anemia of newborns, for congo red at renal and some other diseases [4].

At the same time causes of these abnormalities remained unclear. What are roles of HSA conformation, its occupation with ligands (metabolites, fatty acids), HSA and dye interaction with other plasma components? Optimization of these tests has been questionable: e.g., in the test with congo red one HSA molecule binds 12-19 dye molecules while HSA have only high-affinity 1-5 sites (excluding sites for fatty acids).

At once it has been shown by other methods that some diseases lead to HSA modification and change its physical chemical properties [122, 123]. Bilirubin in HSA is known to reduce ANS fluorescence intensity [96]. In 1989 the study of ANS in pathology has been carried out again. However ANS decay time in hyperbilirubinemia was the same as in normal serum, i.e. there was no fluorescence energy transfer from ANS to bilirubin. Bilirubin adding into normal serum does not reduce fluorescence in the same extent as the same bilirubin level in jaundice patients [103]

In these cases, the deviations of ANS fluorescence intensity from normal one did not correlate with total or indirect bilirubin level. Moreover, ANS fluorescence was decreased in diseases at normal bilirubin level (e.g., miocardial infarction) [124]. Hence, there exist "nonbilirubin" causes of these abnormalities, some other factors affect properties of HSA molecules changing a significant fraction, may be total HSA population in blood.

That kind of ANS sensitivity to HSA anomalies has been overcome by changing experimental conditions [125]; hence, HSA molecule is not dramatically and irreversibly destroyed. At the same time these abnormalities detected by the fluorescent probe are of significant interest. Therefore a new fluorescent probe named K-35 or CAPIDAN has been synthesized specially for study of HSA pathological changes.

Creating Fluorescent Probe CAPIDAN (K-35)

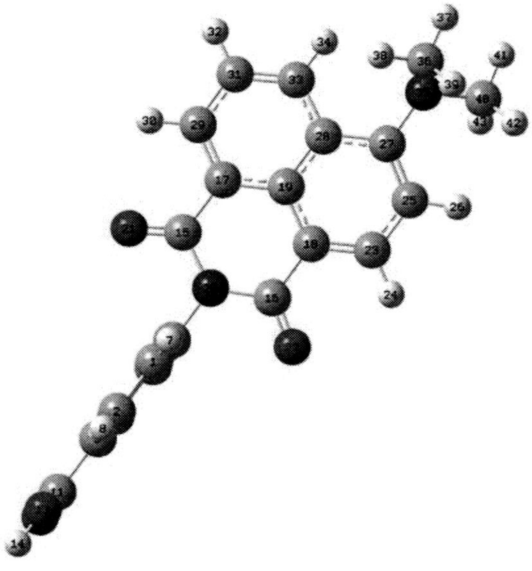

Figure 5. K-35 molecule. Atom numeration is consistent with quantum chemical data.

An attempt to create a probe more sensitive to HSA abnormalities than ANS has been made. B.M.Krasovitsky and colleagues have synthesized a set of fluorescent organic anions. One of them named K-35 or CAPIDAN (Fig. 5) was synthesized by L.I.Kormilova and I.G.Ermolenko. Fluorescence intensity of any probe, ANS or K-35, decreases in HSA solutions at addition of drugs specifically bound with HSA, however K-35 sensitivity to drugs was several times more than ANS (Fig.6). Fraction of HSA-associated K-35 fluorescence in diluted plasma exceeds 95 % (Table 2) [124, 126] while only 87 % in the case of ANS (see above).

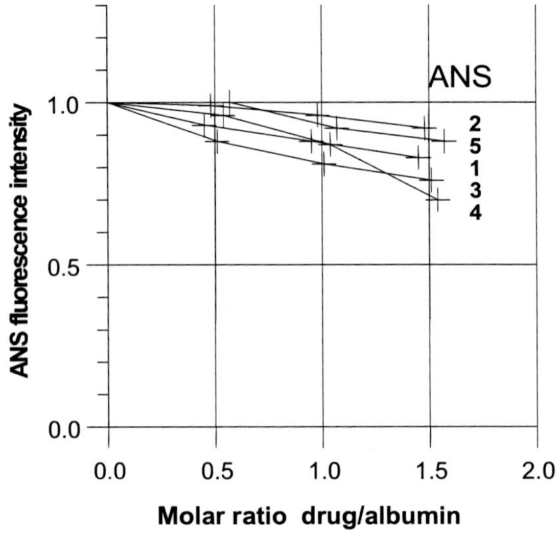

Figure 6. Decrease of ANS and K-35 fluorescence intensity in HSA solutions due to drug addition. 1 - bilignost; 2 - iopanoic acid; 3 - ibuprofen; 4 - phenylbutazone; 5 - sulphadimetoxine.

Table 2.Fluorescence intensity (*F, arbitrary units*) of K-35 at 520 nm (excitation at 440 nm) in different fractions of serum of healthy donors [127]. Globulins are serum minus HSA and minus lipoproteins. Concentration of each fraction corresponds to its concentration in normal serum diluted by 200 times.

	F
None (buffer pH 7.4)	<1
Lipoproteins	2
Globulins	8
HSA	100
HSA + Globulins	102

Quantum Chemistry Calculation

Quantum chemistry calculation of the molecule was performed for better understanding properties of the new fluorescent probe, K-35. Its molecule has carboxyl negatively charged at pH 7.4, phenyl ring, naphthalimide and dimethylamino groups. Calculations *ab initio* show that the phenyl ring is perpendicular to the naphthalimide plane (Fig. 7), therefore their optical conjugation is very low. At the same time rotation barrier of the phenyl about the bond C–N up to angles of 45^0 is rather low – below 3 kcal/mol (Fig. 8). Rate constant for that rotation is of the order of 30 ps when its value is estimated by the common equation: where k_B

is the Boltzmann constant, T is temperature, N_A is the Avogadro number and ΔG is energy from Fig. 8.

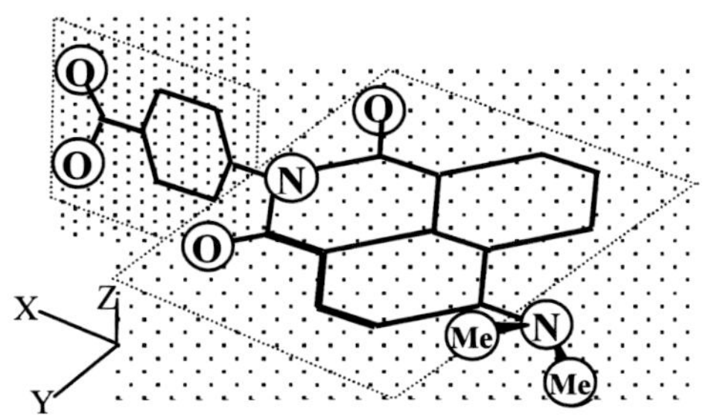

Figure 7. Shape of K-35 molecule as calculated *ab initio* by the Hartree-Fock method.

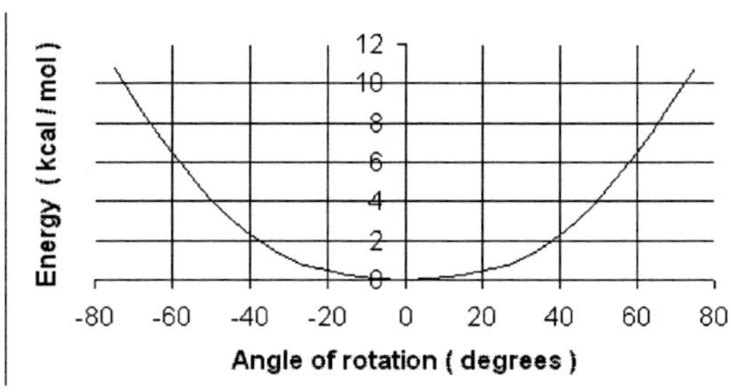

Figure 8. Energy of phenyl ring rotation about C–N bond in K-35 molecule calculated *ab initio* by the Hartree-Fock method.

The molecule in its ground state has rather large dipole moment: μ_g = 7,75 D along X axis (Fig. 7.). It raises to 8.7 D in nonpolar media and to 10.0-10.2 D in polar media. Such values of the dipole moment are a result of intramolecular charge transfer from dimethylamine group to oxigen atoms.

^{1}H NMR Data

Some major peaks of K-35 proton magnetic resonance are presented in Table 3. Six methyl protons of the dimethylaminogroup give a narrow and high peak; doublets and triplets of aromatic protons also are seen well. Solvent polarity does not shift the methyl proton peak while aromatic peaks go to lower fields in polar media.

Table 3. Peaks of K-35 ^{1}H NMR spectrum in deutered solvents The numbers of protons corresponds to Fig 5 dd-doublet, t-triplet Water buffer had pD 7.1.

Proton ## (Fig.5)	37-39, 41-43 $N(CH_3)_2$	30	32	34	26	24	8	7
chloroform	3.15	8.61	8.00 (t)	8.51	7.44	8.51	7.68 (dd)	8.25 (dd)
acetone	3.16	8.62	7.80 (t)	8.42	7.30	8.52	7.55 (dd)	8.18 (dd)
waterPBS	3.15	8.39	7.56 (t)	8.12	7.01	8.25	7.31 (dd)	8.05 (dd)

Binding Sites for CAPIDAN and Their Heterogeneity

Understanding the probe-albumin interaction leads to rising of two principal questions: (i) what are HSA binding sites for and their affinity to the probe and (ii) how do these binding sites form K-35 fluorescence. To answer them we used quantum chemistry calculculation, ^{1}H-NMR, equilibrium dialysis time-resolved fluorescence spectroscopy, resonance energy transfer and special fluorescence titration techniques. ^{1}H-NMR and equilibrium dialysis are described elsewhere. The particularities of other approaches are briefly described below.

HSA Binding Sites and Their Affinity

Fig.6. shows that anionic drugs decrease significantly K-35 fluorescence. Hence, that effect can be due to their competition for the same binding sites, or drug and probe molecules localize very close to each other in HSA. To check the competition idea, a technique has been used basing on fluorescence energy transfer.

Fluorescence Energy Transfer

The method used (described in [128]) is based on fluorescence energy transfer from a label (donor) covalently bound with HSA to a ligand (fluorescent probe CAPIDAN). An isoindole label – the product of albumin reaction with orthophthalic dialdehyde in presence of 2-mercaptoethanol – was proposed to be used as the donor (Fig. 9):

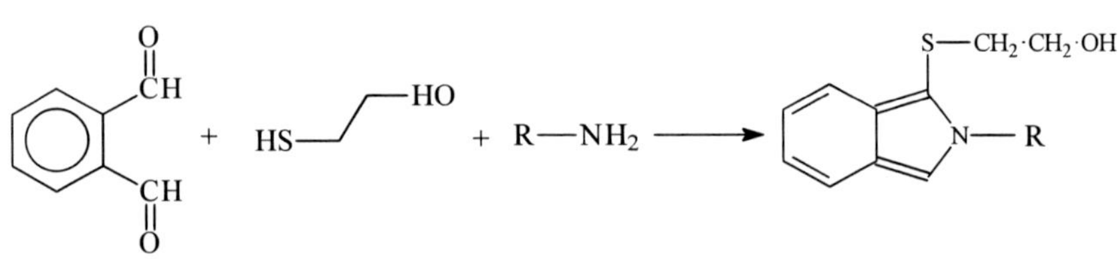

Figure 9. Isoindole labelling of HSA.

Let CA be the HSA molar concentration and r be the concentration of the HSA-bound ligand. Let the protein contain a fluorescent group whose (which) fluorescence emission overlaps the absorption spectrum of the ligand. In this case, the nonradiative energy transfer efficiency is found as

$$-\ln\left(\frac{F}{F_0}\right) = k\frac{r}{C_A},$$

where k is the product of an average transfer rate constant by the instrument constant; and F_0 and F are the fluorescence intensities of the donor before and after addition of the ligand, respectively [129].

The primary data for constructing the transfer efficiency plot is the value –ln(F/F0) as a function of the acceptor (CAPIDAN) added. The titration curves obtained can be processed by any available method, e.g. by the Klotz or Scatchard methods [18] . Practically, we have used the following technique. To determine the parameters of the acceptor binding (bound) to the albumin, the transfer efficiency was measured at several (two, in the simplest case) protein concentration. Then, such values of the binding constant K and number n of binding sites were matched that (1) the transfer efficiency –ln(F/F0) is proportional to the occupation degree, r/CA, of HSA binding sites by the ligand and that (2) $\chi 2$ calculated as a measure of coincidence of a model and experimental data should be minimal; i.e. the –ln(F/F0) versus r/CA plots at different protein concentrations are the best coincide within experimental errors.

"Amplitude Standard": a Technique to Estimate Fluorescent Molecules Absolute Concentration

The problem is that fluorescence of some fraction of a protein-bound fluorescent probe can be strongly quenched, e.g. by protein internal groups, hydrogen bond, etc., and such molecules can not be detected by steady-state fluorescence measurements. Nevertheless, the evaluation of the fraction bound but strongly quenched ligand molecules, as well as evaluation of the concentration of probe molecules fluorescing with different quantum yield, can be useful for diagnostics of protein dysfunctions. The "amplitude standard" approach was developed for quantitative evaluation of fluorescent probe molecules, including completely quenched (non-fluorescent) ones in heterogeneous media, e.g. in proteins, using the analysis of fluorescent decay amplitudes [130, 131].

The principle of the approach can be described as follow.

Usually, the decay of the probe fluorescence in proteins is non-exponential but can be described as a sum of several monoexponential components:

$$F(t) = A_1 \exp(-t/\tau_1) + A_2 \exp(-t/\tau_2) + ,$$

where A_1, A_2 ... are fractional amplitudes and τ_1, τ_2 ... are fractional decay times. The more the concentration, C_i, of the fraction, the more its amplitude is. However its concentration cannot be calculated directly using its amplitude because the amplitude depends also on the instrument scale, m, and on radiative rate constant, k_r, of the probe:

$$c_i = A_i k_r m$$

This value can be estimated using some standard sample of the probe where its fluorescence quantum yield, Q_S , should be close to 1, i.e. the most probe molecules have fluorescence with a decay time τS:

$$k_r = Q_S / \tau_S \;;$$

$$C_S = A_S k_r m \;;$$

$$\text{then } C_i = A_i (C_S / A_S)$$

More over, some quantity of strongly quenched molecules of bound probe with very low τ can exist. This approach allows also to estimate concentration of nonfluorescent molecules. Details are described in [130, 131].

A Special Fluorescent Titration Technique

Fluorescence intensity, *F*, of a probe is microenvironment dependent, and its changes after the probe binding with HSA depends on concentration, *r*, of bound probe. In the simplest case, that difference is proportional to *r*:

$$F(r) = r \bullet const$$

In this case, the value of *F(r)* allows calculating *r* directly. However relation between $\Delta F(r)$ and *r* is, as a rule, more complex, this "*const*" is *r*-dependent. To overcome this difficulty, a new method has been proposed [132]. Two HSA concentrations are titrated stepwise with the probe. Some calculations allow to estimate *r* at any concentration of added probe without any *a priori* assumptions on the relation between *F(r)* and *r*.

This method is illustrated by Fig.10. Indeed, the more of K-35 binds with HSA, the less fluorescence increases, $\Delta F(r)$, per each new portion of bound probe, Δr, is observed.

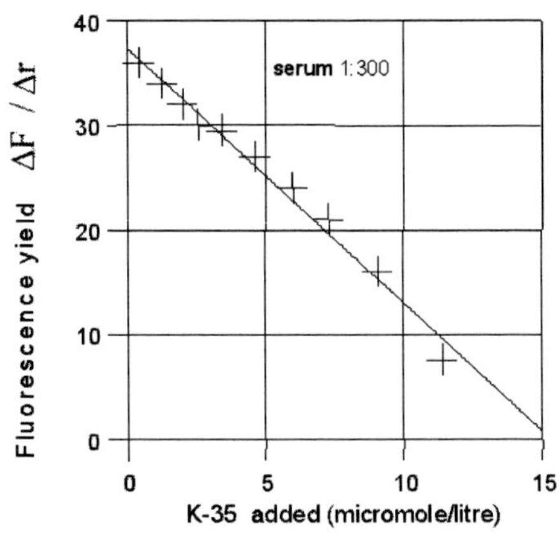

Figure 10 .Lowering fluorescence intensity rise, $\Delta F(r)$, due to bound probe concentration increasing, Δr, at K-35 adding to human serum diluted by 300 times.

Energy of Probe-Medium Interactions

In water solutions, K-35 molecule prefers to bind with HSA while volume ratio probe/HSA is of order of 1/1000 and less. What are causes of such preference? Energy of different interactions between K-35 and environment has been estimated by quantum chemistry methods (Table 3). The estimation shows that K-35 dipoles weakly interact with nonpolar heptane and rather strong - with polar acetone and water. However cavitation energy in water is so high that the probe have to prefer acetone. Perhaps, this is a major cause of redistribution of the probe from water to HSA.

Table 3. Energy (kcal/mol) of K-35 transfer from vacuum to different media calculated *ab initio.*

Heptane	Acetone	Water
-8.7	-27.3	-30.3
29.3	31.5	45.4

These data allow to suggest that if in HSA molecule nonpolar (heptane-like) and rather polar (acetone-like) pockets simultaneously exist, K-35 should prefer the last ones. Pavement of HSA binding sites could be polar.

Ligand Effect on CAPIDAN Binding and Fluorescence

As has been seen above (Fig.6), the addition of drugs leads to K-35 fluorescence lowering in HSA. Two hypothetic causes can be responsible: (i) competition between K-35 and drugs and (ii) indirect (allosteric, conformational) quenching K-35 fluorescence by drugs. For competition, K-35 should occupy the drug binding sites, firstly drug binding site I and site II.

The techniques described above were used to estimate possible localization of CAPIDAN binding sites in the HSA and to evaluate probe-to-HSA binding parameters. We have used markers of HSA binding sites as possible competitors of CAPIDAN for binding to the HSA. The displacement of the probe by a marker was considered as a competition for the same site in albumin. In parallel, the CAPIDAN fluorescence quantum yield in the HSA in absence as well as in presence of markers was measured. The results are shown in Table 4.

Table 4. The effect of HSA binding site markers on CAPIDAN binding with and its fluorescence in HSA

HSA preparation	Ligand	CAPIDAN parameters studied		
		Binding	Fluorescence intensity	Fluorescence quantum yield
Fatty acid free HSA	Palmitic acid	↑ 1.15-1.3	↑ 1.8-2.2	↑ 1.5-2.0
	Phenylbutazone (drug site I)	↓ 1.6-1.8	↓ 2-2.1	↓ 1.2-1.3
	Ibuprofen (drug site II)	N/e	↓ 1.15–1.25	↓ 1.2–1.3
	Bilirubin	N/e	↓ 1.7-1.8	↓ 1.7-1.8
	Cu^{2+}	N/e	N/e	N/e
	NEM	N/e	N/e	N/e
HSA loaded with palmitic acid 1:1 mol/mol	Phenylbutazone (drug site I)	↓ 1.2-1.3	↓ 2	↓ 1.6-1.7
	Ibuprofen (drug site II)	N/e	↓ 1,2-1,4	↓ 1,2-1,4
	Bilirubin	N/e	↓ 2,2-2,4	↓ 2,2-2,4
	Cu^{2+}	N/e	N/e	N/e
	NEM (thiol reagent)	N/e	N/e	N/e

Arrows show: ↑ – increasing by … times, ↓ – decreasing by …times, N/e – No effect.

All the data were obtained at ratio ligand/HSA ≤1 mol/mol. Commercial essential fatty acid free HSA preparation (Sigma) was used. Chemical structure of the ligands is shown above (Fig. 4); NEM – N-ethylmaleimide for HSA-SH blockage.

The results allow to assume that phenylbutazone is the competitor of CAPIDAN and, thus, one of the possible probe binding sites is site I or a region very near this (the) site. Binding of palmitic acid increased both binding of CAPIDAN as well as its quantum yield. Such behavior is typical for a fluorescent marker of site I, dansylimide [133, 134]. Hense, one assumes that K-35 occupy the site I. At the same time ibuprofen, a marker of site II, affect the probe changing its fluorescence quantum yield but not its binding; probably the interaction is allosteric. Cupric ions and N-ethylmameimide (blocking SH group of Cys-34 residue) did not affect CAPIDAN binding and fluorescence. The effect of bilirubin on the probe fluorescence, probably, is due to resonance energy transfer from the probe to the ligand.

Thus, at least one binding site for CAPIDAN is placed in site I region and the probe fluorescence is sensitive to conformational events there.

But how many sites for CAPIDAN are there in albumin? Are they “fluorogenic” and what is their affinity to the probe? The results of binding parameters determination in different HSA preparations by different methods are showed in Table 5.

Table 5. CAPIDAN binding parameters in different HSA preparations (*n* is number of binding sites per one HSA molecule, *K* is the apparent equilibrium constant)

Method	Commercial HSA (Sigma)		Serum	
	n	*K* L/micromole	*n*	*K* L/micromole
Equilibrium dialysis and ultrafiltration [128]	4.2 ± 0.3	0. 009 – 0.012	–	–
Resonance energy transfer [128]	4.0 ± 0.2	0. 01 – 0. 015	0.9 – 1.5	0.012 – 0.1
Conventional fluorometric titration	1.6 ± 0.2	0. 01 – 0. 03	1.8 ± 0.3	0.09 – 0.4
Special fluorometric titration [132]	–	–	≥ 2.3	0.2 – 0.3
NMR*	95 ± 5	0.002 – 0.003	–	–

The data indicate that there are a lot of binding sites for CAPIDAN, and about four of them have rather high affinity. These sites of strong binding have close values of binding constants because of straight line in Scatchard plot (Fig.11). At the same time, NMR data obtained at high HSA-to-probe ratio (more than 10:1) allow to assume that there are a large number of weak affinity sites. On the other hand, the results of fluorometric titrations demonstrates that no more than 2 of binding sites are “fluorogenic”, i.e. CAPIDAN bound to them fluoresces and can be detected fluorometrically.

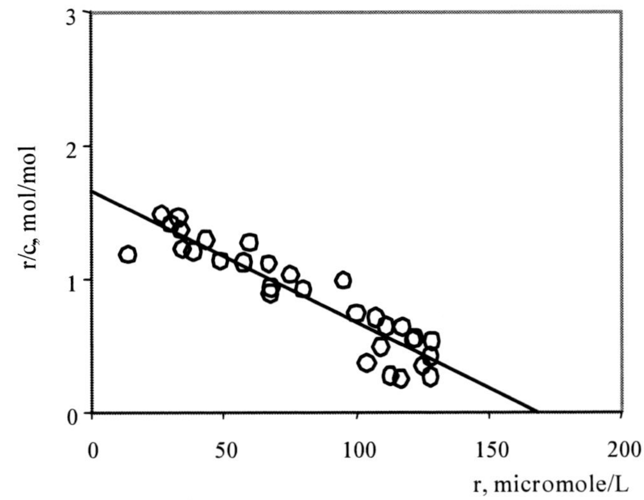

Figure 11. Scatchard plot for equilibrium dialysis data. Preparation of commercial (Sigma) fatty acid free HSA was under study.

Thus, analysis of the binding data allow to suggest that HSA binding sites for CAPIDAN are strongly heterogeneous in their potential to form the probe microenvironment in which fluorescence is not quenched.

CAPIDAN Fluorescence Decay and HSA Binding Sites Heterogeneity

Indeed, K-35 fluorescence decay in HSA preparations as well as in sera can not be described by a model of a single type of binding sites with similar probe environment in all occupied sites; three exponentials are needed to describe the decay:

$$F(t) = A_1\, exp(-t/\tau_1) + A_2\, exp(-t/\tau_2) + A_3\, exp(-t/\tau_3)$$

(see above). This set of three decay times τ_i (Fig.12a) is independent of the occupation degree while ratio of amplitudes $A_1 / A_2 / A_3$ is (Fig.12b). That points on different binding constants K for three types of sites. Hence, three types of binding sites exist where K-35 molecules have different microenvironment. These data allow to estimate their partial fluorescence quantum yields Q_i:

$$Q_i = \tau_i \cdot k_r$$

where k_r is radiation rate constant: $k_r = 7.2 \cdot 10^7$ sec^{-1}. Moreover, there exists a fourth - "nonfluorescent" - type of sites where $\tau_4 << 1$ ns.

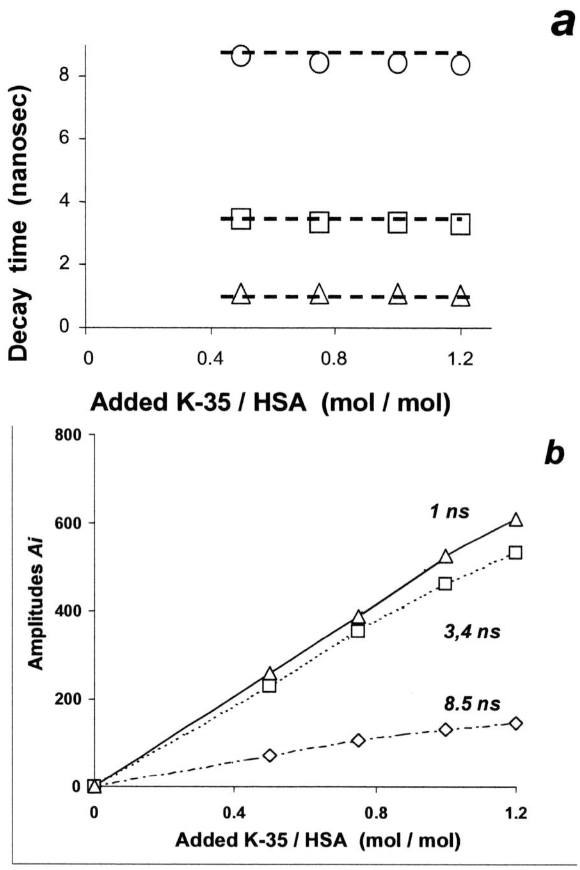

Figure 12. Decay times τ_i (*a*) and amplitudes A_i (*b*) of exponential components of K-35 bound with HSA in human serum. HSA concentration in the diluted serum was 38 microM.

Albumin Fluorescent Test

Test for Variation of HSA Binding Sites Conformation

The studies of CAPIDAN-albumin interaction and high sensitivity of the probe fluorescence to HSA binding sites conformation (see Table 4) lead to creating a routine procedure called "albumin fluorescent test" This test was developed to study conformational flexibility and was adopted – terminologically and technically – for its use in clinical laboratories.

That "conformational" basis of the approach is as follows [135]. The properties of the albumin globule and the state of its binding sites were evaluated using a procedure in which albumin undergoes a conformational transition known as normal-fast (N-F) [71, Peters, 1996 #5]. CAPIDAN fluorescence intensity is measured twice in the same sample. One measurement is made at pH 7.4 and is sensitive to changes of a natural N-conformation. Here the fluorescence intensity, $F(7.4)$, of CAPIDAN is proportional to HSA concentration, C_A:

$$F(7.4) = k \cdot (C_A) \cdot \chi_N ,$$

were χ_N is a conformational factor depending on HSA translational or posttranslational modification, and thus influencing the probe interaction with the globule; k is an instrument constant measured by use of respective standards. Then pH of the albumin sample is adjusted to 4.2 to induce N–F conformational transition and 0.03% of Brij-35 is added to stabilize the globule. In these conditions

$$F(4.2) = \mathrm{k} \cdot (C_A) \cdot \chi_{F},$$

however the factor χ_F is insensitive to conformational changes, in contrast with $F(7.4)$. In other words, $F(4.2)$ depends only on HSA concentration. The ratio of these CAPIDAN fluorescence intensities, $F(7.4)/F(4.2)$, is concentration independent and reflects only modifications of binding sites.

Test for Clinical use. Parameters EAC, TAC and TI

The procedure described above was adopted for clinical use. The conditions of the test at pH 4.2 (see previous section) were developed to make albumin globule insensitive to the presence of ligands and to modifications. This allowed creating a method for determination of albumin mass concentration in whole serum. The method is based on measuring of CAPIDAN fluorescence intensity at pH 4.2 in presence of Brij-35, the value of F(4.2) [126, 136]. The measured HSA concentration was called "Total albumin concentration," TAC (g/l). This parameter is equivalent to HSA concentration determined with biochemical assays, e.g. with bromcresole dyes. The testing of the method showed that the values of TAC determined with bromcresole green assay and with CAPIDAN method are in good agreement; the mean discrepancy does not exceed 3-5% [136]. Moreover, the study of the test has shown that, at least at low albumin concentration, the discrepancy is mainly due to errors of bromcresole than CAPIDAN method. High sensitivity of the fluorescent method for the determination of TAC allowed to work with microvolumes of serum (5-10 microL) and to measure HSA concentration in liquids with very low, by 2 orders comparing with serum, albumin level, e.g. in aqueous humor and in tear [137].

This CAPIDAN parameter $F(7.4)$ after conventional calibration for clinical purposes was called "Effective albumin concentration", EAC (g/l) [138]. The parameter $F(4.2)$ after calibration is the Total albumin concentration, TAC [136] (see Fig.13). It was postulated that in norm (helthy volunteers) EAC=TAC [138, 139].

The ratio EAC/TAC can be used to evaluate conformational alterations of albumin binding sites in diseases. The difference between EAC and TAC can be characterized with relative parameter called as "Toxicity Index", TI [139]:

$$\mathrm{TI} = (\mathrm{TAC}/\mathrm{EAC}) - 1$$

This index is close to 0 in normal sera and raises as rule up to 1 in pathology. This is a simple biophysical express micromethod for evaluation of HSA properties in whole serum.

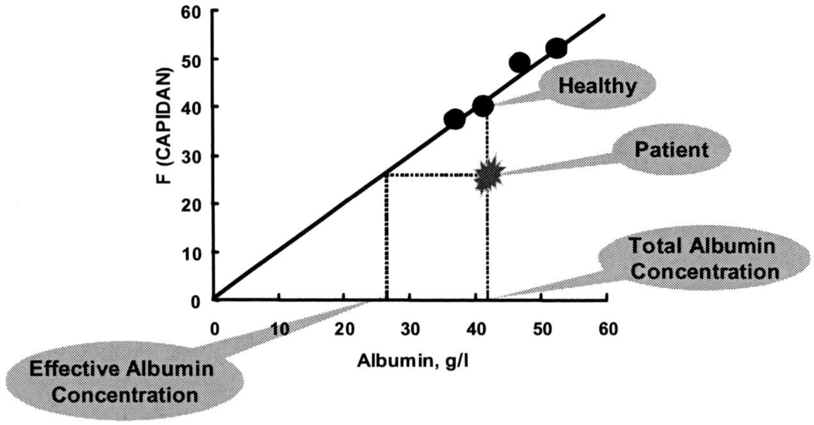

Figure 13. A scheme, illustrating parameters “Total” and “Effective” albumin concentration.

Using the Albumin Fluorescent Test in Diseases

It is well known that concentration of albumin in serum is an important predictor of total and Intensive Care Unit mortality in emergencies and in critical states. Moreover, beyond such urgent states HSA level correlates negatively with total mortality: the low HSA level, the higher probability of death from any reason [1-3].

The result of the research described in previous sections was express micro method for testing HSA parameters dependent not only on the protein concentration (TAC) but sensitive to ligand presence, oxidation and other causes that induce conformational modification of albumin globule as well (EAC, EAC/TAC). The question arises: whether the new albumin parameters can be useful in medicine or the information provided with them has gain no advantage in the face over albumin level (TAC). The aim of clinical observation and studies described in this section was to make the question clear, at least at the stage of pilot steps.

Total Peritonitis

Total peritonitis is a very severe disease with mortality up to 30 % and more. It is very important to evaluate the state of the patient and to make the prognosis of the disease development at the first day after surgical operation. The study was multicenter (in 5 clinics of Moscow and Moscow Region), non-randomized and blind in concern the fluorescent albumin parameters. 220 patients were under observation in the disease dynamics, 180 patients survived. The study protocol comprised the data of clinical examinations, scales SAPS and MPI as well as laboratory findings and albumin parameters EAC, TAC (about 80 parameters in total) [140]. The subsequent regression analysis denoted that a combination of 2 parameters, i.e. EAC and serum urea concentration routinely measured, is highly informative in evaluating the total peritonitis severity and correlates well with the survival rate in this disease. The (the) using these two parameters allows to discriminate three groups of patients: with poor (survival 10%, 95-% CI 1-26%), average (53%, 95-5 CI 30-72%) and good (survival rate 95%, 95-% CI 82-99%) It is interesting, that using TAC instead of EAC lead to notable overlap of confidence intervals that corresponded to worse prognostic significance of this parameter comparing with EAC.

We believe, that the study results could be a basis for an early individual prognostication of total peritonitis complications made on the 1-st postoperative day [140].

Acute Pancreatitis

The diagnostic and prognostic value of fluorescent test in acute pancreatitis was evaluated in a pilot study [141, 142] . There were 140 patients with acute pancreatitis: 80 with necrotic form (acute necrotizing pancreatitis), 60 with edematous form of the disease. The comparison of these two forms by albumin fluorescent test revealed their difference by EAC, TAC and TI parameters. Within first 2-3 days of the disease, all patients with EAC<20 g/l suffered from necrotic form. At the first day of the disease, if the TAC was greater than 28 g/l and/or the EAC was grater than 14 g/l, the risk of shock development was minimal. The results indicate that albumin fluorescent test parameters are linked with the severity of acute pancreatitis.

Acute Poisoning with Psychotropics

Two hundred and five patients with toxicogenic and somatogenic intoxication with psychotropic with psychotropic (PT) drug overdose in phase I (drug in body liquids is detected) and phase II (consequences of poisoning) were examined for the purpose of evaluating the prognostic value of the albumin fluorescence test [143]. An analysis of distributions of laboratory indices in the groups of survived (n = 176) and diseased deceased (n = 29) patients demonstrated the following: the parameter "Efficient albumin concentration" (EAC) measured by CAPIDAN fluorescence is informative in respect to prognosticating an outcome of PT intoxication of both toxicogenic and somatogenic forms (phase I and phase II, correspondingly). EAC provides for isolating a subgroup of patients with the risk of unfavorable outcome that is 3-4-fold higher versus the mean value in the total group. The routine laboratory parameters, e.g. level of creatinine, serum urea are rather of a less prognostic significance versus the EAC fluorescence parameter [143].

Anxious Depression

In critical states, the decrease of TAC as well as EAC was observed. However, the EAC value decreased at more great extent, and this led to greater prognostic significance of EAC comparing with TAC.

Albumin in blood, in fact, is the protein coming from tissues. These are dramatic events in tissues – inflammation, oxidation, etc. – that result in HSA modifications detected in serum samples with the fluorescent test. It was of interest to study albumin test parameters in diseases without such great albumin leakage as in emergencies.

In cooperation with Moscow Research Institute of Psychiatry we have performed a study of albumin properties in patients with anxious depression (F 3. ICD-10) [144]. 26 patients and 24 healthy volunteers were under observation. The TAC parameter did not differ in both groups However, the EAC value was dramatically ($p<0.01$) lower in depressive patients than in controls.. Consequently, EAC/TAC value in patients serum was 0.59±0.03 and 0.96±0.02

in patients and in controls. This allowed us to suggest dramatic changes of HSA binding sites in anxious depression.

To confirm conformational nature of the changes founded, HSA binding sites for CAPIDAN were studied by fluorescence quenching technique [145] , using ninrate anion as a quencher. In this group of patients the quenching constant was significantly lower ($p< 0.01$) than in volunteers, it was equal to 2.7 ± 0.3 l/mol and 4.6 ± 0.3 l/mol, correspondingly. The fraction of fluorescent signal accessible to quenching was significantly ($p<0.01$) lower in the group of patients comparing with donors (M±m): 0.59 ± 0.03 and 0.73 ± 0.03, correspondingly. We consider the results obtained as an additional evidence of conformational albumin changes in diseases.

Can Stress Lead to Albumin Conformational Changes?

The observations performed in clinics let us to assume that conformational changes of albumin binding sites represent, at least in some cases, a non-specific response to pathological process. This response might reflect pathological events in body tissues that result to the HSA modification. One can say, that albumin globule is “a molecular probe” for tissues, whereas CAPIDAN is a probe for HSA binding sites. If so, the non-specific changes in the body, possibly non-related directly with diseases, should induce, in less or more extent, albumin changes detected with CAPIDAN.

We have tested this hypothesis in experiments on rats subjected to acute stress. Although human and rat serum albumins are not identical, the molecular region bound CAPIDAN seems to be generally similar.

We have found that behaviour of rats under the open field test condition and effect of subsequent acute stress relate to conformational properties of albumin.

Behaviour of 78 Wistar mail rats was characterized from the fraction of time taken (spent) for exploratory and ambulatory activity during the open field test. In rats not subjected to stress ($n = 40$), a negative correlation was revealed between open field test activity and CAPIDAN parameter $F(7.4)/F(4.2)=R(N/F)$, the analog of EAC/TAC (see above sections), $r = -0.55$, $p<0.0005$. In the group of rats subjected to acute stress (immobilization plus stochastic cutaneous electrostimulation) the correlation between behavioral activity and R(N/F) parameter was significantly positive ($r = 0.59$, $p<0.0001$): i.e. R(N/F) parameter increased in highly active (high-active) rats and decreased in the low-active animals. It was shown that the mechanism of observed effect may involve (however, not determine completely) differences in non-esterified fatty acid production and their binding to albumin in stress [135].

These data are in a accordance with a clinical observation: the lowering the ratio EAC/TAC in patients that are waiting for surgical operations [146].

Considering the obtained results *in toto*, we believe that albumin conformational changes detected with fluorescent probe, CAPIDAN, are non-specific but related to the severity of diseases or state of the body and can be useful for its evaluation.

Conclusion

Human serum albumin is a participant of many different processes in the body - from transfer of substances to influence on oxidative processes in tissues and blood. These processes affect structure and conformational of albumin molecule. State of the molecule is reflected in state and properties of its binding sites. These sites can be probed by special fluorescent dyes. These dyes (probes) bind with albumin binding sites both in isolated albumin and in whole plasma (serum) where this protein is in its natural environment. About 96 % of fluorescence of the new probe, CAPIDAN (K-35), in plasma can be attributed to albumin-bound probe despite of presence of other plasma components. That approach shows that binding site structure is affected by pathological processes. These structural changes correlate with severity of the disease and its prognosis. That biophysical approach might be a source of additional useful information to usual biochemical and immunological tests for blood plasma and serum.

Acknowledgements

This work was partly supported by the International Scientific and Technical Center, grant #3156, and the Russian Fund for Basic Research, grants # 06-04-08254, Russian Foundation of Humanity (#05-06-06251a).

Authors are very grateful to colleagues that have presented some new data on K-35 properties and its interaction with albumin: T.I. Syrejshchikova for time-resolved data, B.M.Poliak and S.K.Gularian for quantum-chemistry calculations,

T.A.Babushkina and T.M.Klimova for NMR data, N.V.Smolina for quenching data and for the help in manuscript preperation.

References

[1] Nagai, H., Shichita, K., Haga, H., Suyama, Y., Matsuzaki, T., Shibata, H., Koyano, W., and Hatano, S. (1984). Relationship of serum albumin concentration to aging and survivorship in the community volunteers aged 70. *Nippon Ronen Igakkai Zasshi* **21**, 588-592.

[2] Shibata, H., Haga, H., Nagai, H., Suyama, Y., Yasumura, S., Koyano, W., and Suzuki, T. (1992). Predictors of All-Cause Mortality Between Ages 70 and 80 - *The Koganei Study. Arch Gerontol Geriatr* **14**, 283-297.

[3] Tell, G.S., Rutan, G.H., Kronmal, R.A., Bild, D.E., Polak, J.F., Wong, N.D., and Borhani, N.O. (1994). Correlates of blood pressure in community-dwelling older adults. The Cardiovascular Health Study. Cardiovascular Health Study (CHS) *Collaborative Research Group. Hypertension* **23**, 59-67.

[4] Csögör, S.I. (1972). Functia de transport a serumalbuminei (Bucuresti: Ed.Acad.Republic.Socialiste Romania).

[5] Landau, M.A. (1985). Molecular nature of some physiological processes (Moscow: NAUKA).

[6] Peters, T.J. (1996). All about albumin: biochemistry, genetics and medical applications (San Diego: Academic Press).
[7] Otagiri, M. (2005). A molecular functional study on the interactions of drugs with plasma proteins. *Drug Metab Pharmacokinet* **20**, 309-323.
[8] Peters, T.J. (1985). *Serum albumin. Adv Protein Chem* **37**, 161-245.
[9] Nickerson, J.M. (2001). Extra-hepatic expression of serum albumin mRNA in mouse retina. *Current Eye Research* **22**, 182-189.
[10]Horsey, P.J. (2002). The Cochrane 1998 Albumin Review - not all it was cracked up to be. European Journal of Anesthesiology *19*, 701-704.
[11]Takeda, Y., and Reeve, E.B. (1963). Studies of the metabolism and distribution of albumin with aytologous I(131)-albumin in healthy men. *J. Lab. Clin. Med.* **61** 183-202.
[12]Dewey, W.C. (1959). Vascular-extravascular exchange of I(131) plasma proteins in rat. *American Journal of Physiology* **197**, 423-431.
[13]Shimoda, M., Kokue, E., Hayama, T., and Vree, T. (1989). Effect of albumin distribution. A simulation analysis of the effect of altered albumin distribution on the apparent volume of distribution and apparent elimination rate constant of drugs. *Pharm Weekbl Sci* **11**, 87-91.
[14]He, X.M., and Carter, D.C. (1992). Atomic structure and chemistry of human serum albumin. *Nature* **358**, 209-215.
[15]Tanford, C. (1961). Physical Chemistry of Macromolecules (New York: Wiley).
[16]Tanford, C., Swanson, S.A., and Shore, W.S. (1955). Hydrogen Ion Equilibria of Bovine Serum Albumin. *J. Am. Chem. Soc.* **77**, 6414 - 6421.
[17]Stamler, J.S., Jaraki, O., Osborne, J., Simon, D.I., Keaney, J., Vita, J., Singel, D., Valeri, C.R., and Loscalzo, J. (1992). Nitric oxide circulates in mammalian plasma primarily as an S-nitroso adduct of serum albumin. *Proc Natl Acad Sci U S A* **89**, 7674-7677.
[18]Vladimorov, Y.A., and Dobretsov, G.E. (1980). Fluorescent probes in study o fbiological membranes (Moscow: NAUKA).
[19]Petitpas, I., Bhattacharya, A.A., Twine, S., East, M., and Curry, S. (2001). Crystal structure analysis of warfarin binding to human serum albumin: anatomy of drug site I. *J Biol Chem* **276**, 22804-22809.
[20]Babushkina, T.A., Ivanov, A.I., Sarnatskaya, V.V., and V.G., N. (1994). Comparative study of water proton relaxation in serum albumin solutions from healthy volunteers and from patients with renaland liver failure. In Blood serum albumin in clinical medicine, Y.A. Gryzunov and G.E. Dobretsov, eds. (Moscow: IRIUS), pp. 154-161.
[21]Brodersen, R., Honore, B., and Larsen, F.G. (1984). Serum albumin--a non-saturable carrier. *Acta Pharmacol Toxicol (Copenh)* **54**, 129-133.
[22]Benson, J.M., Boudinot, F.D., Pennell, A.T., Cunningham, F.E., and DiPiro, J.T. (1993). In vitro protein binding of cefonicid and cefuroxime in adult and neonatal sera. *Antimicrob Agents Chemother* **37**, 1343-1347.
[23]Wadsworth, S.J., and Suh, B. (1988). In vitro displacement of bilirubin by antibiotics and 2-hydroxybenzoylglycine in newborns. Antimicrob Agents Chemother *32*, 1571-1575.
[24]Luyk, A.I., and V.D., L. (1984). Serum albumin and biotransport of poisons (Moscow: Medicine).
[25]Halliwell, B. (1988). Albumin, an important extracellular antioxidant. *Biochem Pharmacol* **37**, 569-571.

[26]Halliwell, B., and Gutteridge, J.M. (2007). Free radicals in biology and medicine, Forth Edition (Oxford: University Press).

[27]Halliwell, B., and Gutteridge, J.M.C. (1986). Oxygen free radicals and iron in relation to biology and medicine: some problems and concepts. *Archive of Biochemistry and Biophysics* **246**.

[28]Soriani, M., Pietraforte, D., and Minetti, M. (1994). Antioxidant potencial of anaerobic human plasma: role of serum albumin and thiols as scavengers of carbon radicals. *Archive of Biochemistry and Biophysics* **312**, 180-188.

[29]Yan, L.-J., Traber, M.G., Kobushi, H., Matsugo, S., Tritschler, H.J., and Packer, L. (1996). Efficiacy of hypochlorus acid scavengers in the prevenrion of protein carbonyl formation. *Archive of Biochemistry and biophysics* **327**, 330-334.

[30]Quinlan, G.J., Margarson, M.P., Mumby, S., Evans, T.W., and Gutteridge, J.M. (1998). Administration of albumin to patients with sepsis syndrome: a possible beneficial role in plasma thiol repletion. *Clinical Science* **95**, 459-465.

[31]Garaceni, P., Gasbarrini, A., Tiel, D.H.V., and Borle, A.B. (1994). Oxygene free radical formation by rat hepatocytes during postanoxic reoxygenation: scavenging effect of albumin. *American Journal of Physiology* **266**, G451-G458.

[32]Halliwell, B., and Gutteridge, J.M. (1990). The antioxidants of human extracellular fluids. *Arch Biochem Biophys* **280**, 1-8.

[33]Himmelfarb, J., and McMonagle, E. (2001). Albumin is the major plasma protein target of oxidant stress in uremia. *Kidney Int* **60**, 358-363.

[34]Anraku, M., Yamasaki, K., Maruyama, T., Kragh-Hansen, U., and Otagiri, M. (2001). Effect of oxidative stress on the structure and function of human serum albumin. *Pharm Res* **18**, 632-639.

[35]Kashiba-Iwatsuki, M., Miyamoto, M., and Inoue, M. (1997). Effect of nitric oxide on the ligand-binding activity of albumin. *Arch Biochem Biophys* **345**, 237-242.

[36]Oettl, K., and Stauber, R.E. (2007). Physiological and pathological changes in the redox state of human serum albumin critically influence its binding properties. *Br J Pharmacol* **151**, 580-590.

[37]Gryzunov, Y.A., Arroyo, A., Vigne, J.L., Zhao, Q., Tyurin, V.A., Hubel, C.A., Gandley, R.E., Vladimirov, Y.A., Taylor, R.N., and Kagan, V.E. (2003). Binding of fatty acids facilitates oxidation of cysteine-34 and converts copper-albumin complexes from antioxidants to prooxidants. *Archive of Biochemistry and Biophysics* **413**, 53-66.

[38]Garreau, I., Pejoan, C., Bressollier, P., Verneuil, B., Cucumel, K., and Cupo, A. (1994). Purification and characterization of enkephalin-related peptides released by in vitro peptic digestion of bovine plasma proteins. *Peptides* **15**, 1195-1204.

[39]Tavoulari, S., Frillingos, S., Karatza, P., Messinis, I.E., and Seferiadis, K. (2004). The recombinant subdomain IIIB of human serum albumin displays activity of gonadotrophin surge-attenuating factor. *Hum Reprod* **19**, 849-858.

[40]Jorgensen, K.A., and Stoffersen, E. (1979). Heparin like activity of albumin. *Thromb Res* **16**, 569-574.

[41]Jorgensen, K.A., and Stoffersen, E. (1980). On the inhibitory effect of albumin on platelet aggregation. *Thromb Res Suppl* **17**, 13-18.

[42]Simon, D.I., Stamler, J.S., Jaraki, O., Keaney, J.F., Osborne, J.A., Francis, S.A., Singel, D.J., and Loscalzo, J. (1993). Anti-plalelet properties of protein S-nitrosothiols

derivered from nitric oxide and endothelium-derived relaxig factor. *Atherosclerosis Thrombosis* **13**, 791-799.

[43]Lui, L., Murray, D.K., Cameron, C.T., Parker, C.J., and Rodgers, G.M. (1997). Biochemical characterization of procoagulant albumin. *Thromb Res* **85**, 399-411.

[44]Doweiko, J.P., and Bistrian, B.R. (1994). The effect of glycosilated albumin on platelet aggregation. *JPEN* **18**, 516-520.

[45]Iglesias, J., Abernethy, V.E., Wang, Z., Lieberthal, W., Koh, J.S., and Levine, J.S. (1999). Albumin is a major serum survival factor for renal tubular cells and macrophages through scavenging of ROS. *Am J Physiol* **277**, F711-722.

[46]Zoellner, H., Hou, J.Y., Lovery, M., Kingham, J., Srivastava, M., Bielek, E., Vanyek, E., and Binder, B.R. (1999). Inhibition of microvascular endothelial apoptosis in tissue explants by serum albumin. *Microvasc Res* **57**, 162-173.

[47]Arici, M., Chana, R., Lewington, A., Brown, J., and Brunskill, N.J. (2003). Stimulation of proximal tubular cell apoptosis by albumin-bound fatty acids mediated by peroxisome proliferator activated receptor-gamma. *J Am Soc Nephrol* **14**, 17-27.

[48]Qiao, R., Siflinger-Birnboim, A., Lum, H., Tiruppathi, C., and Malik, A.B. (1993). Albumin and Ricinus communis agglutinin decrease endothelial permeability via interactions with matrix. *Am J Physiol* **265**, C439-446.

[49]Ramirez-Vick, J., and Vargas, F.F. (1993). Albumin modulation of paracellular permeability of pig vena caval endothelium shows specificity for pig albumin. *Am J Physiol* **264**, H1382-1387.

[50]Lum, H., Siflinger-Birnboim, A., Blumenstock, F., and Malik, A.B. (1991). Serum albumin decreases transendothelial permeability to macromolecules. *Microvasc. Res.* **42**, 91-102.

[51]Farmer, P.J., Bernier, S.G., Lepage, A., Guillemette, G., Regoli, D., and Sirois, P. (2001). Permeability of endothelial monolayers to albumin is increased by bradykinin and inhibited by prostaglandins. *Am J Physiol Lung Cell Mol Physiol* **280**, L732-738.

[52]Horstick, G., Lauterbach, M., Kempf, T., Bhakdi, S., Heimann, A., Horstick, M., Meyer, J., and Kempski, O. (2002). Early albumin infusion improves global and local hemodynamics and reduces inflammatory response in hemorrhagic shock. *Crit Care Med* **30**, 851-855.

[53]Oncley, J.L. (1942). The investigation of proteins by dielectric measurements. *Chem Revs* **33**, 433.

[54]Oncley, J.L., Scatchard, G., and Brown, A. (1947). Physical-chemical characteristics of certain of the proteins of normal hyman plasma. J. Phys. & Colloid Chem. *51*, 184-198.

[55]Ritland, Y.N., Kaesberg, P., and Beeman, W.W. (1950). An X-ray investigation of the shapes and hydrations of several protein molecules in solution. *J Chem. Phys.* **18** 1237-1242.

[56]Low, B.W. (1952). Preparation and properties of serum and plasma proteins. XXXIV. An X-ray study of crystalline human serum albumin preparations. *J.Am. Chem. Soc.* **74**, 48304834.

[57]Anderegg, J.W. (1952). *Phys Rev* **87**, 186.

[58]Anderegg, J.W. (1955). Investigation of the size, shape and hydration of serum albumin by small-anglr x-ray scattering. *J Am Chem Soc* **77**, 2927.

[59]Loeb, A.J., and Scheraga, H.A. (1956). Hydrodynamic and thermodynamic properties of bovine serum albumin at low pH. *J Chem. Phys*. **60**, 1633-1644.

[60]Tanford, C., Buzzell, J.G., Rands, D.G., and Swanson, S.A. (1955). The Reversible Expansion of Bovine Serum Albumin in Acid Solutions. *J. Am. Chem. Soc.* **77** 6421-6428.

[61]Champagne, M.J. (1957). Contribution a l'йtute de la mophologie de la serumalbumine en solution йtentue. *J Chem. Phys.* **54**.

[62]Luzzati, V., Witz, J., and Nicolaeff, A. (1961). La structute de la serum albumine de boeuf en solution a pH 5.3 et 3.6: e'tudepar diffusion centrale absolue des rayons *X. J Mol. Biol.* **3,** 379-372.

[63]Laggner, P., Kratky, O., Palm, W.H., and Holasek, A. (1971). R6ntgenkleinwinkeluntersuehungen an Liisungen von Schweineserum-Albumin bei verschiedenen pH-Werten. *Monatshefte fur Chemie* **102**, 1729-1744.

[64]Creeth, J.M., and Knight, C.G. (1965). On the estimation of the shape of macromolecules from sedimentation and vscosity measurements. *Biochim. Biophys. Acta* **102**, 549-558.

[65]Moser, P., and Squire, P.G. (1966). Electric polarization in proteins - dielectric dispersion and Kerr effect. Isoionic bovine serum albumin. *J Phys Chem* **70**, 744.

[66]Squire, P.G., Moser, P., and O'Konski, C.T. (1968). The hydrodynamic propreties of bovine serum albumin. *Biochemistry* **7**, 4261-4271.

[67]Hagag, N., Birnbaum, E.R., and Darnall, D.W. (1983). Resonance energy transfer between cysteine-34, tryptophan-214, and tyrosine-411 of human serum albumin. *Biochemistry* **22**, 2420-2427.

[68]Feng, L., Andrade, J.D., and Hu, C.Z. (1989). Scanning tunneling microscopy of proteins on graphite surfaces. *Scan. Microsc.* **3**, 399-410.

[69]Carter, D.C., He, X.M., Munson, S.H., Twigg, P.D., Gernert, K.M., Broom, M.B., and Miller, T.Y. (1989). Three-dimensional structure of human serum albumin. *Science* **244**, 1195-1198.

[70]Carter, D.C., and He, X.M. (1990). Structure of human serum albumin. Science *249*, 302-303.

[71]Carter, D.C., and Ho, J.X. (1994). Structure of serum albumin. *Adv Protein Chem* **45**, 153-203.

[72]Sjoberg, B., and Mortensen, K. (1994). Interparticle interactions and structure in nonideal solutions of human serum albumin studied by small-angle neutron scattering and Monte Carlo simulation. *Biophys Chem* **52**, 131-138.

[73]Sugio, S., Kashima, A., Mochizuki, S., Noda, M., and Kobayashi, K. (1999). Crystal structure of human serum albumin at 2.5 A resolution. *Protein Eng* **12**, 439-446.

[74]Bhattacharya, A.A., Grune, T., and Curry, S. (2000). Crystallographic analysis reveals common modes of binding of medium and long-chain fatty acids to human serum albumin. *J Mol Biol* **303**, 721-732.

[75]Curry, S., Mandelkow, H., Brick, P., and Franks, N. (1998). Crystal structure of human serum albumin complexed with fatty acid reveals an asymmetric distribution of binding sites. *Nat Struct Biol* **5**, 827-835.

[76]Kiselev, M.A., Gryzunov, Y.A., Dobretsov, G.E., and Komarova, M.N. (2001). The size of human serum albumin molecules in solution. *Biophysics* **46**, 402-405.

[77]Ferrer, M.L., Duchowicz, R., Carrasco, B., de la Torre, J.G.a., and Acuna, A.U. (2001). The conformation of serum albumin in solution: a combined phosphorescence depolarization-hydrodynamic modeling study. *Biophys J* **80**, 2422-2430.

[78]Fujiwara, S.-I., and Amisaki, T. (2006). Molecular dynamics study of conformational changes in human serum albumin by binding of fatty acids. *Proteins: Structure, Function, and Bioinformatics* **64**, 730 - 739.

[79]Sorokovoi, V.I., Dobretsov, G., Petrov, V., Nikitina, A., and Vladimirov, I. (1972). [Dimethylamiochalcone as a luminescent dye sensitive to protein conformational changes]. *Dokl Akad Nauk SSSR* **206**, 500-502 (Russian).

[80]Dobretsov, G.E. (1975). The study of protein structure by fluorescent probe method, Volume 6 (Moscow: VINITI).

[81]Dobretsov, G.E., Kharitonenkov, I.G., Mishijev, V.E., and Vlafimirov, Y.A. (1975). Relationship between fluorescence and circular dichroism of the complex of a fluorescent probe 4-dimethylaminochalcone with serum albumin. . *Biophysics* **20**, 581-585.

[82]Grollman, A. (1925). The combination of phenol red and proteins. *J.Biol.Chem.* **64** 141-160.

[83]Goldstein, A. (1949). The interactions of drugs and plasma proteins. *Pharmacol.Rev. 1,* **102**-165.

[84]Lindup, W.E. (1975). Drug-albumin binding. *Biochem.Soc.Transact.* **3**, 635-640.

[85]Kosh-Weser, J., and Sellers, E. (1976). Binding of drugs to serum albumin. New England J.Med. *294*, 311-316.

[86]Richardson, A., Miller, I., Schumacher, C., Jamber, W., Pancy, F., and Lapedes, D. (1946). Physiological disposition of penicilli G and K in dogs. *Proc.Soc.Exper.Biol.Med.* **63**.

[87]Klotz, I.M. (1946). Spectrophotometric investigations of the interactions of proteins with organic anions. *J.Amer.Chem.Soc.* **68**, 2299-2304.

[88]Klotz, I.M., Burkhard, R., and Urquhart, J. (1952). Structural specificities in the interaction of some organic ions with serum albumin. *J.Amer.Chem.Soc*. **74**, 202-208.

[89]Klotz, I.M., and Walker, F.M. (1947). Binding of organic ions by proteins - charge and pH effects. *J.Amer.Chem.Soc*. **69**, 1609-0612.

[90]Frederichsen, P., and Kierulf, P. (1979). A more accurate dye-binding method for the routine determination of serum albumin. *Clin.Chem.* **25**, 1180.

[91]Doumas, B.T., Watson, W.A., and Biggs, H.G. (1971). Albumin standards and the measurement of serum albumin with bromcresol green. *Clin.Chim.Acta* **31**, 87-96.

[92]Melnik, I.A., Baranovsky, P.V., and Nesterenko, L.I. (1985). A new method for estimation of serum albumin transport function. Laboratornoe Delo (Moscow), 202-204 (Russian).

[93]Salomatin, V.V., and Lifshitz, R.I. (1987). Binding capacity of serum albumin at thermal burns. Voprosy Medic. Chimii (Moscow) 33, 73-77 (Russian).

[94]Laurence, D.J.R. (1952). Adsorbtion of dyes on bovine serum albumin by the method of polarization of fluorescence. *Biochem.J.* ***51***, 168-180.

[95]Weber, G., and Laurence, D.J.R. (1954). Fluorescent indicators of adsorption in aqueous solution and on the solid phase. *Biochem.J.*, 1954, v.56, p.XXXI *56*, XXXI

[96]Rees, V.H., Fildes, J.E., and Laurence, D.J.R. (1954). The dye-binding capacity of human plasma determined fluorimetrically and its relation to the determination of plasma albumin. *J Clin Path* **7**, 336-340.

[97]Weber, G., and Young, L. (1964). Fragmentation of bovine serum albumin by pepsin. *J.Biol.Chem.* **239**, 1415-1423.

[98]Daniel, E., and Weber, G. (1966). Cooperative effects in binding by bovine serum albumin. II. Fluorimetric titrations. *Biochemistry* **5**, 1893-1900.

[99]Anderson, S. (1969). Circular dichroism spectra of the complexes of 1-anilino-8-naphthalenesulfonate with bovine serum albumin. Evidence for heterogeneity of binding. *Biochemistry* **8**, 4838-4842.

[100] Anderson, S.R., and Weber, G. (1969). Fluorescence polarization of the complexes of 1-anilino-8-naphthalenesulfonate with bovine serum albumin. Evidence for preferential orientation of the ligand. *Biochemistry* **8**, 371-377.

[101] Stryer, L. (1965). The interaction of a naphthalene dye with apomyoglobin and apohemoglobin. *J.Mol.Biol.* **13**, 482-495.

[102] Kolb, D.A., and Weber, G. (1975). Cooperativity of binding of anilinonaphthalenesulfonate to serum albumin by a second ligand. *Biochemistry* **14**, 4476-4481.

[103] Miller, Y.I., Dobretsov, G.E., and Kolchin, Y.A. (1991). Anion-binding sites of human serum albumin at the N-F conformational transition in isolated albumin and blood serum. *Biophysics* **36**, 505-508.

[104] Aidiraliev, R.K., Dobretsov, G.E., Lapshin, E.N., Krasovitsky, B.M., Shevchenko, E.A., and Distanov, V.B. (1988). Interaction of fluorescent probes with blood plasma. Aidiraliev R.K., Dobretsov G.E., Lapshin E.N., Krasovitsky B.M., Shevchenko E.A., Distanov V.B. Interaction of fluorescent probes with blood plasma. *Biophysics*, 1988, v.33, *33*, 378.

[105] Dobretsov, G.E., Lapshin, E.N., Kurek, N.K., Kosnikov, V.V., and Aidiraliev, R.K. (1991). Application of fluorescent probes in medical diagnosis. Part 1: Theoretical and experimental background. *Sov.Medical Reviews.Sect.B: Physicochemical Aspects of Medicine* **3**, 1-33.

[106] Miller, Y.I. (1989). The use of a fluorescent probe for estimation of human serum albumin binding capacity in hepatic insufficiency. Laboratornoe Delo (Moscow), 20-23 (Russian).

[107] Sklar, L.A., Hudson, B.S., and Simoni, R.D. (1977). Conjugated polyene fatty acids as fluorescent probes: binding to bovine serum albumin. *Biochemistry* **16**, 5100-5008.

[108] Tatarinov, B.A., Cherenkevich, S.N., and Kolian, A.I. (1978). The study of temperature-dependent conformational transitions of serum albumin by using an adsorbed dye. *Molecular Biology* **12**, 83-90.

[109] Schulman, S.G., and Rutledge, J.M. (1986). Enhancement of the fluorometric properties of some carboxyl substituted anthracenes by complexation with bovine serum albumin. *Anal.Letters* **19**, 2141-2145.

[110] Seki, T., Akimoto, T., and Tijima, T. (1987). The effect of the N-B transition of bovine and human serum albumins on the binding sites of fluorescent probes. *Colloid Polymer Sci.* **265**, 148-154.

[111] Wilton, D.C. (1990). The fatty acid analogue 11-(dansylamino)undecanoic acid is a fluorescent probe for the bilirubin-binding sites of albumin and not for the high-affinity fatty acid-binding sites. *Biochem.J.* **270**, 163-166.

[112] Mazzini, A., Cavatorta, P., Iori, M., Favilla, R., and Sartor, G. (1992). The binding of 4',6-diamidino-2-phenylindole to bovine serum albumin. *Biochem.J.* **42**, 101-109.

[113] Kessler, M., and Wolfbeis, O. (1992). Laser-induced fluorometric determination of albumin using longwave absorbing molecular probes. *Anal Biochem.* **200**, 254-259.

[114] Altamentova, S.M., Shaklai, N., Arav, R., and Miller, Y.I. (1998). A fluorescence method for estimation of toxemia: binding capacity of lipoproteins and albumin in plasma. *Clini.Chim.Acta* **271**, 133-149.

[115] Ribou, A.C., Vigo, J., Viallet, P., and Salmon, J.M. (1999). Interaction of a protein, BSA, and a fluorescent probe, Mag-Indo-1, influence of EDTA and calcium on the equilibrium. *Biophys.Chem.* **81**, 179-189.

[116] Elcelen, S., Klymchenko, A.S., and Demchenko, A.P. (2003). Novel two-color fluorescence probe with extreme specificity to bovine serum albumin. *FEBS Lett.* **538**, 25-28.

[117] Elcelen, S., Klymchenko, A.S., and Demchenko, A.P. (2005). The binding of novel two-color fluorescence probe FA to serum albumins of different species. *Int.J.Biol.Macromol.* **35**, 231-242.

[118] Zhu, X., Hu, Y., and Gong, A. (2007). Investigation on the interaction of tetrachloride fluorescein-bovine serum albumin-beta-cyclodextrin and the determination of protein by flow injection analysis. *Anal.Chim.Acta* **592**, 24-29.

[119] Westphal, U., and Gedigt, P. (1951). Azorubin-binding capacity of normal and pathological sera. *Proc Soc Exp Biol Med.* **76**, 838-843.

[120] Webster, D. (1974). A study of the interaction of bromocresol green with isolated serum globulin fractions. *Clin Chim Acta*, 109-115.

[121] Gustaffson, J.E.C. (1976). Improved specifity of serum albumin determination and estimation of acute phase by use of bromcresol green reaction. *Clin Chem* **22**.

[122] Troitskii, G., Kasymova, G., and Borisenko, S. (1987). [The degree of modification of serum albumin as a toxicological test]. *Vopr Med Khim* **33**, 38-41(Russian)

[123] Troitskii, G. (1987). [Post-synthetic modification of proteins in pathology (defective proteins)]. *Ukr Biokhim Zh.* **59**, 91-117 (Review. Russian).

[124] Dobretsov, G.E., and Miller, Y.I. (1994). Biochemistry and physical chemistry of serum albumin. Binding sites for organic molecules. In Serum albumin in clinical medicine, Y.A. Gryzunov and G.E. Dobretsov, eds. (Moscow: IRIUS), pp. 13-28.

[125] Aidyraliev, R.K., and Dobretsov, G.E. (1993). Assay for total plasma protein, globulins and albumin on the base of the fluorescent probe ANS and intrinsic plasma fluorescence. *Physical Chemical Biology and Medicine* **1,** 45-51.

[126] Gryzunov, I., Miller, I., Dobretsov, G., and AB, P. (1994). [Fluorescent method of determining mass concentration of human serum albumin]. *Klin Lab Diagn*. 1994 Sep-Oct;(5):27-31., 27-31 (Russian).

[127] Miller, I., and Dobretsov, G. (1994). [Molecular bases of the fluorescence method in the determination of binding capacity of serum albumin]. *Klin Lab Diagn* 20-23 (Russian).

[128] Gryzunov, Y., and Komarova, M.N. (2004). Study of ligand binding to protein using the energy transfer from an isoindole fluorescent label to a ligand. *Biphysics* **49**, 880-887.

[129] Dobretsov, G.E. (1989). Fluorecent probes in study of cells, membranes and lipoproteins (Moscow: NAUKA).

[130] Dobretsov, G.E., Gryzunov, Y.A., Komarova, M.N., Syrejshchikova, T.I., and Yakimenko, M.N. (1998). Determination of the number of fluorescence emitting molecules in heterogeneous biological objects by use of fluorescence decay function

and synchrotron pulse excitation. *Nuclear Instruments and Methods in Physics Research, Section A* **A-405**, 344-347.

[131] Dobretsov, G.E., Syrejshchikova, T.I., Gryzunov, Y.A., and M.N., Y. (1998). Quantification of fluorescent molecules in heterogeneous media by use of the fluorescence decay amplitude analysis. Journal of Fluorescence **8**, 27-34.

[132] Dobretsov, G.E. (1996). Method for determining concentration of fluorescent probes binding to biological structures. *Biophysics* **41**, 1062-1067

[133] Birkett, D.J., Myers, S.P., and Sudlow, G. (1977). Effects of fatty acids on two specific drug binding sites on human serum albumin. *Mol Pharmacol* **13**, 987-992.

[134] Birkett, D.J., and Wanwimolruc, S.V. (1985). Protein binding and drug transport, Tillement J.P. L Edition. (Stuttgard), pp. 11-23.

[135] Gryzunov, Y., Koplik, E., Smolina, N., Kopaeva, L., Dobretsov, G., and Sudakov, K. (2006). Conformational properties of serum albumin binding sites in rats with different behaviour in the open field test.*Stress*. 2006 Mar;9(1):53-60. . Stress *9*, 53-60.

[136] Gryzunov, Y., and Lukicheva, T. (1994). [Accuracy and reproducibility of fluorescent method for determination of mass serum albumin concentration]. *Klin. lab. Diagn*, 25-27(Russian).

[137] Gryzunov, I., Deev, A., Kurysheva, N., and Komarova, M. (1999). [Fluorescent analysis of albumin in aqueous humor and tears]. *Biull Eksp Biol Med.* **128**, 597-600.

[138] Gryzunov, Y., and Dobretsov, G.E. eds. (1998). *Serum albumin in clinical medicine*, Volume 2 (Moscow: GEOTAR Publ.).

[139] Gryzunov, Y., and Dobretsov, G.E. eds. (1994). *Serum albumin in clinical medicine*, Volume 1 (Moscow: IRIUS Publ.).

[140] Gryzunov, I., Grinberg, A., Stupin, V., Rodoman, G., Musselius, S., Fedorovskii, N., Dobretsov, G., Chernysh, T., Shalaeva, T., and VI., P. (2003). [Informative value of the parameter "effective concentration of albumin" in diffuse peritonitis: data from a multi-center study]. *Anesteziol Reanimatol.*, 32-35 (Russian).

[141] Rodoman, G., Shalaeva, T., Dobretsov, G., and AL., K. (1999). [Concentration and properties of albumin in blood and exudate during acute pancreatitis]. *Biull Eksp Biol Med.* **128**, 660-662 (Russian).

[142] Kuznetsov, N.A., Rodoman, G.V., Dobretsov, G.E., and Shalaeva, T.I. (2003). Prognosis of severity and termination of acute pancreatitis. *Russian Medical Journal*, 20-25 (Russian).

[143] Gryzunov, I., Syromiatnikova, E., and KK., I.i. (2004). [The prognostic value of albumin fluorescence test in the evaluation of outcome of acute intoxication with psychotropics]. *Klin Lab Diagn*. 2004 Nov;(11):39-42. Russian. , 39-42 (Russian)

[144] Smolina, N., Gryzunov, Y., Maximova, N., Dobretsov, G., Uzbekov, M., Misionzhnik, E., and Vertogradova, O. (2007). The properties of albumin binding sites in anxious depressed patients: the study with fluorescence quenching technique. Buill.Exp.Biol.Med, (in press).

[145] Eftink, M.R. (1991). Fluorescence Quenching Reactions. In Biophysical and Biochemical Aspects of Fluorescence, T.G. Dewey, ed. (New York: Plenium Press).

[146] Rodoman, G., Dobretsov, G., Shalaeva, T., and Naumov, E. (2001). Albumin fluorescent test in surgical stress. *Bull Exp Biol Med.* **131**, 302-304

In: Protein Conformation: New Research
Editor: Linda B. Roswell, pp. 161-199
ISBN: 978-1-60456-219-4

Chapter 4

Domain Motions Associated with Substrate and Inhibitor Binding: The Relevance of Distance Matrixes

Carla S. Alves, Visvaldas Kairys and Miguel X. Fernandes
Centro de Química da Madeira, Departamento de Química
Universidade da Madeira, Campus da Penteada
9000-390 Funchal, Portugal

Abstract

In this work, a difference matrix based method for analyzing and comparing conformational changes associated with substrate or inhibitor binding was implemented. The method was applied on a series of protein structures: aquaporin (AQP0, AQP1, AQPGlpF, AQPZ), calmodulin (CaM), chloride channel (ClC), cyclin-dependent kinase 2 (CDK2), cyclooxygenases (COX-1 and COX-2), factor Xa and the human immunodeficiency virus type 1 (HIV-1) protease. Various receptor conformations were analyzed, where applicable: bound vs. unbound, open vs. closed, active vs. inactive forms. This approach allowed for the comparison of conformational changes upon binding of substrate versus inhibitor or activators. Clear trends of domain motions associated with substrate binding were detected in the various conformations of each protein analyzed. The application of this tool between the structural pairs is more informative compared to the single number reported by root-mean-square-deviation (RMSD) as it provides detailed information to the mechanisms associated with ligand binding. Identification of "hot spots" in conformations can provide a better understanding of molecular recognition phenomenon and its implications on resistance, specificity and drug design.

Introduction

The functioning of proteins is very often accompanied by a movement of regions (domains) inside of the protein. An example is the opening of the flaps in human immunodeficiency virus (HIV) protease, in order to cleave viral polyproteins into functional products necessary for the activity of the virus.

Distance matrix method has been commonly used in various fields of science. In the area of protein conformation research, distance matrixes have been used, for example, for protein structure alignment [1] and for protein fold comparison [2]. Distance matrixes and difference distance matrixes have been used to identify protein secondary structure elements [3]. Difference distance matrixes are a more objective means of comparing structures as they do not depend on a particular superposition scheme [4]. In order to investigate changes in the three-dimensional structure of the protein upon substrate or inhibitor binding, use of difference between two distance matrixes proved to be useful. Using difference distance matrix enabled Rose *et al.* [5] to identify domains in HIV protease associated with substrate binding. This analysis allowed a new interpretation of anti-drug mutation resistance for mutations far removed from the binding site.

In this Chapter, we will use the difference distance matrix approach applied to a series of different proteins and protein families: HIV protease, factor Xa, aquaporins, chloride channel proteins, calmodulin, cyclin dependent kinase 2, and cyclooxygenase.

Methods

The ligand binding patterns and associated domain movements in seven protein families were analyzed using difference distance matrixes. The families investigated were: aquaporin (AQP0, AQP1, AQPGlpF, AQPZ), calmodulin (CaM), chloride channel (ClC), cyclin-dependent kinase 2 (CDK2), cyclooxygenases (COX-1 and COX-2), factor Xa and the human immunodeficiency virus type 1 (HIV-1) protease. A program was written in Fortran 90 determining the distances between the amino acid alpha carbon ($C\alpha$) atoms within individual structures. This method was applied on a series of structures from each protein family, chosen from the Protein Data Bank (PDB) [6]. Swiss-PdbViewer [7] and PDBsum [8] were used to aid in the compilation of the data for each protein family. The root-mean-square-deviation (RMSD) values of the $C\alpha$-atoms of the test and reference structures being compared were calculated using Swiss-PdbViewer [7]. The binding pocket volume for the HIV-1 protease structures were obtained from PDBsum [8]. This was done in order to determine the differences in binding pocket volumes between the test and reference HIV-1 proteases. The data generated was then used to determine the difference in distances between equivalent amino acid $C\alpha$-atoms for specific protein structural pairs. Where applicable, structural comparisons were made for the (1) open unliganded vs. closed unliganded, (2) bound vs. unbound, (3) bound mutant vs. bound wild-type and, (4) active vs. inactive forms of the protein pairs. The data generated for the structural comparisons were displayed graphically using SigmaPlot v9.0 (Systat Software, Inc.). Domains of contacting residues were identified as follows. For the open unliganded vs. closed unliganded structural comparison, residues were assigned to the same domain if the difference in interatomic distances between them varied by less than 1.0 Å. For the remaining structural comparisons a cut-off of 0.5 Å was used. Outliers that deviated by more than 0.5 Å from the distance differences of the rest of the domain were assessed separately.

Conformational Analyses of Proteases

HIV-1 Protease

The HIV-1 protease is an essential enzyme involved in the replication of HIV-1 [9]. The protease cleaves the Gag and Pol polyproteins of the virus into functional protein products that are essential for viral assembly and subsequent activity. This step, which is essential for virus maturation and subsequently infectious particle production, makes the HIV-1 protease an important target of antiretroviral drug design [10-11].

Crystallography plays an important role in the search for an effective drug against HIV. This is due to the fact that the greater the knowledge of the atomic structure, the easier it is to design specialized novel drugs against the pathogen. During the last decade the amount of X-ray crystallographic data characterizing the HIV-1 protease in the apo-form as well in complex with many different drugs and inhibitors has virtually exploded [12]. Over 200 structures of the HIV-1 protease in various forms have been deposited in the PDB [6], allowing to gain a deeper understanding into mechanisms of drug action and drug resistance.

The HIV-1 protease is a small homodimeric protein made up of two identical protein chains, each 99 amino acids long. In the apo-form, the enzyme has two flaps which are about 7 Å distant from the active site and in a very open conformation [9, 12]. Binding of substrate or inhibitor to the active site induces substantial conformational changes in the enzyme. The two flaps close and lose mobility, and the cavity of the active site narrows slightly [12].

To date, nine FDA approved drugs which are available for administration to HIV-1 infected individuals exist. The drugs, amprenavir, atazanavir, darunavir, indinavir, lopinavir, nelfinavir, ritonavir, saquinavir and tipranavir [13-14] are all competitive active site inhibitors. They bind the HIV-1 protease in the extended conformation like substrate does [15]. The protease cannot cleave these peptidomimetic inhibitors and thus they remain lodged in the active site, blocking the normal functioning of the enzyme.

On the whole, the peptidomimetic inhibitors are chemically different [13, 16-17], relatively low molecular weight compounds. An important feature of inhibitor binding involves the aspartic acids constituting the catalytic triad. A hydroxyl group present in the inhibitors positions itself between the Asp25/Asp25' carboxyls of the protease, within hydrogen bonding distance to at least one carboxylate oxygen of each catalytically active aspartic acid [10, 12, 18]. Another common feature to most complexes of the HIV-1 protease is a buried water molecule in the active site. This water molecule bridges the NH groups of Ile50 and Ile50' present in each of the flaps and two hydrogen bond acceptor atoms of the inhibitor. The water molecule is thus tetrahedrally coordinated and shielded from bulk solvent [12, 18].

For the currently approved protease inhibitors, the emergence of mutants that confer multidrug resistance is disturbing [13, 19]. Several of the major resistance mutations in the protease associated with reduced protease inhibitor sensitivity lie within the protease active site. Thus there is a rationale for understanding their action (i.e. they may interfere directly with inhibitor binding). Many other mutations, however, occur far from the active site and do not easily fit this model. Rose *et al.* [5] identified five domains within the HIV-1 protease dimer that move as rigid bodies in accommodation to ligand binding. In the context of these domain motions, it was found that almost all of the drug resistant mutations that are far from

Table 1. HIV-1 Protease structures used for the computation of the difference distance matrixes. The PDB identification number of the structure, name of the ligand bound to the structure, RMS deviations from the reference structure 1g6l, and change of binding pocket volume as reported by PDBsum are listed.

	PDB code	Ligand	Mutations	RMSD(Å)[a]	ΔVol(Å^3)[b]
Closed unliganded	1hhp	none	-	1.07	-369.99
	1rpi	none	D25N, M36V, I84V	2.44	2337.98
	1tw7	none	L10I, D25N, M36V, M46L, I54V, I62V, L63P, A71V, V82A, I84V, L90M	2.36	2640.94
	3hvp	none	C95ABA[c]	1.11	-39.23
	3phv	none	-	1.23	-249.33
Natural substrates	1f7a	CA-p2	Q7K, D25N	0.62	556.87
	1kj4	MA-CA	D25N, Q7K	0.82	-[d]
	1kj7	p2-NC	D25N, Q7K	0.64	631.12
	1kjf	p1-p6	D25N, Q7K	0.56	450.56
	1kjg	RT-RH	D25N, Q7K	0.55	178.87
	1kjh	RH-IN	D25N, Q7K	0.73	332.01
	1mt7	MA-CA	Q7K, D25N, L63P, V82A	0.55	213.05
	1mt8	CA-p2	Q7K, D25N, L63P, V82A	0.64	966.51
	1mt9	p1-p6	Q7K, D25N, L63P, V82A	0.50	473.34
Symmetric inhibitors	1ajv	NMB	-	0.71	-0.84
	1ajx	AH1	-	0.61	149.77
	1bv7	XV638	V82F	0.33	343.41
	1bv9	XV638	I84V	0.30	-227.81
	1bve	DMP323	C95A	1.55	1481.20
	1bvg	DMP323	C95A	1.40	1390.08
	1bwa	XV638	V82F, I84V	0.25	270.42
	1bwb	XV638	V82F, I84V	0.22	402.05
	1dmp	DMP450	-	1.57	167.48
	1g2k	NM1	-	0.70	114.75
	1g35	AHF	-	0.68	9.70
	1hvh	Q8261	-	0.43	99.14
	1hvr	XK263	-	0.36	365.34
	1hwr	XK216	-	0.40	86.84
	1mer	DMP450	I84V	0.25	103.78
	1mes	DMP323	I84V	0.29	378.00
	1met	DMP323	V82F	0.30	685.12
	1meu	DMP323	V82F, I84V	0.27	338.34
	1qbr	XV638	-	0.34	-61.59
	1qbs	DMP323	-	1.55	155.25
	1qbt	SD146	-	0.32	71.30
	1qbu	Q8467	-	0.45	419.34
	1rq9	DMP450	D25N, M36V, I84V	2.16	2676.37
	1t7k	BH0	-	0.78	-259.45
	2a4f	AAU	-	0.78	718.03

Table 1. Continued

Asymmetric inhibitors	1c70	L75[e]	-	0.75	439.17
	1c6y	Indinavir	-	0.79	613.83
	1c6z	Saquinavir	-	0.71	998.58
	1d4s	Tipranavir	V82F, I84V	0.62	-113.49
	1d4y	Tipranavir	Q7K, L33I, L63I	0.57	280.55
	1fb7	Saquinavir	G48V, L90M	0.45	-469.55
	1hpv	Amprenavir	-	0.26	-239.20
	1hsg	Indinavir	-	0.65	236.67
	1hsh	Indinavir	-	1.04	236.67
	1hxb	Saquinavir	-	0.28	288.56
	1hxw	Ritonavir	-	0.70	250.59
	1k6c	Indinavir	Q7K, K14R, V82T, I84V	0.52	332.44
	1k6p	XN3[e]	Q7K, K14R, V82T, I84V	0.53	333.00
	1k6t	XN1[e]	Q7K, K14R, V82T, I84V	0.53	342.98
	1k6v	XN2[e]	Q7K, K14R, V82T, I84V	0.51	354.37
	1mtb	Saquinavir	Q7K, D25N, L63P, V82A	0.53	1069.45
	1mui	Lopinavir	N37S	0.39	94.92
	1n49	Ritonavir	Q7K, D25N, V82A	0.63	-[d]
	1ohr	Nelfinavir	-	0.74	629.86
	1rl8	Ritonavir	V82A	0.57	1022.62
	1rv7	Lopinavir	D25N, M36V, I84	2.34	2789.86[f]
	1s65	Darunavir	V82A	0.57	704.53
	1s6g	Darunavir	Q7K, L33I, L63I, C67A, C95A	0.69	62.01
	1s6s	Darunavir	I84V	0.67	457.31
	1sdt	Indinavir	-	0.77	470.81
	1sdu	Indinavir	L90M	0.58	219.37
	1sdv	Indinavir	V82A	0.75	860.20
	1sh9	Ritonavir	K20R, V32I, L33F, M36I, I54V, L63P, A71V, V82A, I84V, L90M	0.46	966.51
	1t3r	Darunavir	Q7K, L63P	0.59	377.58
	1t7i	Darunavir	Q7K, K14R, V82T, I84V	0.64	569.53
	1t7j	Amprenavir	Q7K, K14R, V82T, I84V	0.58	553.92

[a] The RMSD-values were calculated by superimposition of the $C\alpha$-atoms of the PDB structure onto 1g6l and calculating the root-mean-square-deviation of the $C\alpha$-atoms from 1g6l. This was done using Swiss-PdbViewer [7].

[b] The binding pocket volume for the 1g6l structure obtained from PDBsum is 1851.61 Å^3. The binding pocket volume for the test structures were obtained from PDBsum. The differences in binding pocket volumes between the test structures and 1g6l were calculated using these values. This was done using PDBsum [8].

[c] Cysteine amino acid residue replaced by α-amino-n-butyric acid.

[d] Volumes of the 1kj4 and 1n49 test structures were unavailable from PDBsum.

[e] Indinavir analogues.

[f] The 1rv7 HIV-1 protease structure is in the open conformation. Hence the large difference in binding pocket volume between this structure and the 1g6l HIV-1 protease in the closed conformation.

the active site fall on the boundaries between the moving surfaces of these domains. Thus, the investigation of domain motions can help understand how structural changes impact upon drug resistance.

The approach by Rose *et al.* [5] was more extensively explored in this work. Overall, 71 HIV-1 protease structures in different conformations were chosen from the PDB (Table 1). These included 6 structures of the HIV-1 protease in the apo-form, 9 substrate complexes and 56 different inhibitor complexes. The effect of 14 mutations on inhibitor binding was also explored.

Domain Motions of the HIV-1 Protease

Domain motions occurring within the HIV-1 protease in the apo-form were identified using the $C\alpha$-difference distance matrix approach. Figure 1 shows the representative $C\alpha$-difference distance matrix for the open unliganded and closed unliganded HIV-1 protease structural comparison. Residues 1-99 and 100-198 along either axis of the matrix correspond to the two monomers of the protease dimer. Linear sequences of amino acids displaying interatomic distances varying by less than 1 Å constitute rigid subdomains of the HIV-1 protease. These regions appear along the diagonal of the matrix and are delineated as **a** to **e** and **a'** to **e'**. The ten subdomains form five rigid domains in the HIV-1 protease (Figure 2). Subdomains **a**, **e**, **a'** and **e'** encompass the *terminal* domain. This domain consists of two antiparallel β-sheets (residues 1-3 and 95-99), the turn (residues 4-7) and the α-helix (residues 88-94) of each monomer. The two *core* domains, encompassing subdomains **b** and **d** or **b'** and **d'**, consist of

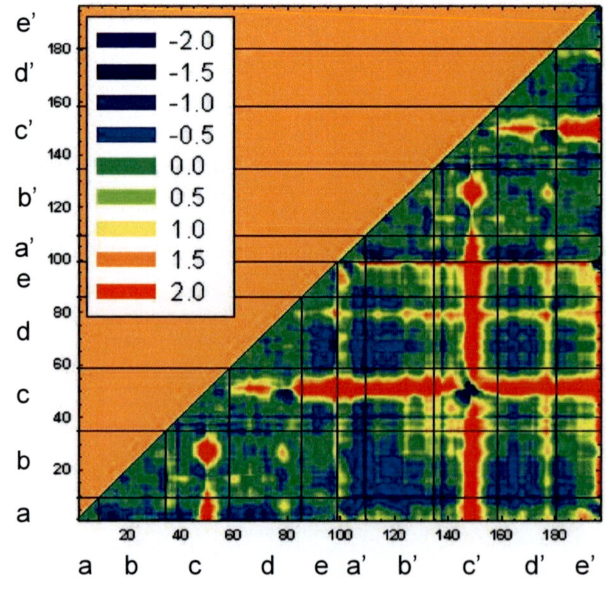

Figure 1. $C\alpha$-difference distance matrix between the open (1hhp) and closed (1g6l) unliganded HIV-1 protease structures. Residue numbers are labelled along both axes: residues 1-99 belong to one monomer and residues 101-199 belong to the other monomer. Letters **a-e** and **a'-e'** indicate the subdomains of the protease with distance differences less than 1.0 Å.

six primarily β-sheets (residues 9-36 and 59-84). Situated at the interface of the two core subdomains in each monomer is the highly conserved catalytic triad sequence, Asp25-Thr26-Gly27. Separating the terminal and core domains is an interface composed residues Arg8, Ile85, Gly86 and Arg87. The two *flap* domains, encompassing subdomains **c** and **c'**, consist of the mostly solvent-exposed β-hairpin loop (residues 43-46) preceding the flap tips, the flap tips (residues 47-52), and the continuation of the β-sheet of the flap (residues 53-56). Separating the flap domain from the core domain in each monomer is an interface. This interface is formed by residues 37-42 of the hinge preceding the flap tips and residues 57-58 of the β-sheet following the flap tips. The residues constituting the interfaces between either the terminal and core domains or the core and flap domains are illustrated in the Figure in grey. Overall, the ten subdomains forming the five rigid domains described here are comparable to those identified previously [5].

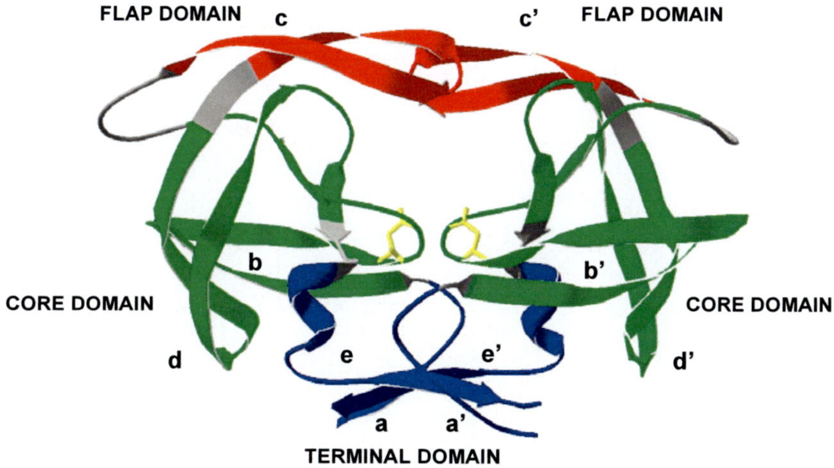

Figure 2. Domain motions of the HIV-1 protease in the closed conformation. The five rigid body domains identified: the terminal domain (residues 1-7 and 88-99 of both monomers) shaded in blue; the two core domains containing the catalytic aspartic acids (residues 9-36 and 59-84 of a single monomer) shaded in green; and the two flap domains (residues 43-56 of a single monomer) shaded in red. Residues comprising the interdomain interfaces are shaded in grey (Figure generated using Swiss Swiss-PdbViewer [7]).

Significant movements between the open and closed unliganded protease monomers are represented on the matrix in Figure 1 by distance differences greater than 1.0 Å. The principal movements are observed between (i) the terminal domain of one protease monomer with the flap domain of the equivalent protease monomer (e.g. **a** with **c**) and (ii) the terminal domain of the one monomer with the flap domain of the opposite monomer (e.g. **a** with **c'**). In the matrix, these movements are illustrated in red. The frequency distribution of the distance differences was then determined for the matrix. The data here indicated expansion of the open HIV-1 protease structure relative to the closed structure. The difference in volume of the HIV-1 protease active site for the open and the closed structures, calculated using values obtained from PDBsum, confirmed this. The RMSDs of the $C\alpha$-atoms of the open protease structure onto the closed structure also indicated differences in conformation between the two structures (Table 1). The domain movements recorded using the *Cα*-difference distance matrix approach, as well as the frequency distribution of the distance differences, however, are more informative and accurate relative to looking at individual atomic deviations. Overall, the degree of extension of the flaps in the HIV-1 protease structure in the open conformation relative to the structure in the closed conformation accounts for the above observations.

Upon ligand binding to the open HIV-1 protease, a 7 Å motion of the flap region [9, 12], followed by narrowing of the active site is observed. Rose *et al.* [5] showed that ligand binding results in a 4° rotation of the two core domains relative to the terminal domain. The catalytic aspartic acids located at the interface of either core domain were shown to move very little (2.5 Å) while the surface residues moved as much as 4 Å. From this it is evident that the highly flexible flap domains [10] play a key role in gating the ligand binding mechanism.

Domain Motions Associated with Substrate/Inhibitor Binding in the HIV-1 Protease

The PDB contains many crystal structures of the HIV-1 protease in the complexed form. We employed the *Cα*-difference distance matrix approach to investigate the conformational differences between the protease in the bound form relative to the protease in the unbound form. These differences were assessed for the protease bound to a natural substrate (CA-p2, MA-CA, p1-p6, p2-NC, RT-RH and RH-IN), or to an inhibitor (asymmetric: amprenavir, darunavir, indinavir, lopinavir, nelfinavir, ritonavir, saquinavir, tipranavir; symmetric: cyclic sulfamide and cyclic urea inhibitors).

Superimposition of the bound HIV-1 protease structures onto the unbound structure resulted in RMSDs for the *Cα*-atoms ranging from 0.22 Å to 2.34 Å (see Table 1). These deviations indicate overall differences in conformation between the bound and unbound HIV-1 protease structures. The domain movements recorded using the *Cα*-difference distance matrix approach however, is more informative and accurate relative to looking at the RMSDs.

Figures 3 (A) and 3 (B) show the *Cα*-difference distance matrixes for the substrate-bound versus unbound and inhibitor-bound versus unbound HIV-1 protease structural comparisons, respectively. The domains were identified here as spatially connected regions with interatomic distances varying by less than 0.5 Å. As with the open versus closed structural comparison, five rigid body domains composed of ten subdomains were identified. The rigid subdomains are delineated as **a** to **e** and **a'** to **e'** in each matrix. The interdomain interfaces identified here however vary slightly from those identified for the open versus closed structural comparison. Separating the terminal and core domains in each monomer is an interface composed mainly of residues 8, 65-72, 85-87. Separating the flap domain from the core domain in each monomer is an interface formed by residues 34-42 of the hinge preceding the flap tips and residues 57-58 of the *β*-sheet following the flap tips.

Subdomains displaying distance differences less than 0.5 Å are considered to move concertedly with respect to one another. For the natural substrate-bound and unbound protease structural comparison, concerted movements between the equivalent subdomains of each monomer pair are observed (i.e. **a** with **a**, **b** with **b**, etc.). Concerted movements between the two terminal subdomains (i.e. **a** and **e**, **a** and **e'**, etc.) of the equivalent protease monomers are also observed. A similar pattern of concerted movements for the inhibitor-bound and unbound protease structural comparison was recorded.

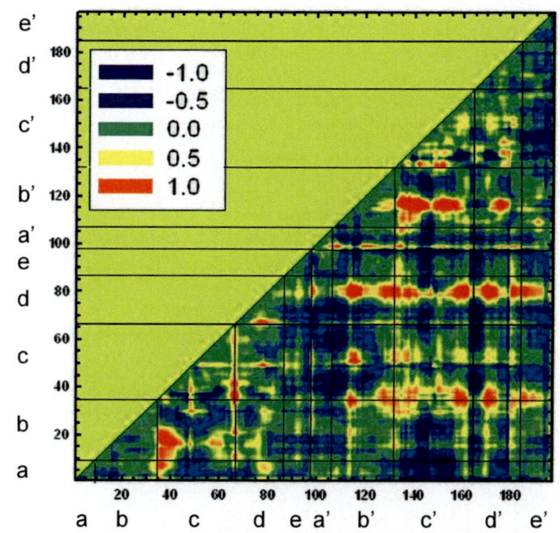

A.

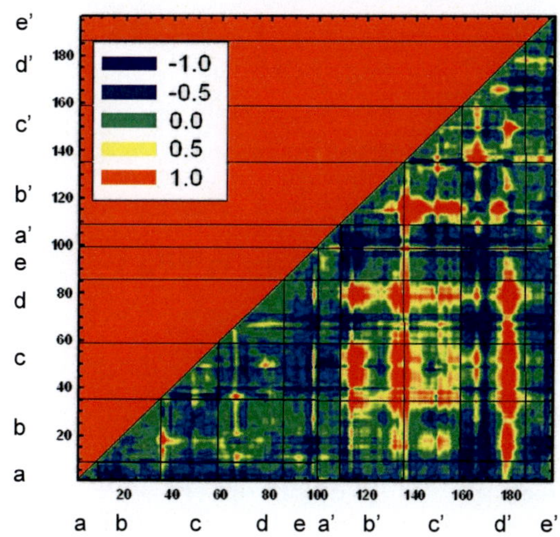

B.

Figure 3. $C\alpha$-difference distance matrix between (A) the substrate CaP2 bound and unbound HIV-1 protease structures and, (B) inhibitor nelfinavir bound and unbound HIV-1 protease structures. Residue numbers are labelled along both axes: residues 1-99 belong to one monomer and residues 101-199 belong to the other monomer. Letters **a-e** and **a'-e'** indicate the subdomains of the protease with distance differences less than 0.5 Å. The asymmetry of movements in (A) and especially in (B) is apparent.

Evaluation of the $C\alpha$-difference distance matrixes in Figures 3 (A) and 3 (B) show several domain motions associated with substrate or inhibitor binding. The key motions recorded for the opposite monomers in each structural pair under investigation occur between the core and the terminal domains of one monomer set with the flap domains of the opposite monomer set (e.g. **a** with **c'**). Three principal motions are observed between the equivalent monomers in each structural pair under investigation. These motions occur between (i) the terminal domain and the core domain (i.e. **a/e** with **b**, **a/e** with **d**), (ii) the terminal domain and the flap domain and, (iii) the core domain and the flap domain (i.e. **b/d** with **c**). Depending on the monomer set under scrutiny, these motions vary slightly. This can be seen as the asymmetric regions in each $C\alpha$-difference distance matrix (see regions **a-e** with **a-e** and **a'-e'** with **a'-e'** along either axis of the matrixes). The asymmetry may be a result of the different extension and interactions of the symmetric inhibitor into either of the monomers. This is consistent with previous observations that different conformations are adopted by the two protease monomers in order to accommodate ligand [11].

The volume of the active site of the complexed HIV-1 protease structures and that of the uncomplexed structure in the closed conformation were also compared. The difference in volume between the complexed and uncomplexed HIV-1 protease structures were computed (see Table 1). The frequency distribution of the distance differences recorded in each matrix was also determined. Generally, the binding pocket volume of the complexed HIV-1 protease structure is slightly expanded relative to the closed uncomplexed structure. The binding pocket volume of some HIV-1 protease complexes (amprenavir 1hpv; saquinavir 1fb7; tipranavir 1d4s; cyclic ureas 1qbr and 1t7k) did indicate slight contraction of the active site. The reason for the decrease in binding pocket volume is unclear. The presence of a relatively small ligand in the protease active site may contribute to this decrease in volume. In some cases it may be related to the influence of residue mutation upon the volume of the binding pocket. For example, mutation of Gly48 to Val in the binding site is accompanied by an increase in side chain volume, and this can explain the observed shrinking of the binding pocket in the 1fb7 protease structure (saquinavir bound to a mutant protease).

Drug Resistance Mutations Related to Domain Motions in the HIV-1 Protease

A major difficulty in the development of anti-HIV drugs is the increased incidence of drug-resistance mutations within the HIV-1 protease that alter the susceptibility of the enzyme to inhibition by specific inhibitors. A recent survey identified 41 residue modifications in the HIV-1 protease associated with drug resistance [19]. Of these modifications, 14 were observed in the HIV-1 protease complexes investigated in this study. Figure 4 locates the sites of these resistance mutations on the HIV-1 protease structure. The mutations that are in direct contact with the inhibitor (e.g. Val32, Gly48, Val82, Ile84) are the easiest to explain. Changes in these specificity-determining residues may directly interfere with the binding of the inhibitor to the enzyme. In contrast, mutations that do not directly interact with inhibitor can influence binding via long-range structural perturbations of the active site, or they can change the efficiency of catalysis and the stability of the enzyme [12, 18]. Using the $C\alpha$-difference distance matrix approach may help rationalise the location of drug-resistance mutations far from the protease active site.

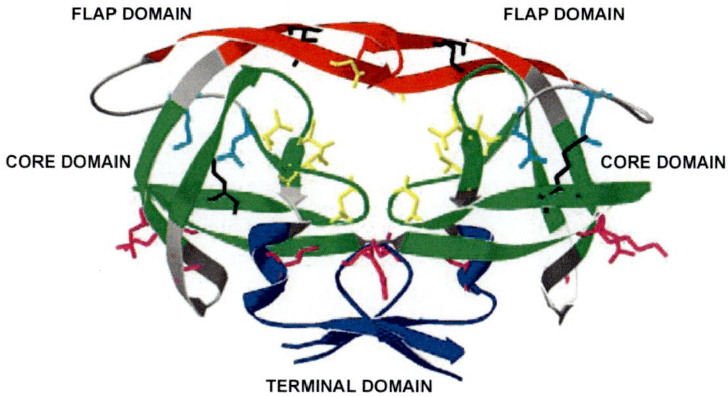

Figure 4. Resistance mutations mapped onto the HIV-1 protease in the closed conformation. The five rigid body domains of the protease are shaded: the terminal domain in blue, the two core domains in green and the two flap domains in red. Residues comprising the interdomain interfaces are shaded in grey. The residues associated with drug resistance are indicated: residues in direct contact with ligand residues are in yellow, residues located at the interface between the core and terminal domains of the HIV-1 protease are shaded in pink, and residues located at the interface between the core and flap domains are shaded in teal. The residues in black are not observed at any interdomain interface (Figure generated using Swiss-PdbViewer [7]).

The distance differences between the $C\alpha$-atoms of variant HIV-1 protease in complex with inhibitor and wild-type protease in complex with the same inhibitor were calculated. The HIV-1 protease in complex with either amprenavir, darunavir, indinavir, lopinavir, nelfinavir, ritonavir, saquinavir, tipranavir, cyclic sulfamide or cyclic urea were analysed. The same rigid body domains and the same domain movements associated with ligand binding identified for the bound versus unbound HIV-1 protease structural comparison were observed here (data not shown). The interdomain interfaces between the terminal and core domains and between the core and flap domains identified here were also equivalent to those identified in the previous section. From the matrixes generated here, 5 sites of resistance mutations were located at or in the vicinity of the interface between the core and terminal domains of the HIV-1 protease (Gln7, Lys14, Leu63, Ala71, Leu90), while 2 were located at the interface between the core and flap domains (Leu33, Met36). Residues Lys20 (located on the core domain) and Ile54 (located on the flap domain) are neither a part of the protease active site nor are they in direct contact between the domains at an interface (see Figure 4). The location of the Gln7Lys, Lys14Arg, Leu33Phe, Met36Ile, Leu63Ile, Ala71Val and Leu90Met mutations at the two interdomain interfaces suggests that domain orientation or movement may be a factor in the development of resistance to protease inhibitors. No differences in orientation of the rigid body domains were observed in the protease-ligand complexes investigated here. Thus it is apparent that mutations at the interdomain interfaces affect the movements of the domains, indirectly affecting the interactions between ligand and protein.

Of the 14 drug resistance mutations considered in this investigation, 9 are not in contact with inhibitor and are far from the HIV-1 protease active site. Many of these mutations exhibit cross-resistance to the different structurally diverse protease inhibitors, implying the fixation of these drug-resistant mutations. For example, saquinavir treatment *in vivo* strongly selects for the Leu90Met mutation, and can also select for the Ala71Val/Thr mutation [19,

20]. For indinavir, variable patterns of multiple substitutions have been reported involving the alteration of residues Lys20 (to Met or Arg), Met36 (to Ile), Ile54 (to Val), Ala71 (to Val or Thr) and/or Leu90 (to Met) [19]. *In vitro* selection with ritonavir produced a resistant variant with five substitutions of which the Leu63Pro and Ala71Val mutations were included [21]. For a more extensive look at the mutations exhibiting cross-resistance to the different protease inhibitors see [19]. In general, the observations recorded using the $C\alpha$-difference distance matrix approach indicate a mechanism for drug resistance that is not specific to each inhibitor. This mechanism may be necessary for the dynamic process of protease closure and as a result may show subtle differences in the various protease-ligand complexes.

Factor X and Associated Domain Motions

Factor X is a trypsin-like serine protease which plays a pivotal role in the blood coagulation cascade [22-25]. It is synthesized in the liver as a single-chain molecule and following post-translational modifications is secreted into the blood as a zymogen [23-24]. The conversion of factor X to the functional factor Xa occurs via the intrinsic (activated by surface contact) or extrinsic (activated by vessel injury) pathways [22-25]. Combined with factor Va and Ca^{2+}, factor Xa converts prothrombin to thrombin on a phospholipid membrane [22-24]. Factor X is thus the first member of the final common pathway or thrombin pathway in the blood coagulation cascade.

Factor X is a glycoprotein consisting of a 139 residue light chain and a 306 residue heavy chain [23-24]. The N-terminal light chain contains three characteristic structural domains: a gamma-carboxyglutamic acid (GLA) rich domain followed by two epidermal growth factor (EGF) domains [23-24, 26]. The first EGF domain, EGF1, corresponds to residues Asp46 to Phe84 while the second EFG domain, EGF2, corresponds to residues Thr85 to Gly128. Each domain possesses distinct functional properties. The GLA rich domain which is critical for Ca^{2+} and phospholipid binding induces the necessary conformational changes required for optimal factor X activation, while the two EGF domains are considered important for protein-protein interactions [26]. It is however unclear whether the three domains interact directly with factor Va in the prothrombin activation complex, or if they position the catalytic domain at a correct distance above the phospholipid membrane [26]. The C-terminal heavy chain, which is joined to the light chain by a disulfide bond, harbours a 52 residue activation peptide (Ser143 to Arg194) and a 254 residue trypsin-like serine protease domain [24, 26]. The catalytic specificity of the protein is supported by this domain. Situated within a β-barrel fold in this domain are two subsites (S1 and S4) and a neighbouring active triad (Ser195-His57-Asp102), which are responsible for the blood coagulation specificity of the protease [23-24, 26].

The proteolytic activation of factor X is achieved by either the tissue factor (TF)-VIIa complex (extrinsic pathway) or by the IXa-VIIIa complex (intrinsic pathway) [23-24, 26]. The selective activation of factor X by either complex results in cleavage at Arg194-Ile195 of the heavy chain and the subsequent release of the activation peptide [24]. Factor Xa then serves as the enzyme component of the prothrombinase complex which is responsible for the rapid conversion of prothrombin to thrombin [26]. Thrombin subsequently converts fibrinogen to fibrin, inducing clot formation and platelet aggregation [22, 26]. An imbalance in these processes can thus lead to thrombotic or bleeding disorders. Thus by targeting factor

Xa, propagation of the anticoagulation cascade can be blocked. A plethora of antithrombotic therapies selective for factor Xa inhibitors exist. These have been divided into two groups: the parenteral Xa inhibitors and oral factor Xa inhibitors [27]. Research on the latter group is well advanced leading the way for the oral antithrombotic therapies targeting factor Xa. More details on the mode of inhibitor action on factor Xa can be found in the review by Yavin *et al.* [27].

Table 2. Factor Xa structures used for the computation of the difference distance matrixes. The PDB identification number of the structure, name of the ligand bound to the structure and RMS deviations from the reference structure 1c5m.

		PDB code	Ligand	RMSD(Å)[a]
Inhibitors	Amino-benzyl-carboxyl-amides	1lpg	IMA	1.69
		1lpk	CBB	1.75
		1lpz	CMB	0.66
		1lqd	CMI	1.71
	Pyridinyl	1ioe	XMA	1.73
		1iqe	XMB	1.69
		1iqf	XMD	1.72
		1iqg	XME	1.69
		1iqh	XMF	1.73
		1iqi	XMG	1.69
		1iqj	XMH	1.69
		1iqk	XMI	1.72
		1iql	XMJ	1.72
		1iqm	XMK	1.69
		1iqn	XMC	1.71
	Sulfonyl-piperazinone	1f0r	815	1.73
		1f0s	PR2	1.74
		1nfu	RRP	1.69
		1nfw	RRR	1.71
		1nfx	RDR	1.70
		1nfy	RTR	1.70
		1v3x	D76	1.72

[a] The RMSD-values were calculated by superimposition of the $C\alpha$-atoms of the PDB structure onto 1c5m and calculating the root-mean-square-deviation of the $C\alpha$-atoms from 1c5m. This was done using Swiss-PdbViewer [7].

The conformational changes associated with factor Xa inhibition were assessed using the $C\alpha$-difference distance matrix approach. Overall, 23 factor Xa structures in different conformations were chosen from the PDB (Table 2). These included 1 factor Xa structure in the apo-form and 22 factor Xa-inhibitor complexes. Several of the factor Xa X-ray crystal structures were invariably solved with the GLA domain cleaved. This is a result of experimental limitations in controlling auto-proteolytic activity [24, 28-30]. Also, in several of these structures, the first EGF domain and the region between the two EGF domains were reported to be disordered [28-30]. For the purpose of this study, the residues selected for comparison of the factor Xa structures were: (1) residues Ile16 to Lys243 encompassing all domains of the heavy chain and (2) residues Cys1 to Leu49 encompassing only the EGF2 domain of the light chain. Due to variations in the numbering used within all the reported factor Xa structures, the residue numbering used here is based on the 1c5m structure.

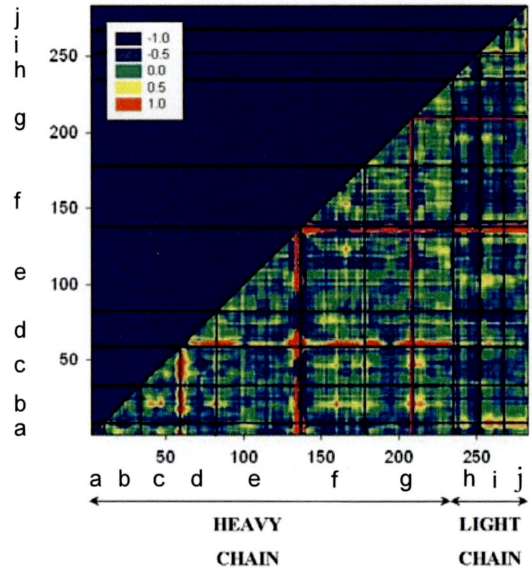

Figure 5. $C\alpha$-difference distance matrix between the inhibitor bound (1lpz) and unbound (1c5m) factor Xa structures. The heavy chain of factor Xa corresponds to residues 1-234 while the light chain of factor Xa corresponds to residues 235-283.

Figure 5 shows the $C\alpha$-difference distance matrix for the inhibitor bound versus unbound structural comparison. Equivalent residues in both the heavy and light chains of each factor Xa structural pair were compared. Residues 1-234 along either axis correspond to residues 16 to 243 of the factor Xa heavy chain, while residues 235-283 along either axis correspond to residues 1 to 49 of the factor Xa light chain. Each chain is identified in the matrix. In some of the structures compared, residues were missing from the factor Xa heavy chain polypeptide. For example, in the 1c5m structure residues 149 and 219 are missing, and in the 1lpz structure residues 146 and 218 are missing. These missing residues in the factor Xa heavy chain were excluded from all analyses and can thus account for the apparent incongruence between the number of residues found in the matrix and the actual heavy chain sequence length.

The subdomains delineated in the $C\alpha$-difference distance matrix in Figure 5 were identified as spatially connected regions with interatomic distances varying by less than 0.5 Å. Subdomains **a-g** comprise two rigid body domains in the factor Xa heavy chain. Domain 1 consists of residues 16-25, 28-52, 53-73, 79-101 and 156-191 of the factor Xa heavy chain, while domain 2 is made up of residues 102-144 and 195-251 of the factor Xa heavy chain. Subdomains **h-j** constitute three domains in the EGF2 domain of the factor Xa light chain. Domain 3 constitutes residues 0-15 of EGF2, domain 4 consists of residues 18-30 of EGF2 and domain 5 of residues 34-51 of EGF2. The domains identified in the heavy and light chains of the factor Xa structure are illustrated in Figure 6.

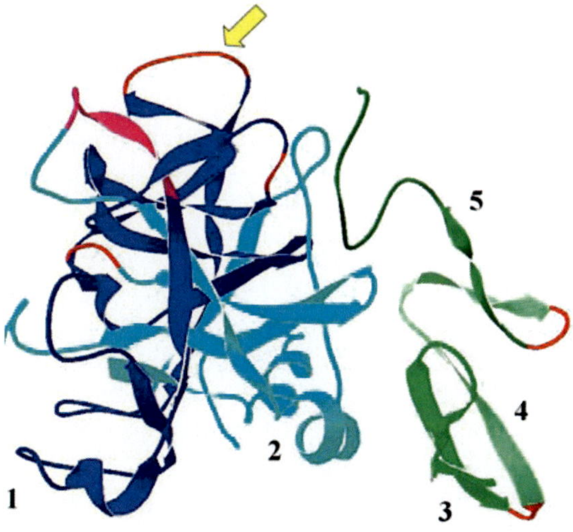

Figure 6. Domain motions of factor Xa. The factor Xa heavy chain consists of two domains made up of seven subdomains: domain 1 (16-25; 28-52; 53-73; 79-101; 156-191) shaded in blue; and domain 2 (102-148; 195-251) shaded in teal. The EGF2 domain of the light chain consists of three domains: domain 3 (0-15), domain 4 (18-30), and domain 5 (34-51), each in a shade of green. Residues comprising the interdomain interfaces are shaded in red. The mobile fragment of the factor Xa heavy chain activation peptide corresponding to residues 149 to 156 are indicated in pink. The most flexible hinge in factor Xa glycoprotein is located in the heavy chain and is indicated with a yellow arrow (Figure generated using Swiss-PdbViewer [7]).

Subdomains displaying distance differences less than 0.5 Å essentially move concertedly with respect to one another. These concerted movements occur between all the equivalent subdomains of the factor Xa heavy chain (i.e. **a** with **a**, **b** with **b**, etc.). Concerted movements are also observed within and between the three domains of the EGF2 domain of the factor Xa light chain. Evaluation of the matrix in Figure 5 does however indicate several conformational differences within the factor Xa heavy and light chains. These differences are illustrated in the matrix in red and correspond to difference distances greater than 0.5 Å. The principal movements between factor Xa in the bound form relative to factor Xa in the apo-form were observed mainly in the heavy chain. The most prominent movement here corresponds to residues 149 to 156 in the factor Xa heavy chain (see positions 133-140 on the x-axis of the matrix with respect to positions 1-140 on the y-axis). These residues, indicated in Figure 6, form part of the 52 residue activation peptide which is located on a disulfide loop in the factor Xa heavy chain. This peptide essentially connects Arg139 at the C-terminal end of the light chain with the N-terminal Ile195 of the heavy chain [24]. Analysis of other serine proteases such as chymotrypsinogen, prethrombin, and proproteinase E indicate that significant reorientation of the N-terminus of the heavy chain occurs upon activation of the respective zymogens [24]. Thus, the $C\alpha$-difference distance matrix approach is capable of detecting this movement in the factor Xa heavy chain. An additional flexible hinge (residues 74-78) was identified in the factor Xa heavy chain. This hinge was found to be responsible for movement between subdomains **c** and **d** of domain 1 in the heavy chain (see Figure 6). In the factor Xa light chain, some movements were detected in domain 2. The conformational differences detected here, in the example represented in Figure 5, were recorded for all the comparisons made in this protein family. Superimposition of the bound structures onto the unbound structure resulted in RMSDs for the $C\alpha$-atoms ranging from 0.66 Å to 1.75 Å (see Table 2). While these deviations indicate some conformational differences between the structural pairs under investigation, the $C\alpha$-difference distance matrix approach provided more details as to these differences. This tool is thus more informative and accurate relative to looking at overall RMSDs between the proteins.

Conformational Analyses of Membrane Channel Proteins

Aquaporins and Associated Domain Motions

Aquaporins (AQPs) are a class of membrane channel proteins selective for the permeation of specific small molecules across cell membranes [31-32]. Members of this superfamily constitute two functionally distinct subgroups: (1) the aquaporins which are water selective and (2) the aquaglyceroporins, which are permeable to water, glycerol and other small neutral solutes [33].

AQPs are found in a wide variety of organisms including bacteria, plants and mammals. In the human genome, more than ten different genes encoding these membrane channel proteins have been identified [33]. In all instances they play an essential role in water uptake and release. It is thus not surprising that physiological disorders including nephrogenic diabetes insipidus, congenital cataract formation and cerebral oedema are connected to the impaired function of these channels [34].

Crystallographic studies have confirmed the structural organization of AQPs in the cell membrane. Based on these studies the functional AQP unit is tetrameric in nature [35-36], with each monomer contributing an independent water channel [35-36]. The AQP subgroups all display the same general folding topology. Each AQP monomer contains six tilted transmembrane α-helices packed together to form a right-handed helical bundle [36-39]. The N- and C- terminal halves of the monomer, each formed from three transmembrane helices, are tandem repeats connected by a long loop spanning the extracellular face of the protein [35-40]. Passing through the centre of the monomer is an individual hour-glass shaped channel outlined mostly by hydrophobic residues and comprising the two highly conserved Asn-Pro-Ala (NPA) motifs [35-38]. The narrow constriction in this channel acts as a size-selective pore which allows small molecules to traverse the channel in single file. Differences in peptide sequence and length amongst the AQP subfamilies contribute to the observed variation in pore size and specificity. Generally, the selectivity filter of aquaglyceroporins is larger and more hydrophobic than that of aquaporins thereby influencing AQP specificity [36, 39].

Table 3. AQP structures used for the computation of the difference distance matrixes. The PDB identification number of the structure, residues compared, name of the ligand bound to the structure and RMS deviations from the reference structure.

	PDB codes[a]	Residues compared[b]	Structural comparison	Ligand	RMSD(Å)[c]
AQP0	1sor_1ymg	6-239	unbound vs. unbound	-	0.90
	2b6o_1ymg	6-239	bound vs. unbound	MC3	0.34
	2b6p_1ymg	6-239	unbound vs. unbound	-	1.33
	2c32_1ymg	6-239	unbound vs. unbound	-	0.00
AQP1	1fqy_1h6i	9-232	unbound vs. unbound	-	3.34
	1ih5_1h6i	9-232	unbound vs. unbound	-	3.44
AQPGlpF	1fx8_1lda	6-259	bound vs. unbound	glycerol	0.17
	1fx8_1ldi	6-259	bound vs. unbound	glycerol	0.16
	1lda_1ldi	6-259	unbound vs. unbound	-	0.13
	1ldf[d]_1fx8	6-259	bound vs. bound	glycerol	0.23
AQPZ	2abm_1rc2	1-227	unbound vs. unbound	-	0.45

[a] The AQP structural pairs, each in different conformations, compared using the $C\alpha$-difference distance matrix approach. For all structural comparisons, the first PDB code corresponds to the test AQP structure and the second PDB code corresponds to the reference AQP structure.

[b] The AQP subgroups differ in peptide sequence. For each subgroup, only the residues common to each structural pair were compared.

[c] The RMSD-values were calculated by superimposition of the $C\alpha$-atoms of the test PDB structure onto the reference structure and calculating the root-mean-square-deviation of the $C\alpha$-atoms from the reference structure. This was done using Swiss-PdbViewer [7].

[d] The 1ldf AQPGlpF structure is a mutant protein, with the W48F and F200T mutations.

The $C\alpha$-difference distance matrix approach was employed to investigate the domain motions occurring within the following AQP structures: AQP0, AQP1, *Escherichia coli* glycerol facilitator (AQPGlpF) and AQPZ. Overall, 14 structures in various conformations were chosen from the PDB (Table 3). Most of the structures reported for the different AQP subgroups start and end with specific residues. As a result, the structural comparisons made for each subgroup took different polypeptide chain sequences into account (see Table 3). The structural comparisons made for this group of proteins included the (1) unbound vs. unbound, (2) bound vs. unbound and (3) bound mutant vs. bound wild-type conformations.

Superimposition of the test AQP structures onto the reference AQP structure resulted in RMSDs for the $C\alpha$-atoms ranging from 0.00 Å to 3.44 Å (see Table 3). These deviations indicate overall differences in conformation between the studied AQP structures. The domain movements recorded using the $C\alpha$-difference distance matrix approach however, is more informative and accurate relative to looking at overall RMSDs between the proteins.

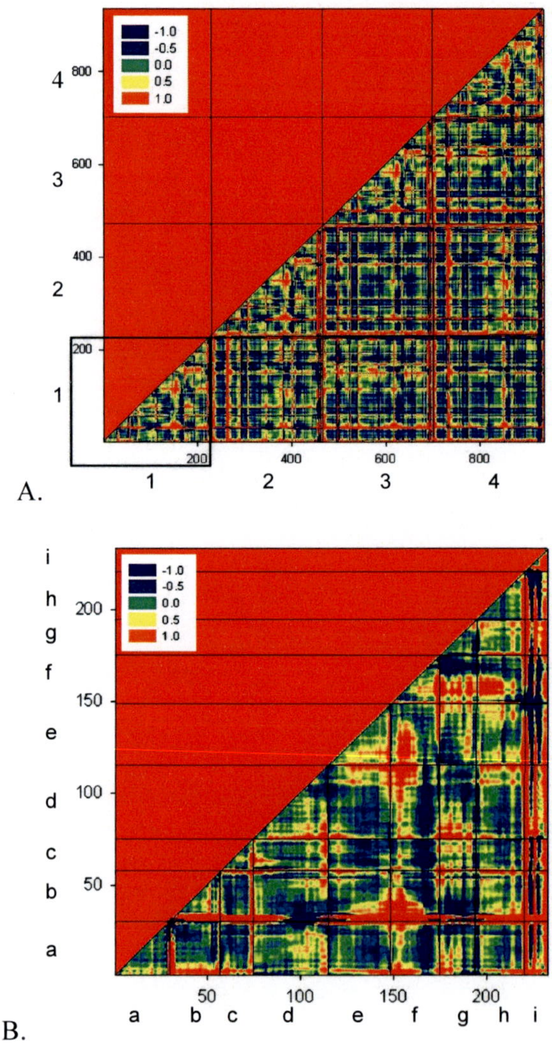

Figure 7. $C\alpha$-difference distance matrix between the two unbound AQP0 structures, 1sor and 1ymg. (A) Corresponds to the complete AQP0 tetramer: residues 1-234, 235-468, 469-702 and 703-936 along either axis of the matrix correspond to each subunit of the tetramer. The boxed area in the graph, representative of a single AQP0 subunit, is amplified in (B) to illustrate the domain and subdomain movements occurring within an AQP0 subunit. Letters **a-i** indicate the subdomains of AQP0 with distance differences less than 0.5 Å.

Figure 7 shows the representative $C\alpha$-difference distance matrix for the unbound versus unbound AQP0 structural comparison. In Figure 7 (A) the $C\alpha$-difference distance matrix for the complete AQP0 tetramer is illustrated. Keeping in mind that the reported AQP0 structures start with residue 6 and end with residue 239 [41], residues 1-234, 235-468, 469-702 and 703-936 along either axis of the matrix correspond to each subunit of the AQP0 tetramer. The subunits, delineated in Figure 7 (A) as 1, 2, 3 and 4, can be discerned along the diagonal of the matrix and by the red horizontal and vertical regions of the matrix. The red regions here correspond to distance differences greater than 0.5 Å.

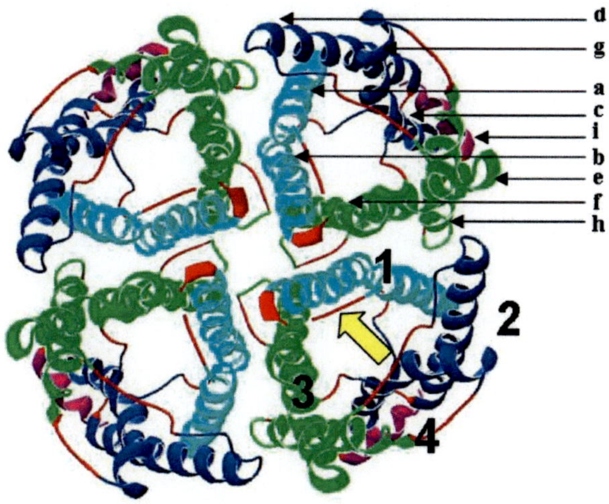

Figure 8. Domain motions of the AQP tetramer. Each AQP subunit consists of four domains made up of nine subdomains: domain 1 (**a**: 6-33; **b**: 41-60) shaded in teal; domain 2 (**c**: 65-78; **d**: 83-114; **g**: 182-196) shaded in blue; domain 3 (**e**: 121-152; **f**: 156-174; **h**: 202-224) shaded in green; and domain 4 (**i**: 230-239) shaded in purple. Residues comprising the interdomain interfaces are shaded in red. The most flexible hinge in the aquaporin subunit is indicated with a yellow arrow (Figure generated using Swiss-PdbViewer [7]).

From Figure 7 (A) it is clear that the domain motions within each AQP0 subunit are equivalent. Thus, in order to extensively analyse the domain movements within an individual AQP subunit, the $C\alpha$-difference distance matrix for a single AQP0 monomer was examined (see Figure 7 B). The domains in the AQP monomer were identified as spatially connected regions with interatomic distances varying by less than 0.5 Å. Nine subdomains, appearing along the diagonal of the matrix, were identified in the AQP0 monomer. These are delineated in the matrix as **a-i**. Combined, these subdomains form four rigid domains (see Figure 8). Domain 1 is formed by subdomains **a** and **b**, domain 2 by subdomains **c**, **d** and **g**, domain 3 by subdomains **e**, **f** and **h** and domain 4 by subdomain **i**. Positioned on subdomains **c** and **g** of domain 2 are the highly conserved NPA motifs. Loops B and E which are equivalent to subdomains **c** and **g**, respectively, are associated by van der Waals interactions between the NPA motifs. It is the interaction of the asparagine hydrogen-bond-forming groups in this motif that readily allow the movement of water molecules through the AQP pore [36-37, 40].

Evaluation of the $C\alpha$-difference distance matrix indicates several conformational differences in the two AQP0s under investigation. The most prominent difference however is observed between subdomain **i** and subdomains **a-g**. Subdomain **i** corresponds to the amino terminus of the AQP molecule which is located at the cytoplasmic side of the cell membrane. This data is indicative of the N-terminal domain not being well ordered in the structure. In fact, Harries *et al.* [41] reported that the N-terminus of the 1ymg AQP0 structure was not well ordered. Also evident from the $C\alpha$-difference distance matrix is the presence of a flexible hinge in the AQP0 structure. Residues 34-38 of AQP0 are observed to move by more than 0.5

Å with respect to residues 39-239. This movement is illustrated by the horizontal red region across the matrix at positions 38-39 on the y-axis. Similar movements for the AQP1, AQPGlpF and AQPZ were recorded, even when comparing the bound vs. unbound or mutant vs. wild-type structural conformations. Essentially, AQPs in the unbound or bound conformations display movement within, as well as between all domains, with domain 4 being the most mobile.

Chloride Channel Proteins and Associated Domain Motions

The ClC chloride channels (ClC) constitute a large family of selective ion channel proteins, found throughout biology in both prokaryotic and eukaryotic cells [42-44]. The conduction of Cl^- ions across the cell membrane into the cytoplasm by the ClCs is vital for the regulation of electrical impulses, transport of salts across cells, and stabilizing the voltage of excitable muscle cells [44-47]. Genetic defects of human ClC Cl^- channels underlie several inherited diseases (e.g. Bartter's syndrome and Dent's disease) [42, 48]. The wide array of physiological functions performed by the ClCs, as well as their link to various diseases, highlights the importance of further investigating this family of proteins.

Crystallographic data of the ClC Cl^- channels depict a class of membrane channel proteins fundamentally different from that of the well-studied cation-selective ion channels [49]. In contrast to the one-pore oligomeric channels, where the pore occurs at the axis of symmetry of the oligomer, the ClC Cl^- channel displays a homodimeric architecture, with each subunit containing a Cl^- conducting pore [48-49]. The ClC Cl^- channel subunit on its own exhibits a complex topology. Each subunit is formed by two roughly repeated, inverted N- and C- terminal halves [42-43]. The two halves wrap around a common centre bringing important amino acids from four highly conserved regions together to form the ClC selectivity filter [42]. This architecture places the selectivity filter at the pseudo two-fold axis of the ClC subunit, with the positive N-termini of the α-helices positioned to stabilize the Cl^- ions inside the membrane [42-44].

In the ClC channel, selective ion conduction and 'fast gating' seem to be intimately coupled [44]. Chloride ions conduct rapidly through the pore thereby influencing the probability that the fast gate will open [44]. Dutzler *et al.* [44] found that gating of the ClC Cl^- selectivity filter of each subunit pore is conducted by the conserved Glu148 side chain. For Cl^- conduction to occur the glutamate carboxyl group within the filter swings out of the way, permitting the entry of a Cl^- ion in its place. Several pieces of evidence do however indicate that this 'fast gating' mechanism requires more than a simple movement of a glutamate residue [49].

The domain motions occurring within the ClC Cl^- channel were investigated using the $C\alpha$-difference distance matrix approach. Of the 10 ClC structures taken from the PDB, 1 was of ClC in the apo-form and 9 were of ClC-Fab complexes in the presence/absence of bound ions (Table 4). Conformational differences between the ClC channel in the bound form relative to the channel in either the unbound form or the bound form were assessed. The RMSDs obtained following superimposition of the test structures onto the reference structure ranged from 0.34 Å to 0.89 Å (see Table 4) indicating minor conformational differences between the test and reference structures.

Table 4. ClC structures used for the computation of the difference distance matrixes. The PDB identification number of the structure, name of the ligand bound to the structure and RMS deviations from the reference structure.

	PDB codes[a]	Ligand	Mutations	RMSD(Å)[b]
Bound vs. unbound	1ots_1kpk	Fab, Cl^-	-	0.85
	1ott_1kpk	Fab, Cl^-	E148A	0.87
	1otu_1kpk	Fab, Cl^-	E148Q	0.89
	2exw_1kpk	Fab	-	0.81
	2exy_1kpk	Fab	E148Q	0.86
	2ez0_1kpk	Fab, Br^-	S107A, E148Q, Y445A	0.84
	2fec_1kpk	Fab	E203Q	0.79
	2fed_1kpk	Fab	E203Q	0.78
	2fee_1kpk	Fab	-	0.81
Bound vs. bound	1ots_2exw	Fab, Cl^- vs. Fab	-	0.51
	1ott_1ots	Fab, Cl^- vs. Fab, Cl^-	E148A vs. wt	0.46
	1otu_1ots	Fab, Cl^- vs. Fab, Cl^-	E148Q vs. wt	0.57
	2exy_2exw	Fab vs. Fab	E148Q vs. wt	0.38
	2ez0_2exw	Fab, Br^- vs. Fab	S107A, E148Q, Y445A vs. wt	0.53
	2fec_2exw	Fab vs. Fab	E203Q vs. wt	0.38
	2fed_2exw	Fab vs. Fab	E203Q vs. wt	0.39
	2fee_2exw	Fab vs. Fab	-	0.34

[a] The ClC structural pairs, each in different conformations, compared using the $C\alpha$-difference distance matrix approach. For all structural comparisons, the first PDB code corresponds to a test, ligand bound ClC structure. The second PDB code for each structural comparison corresponds to a reference unbound (top section of Table) or reference bound ClC structure (bottom section of Table).

[b] The RMSD-values were calculated by superimposition of the $C\alpha$-atoms of the test PDB structure onto the reference structure and calculating the root-mean-square-deviation of the $C\alpha$-atoms from the reference structure. This was done using Swiss-PdbViewer [7].

The representative $C\alpha$-difference distance matrix illustrated in Figure 9 is for the bound Glu148Gln mutant ClC versus the unbound wild-type ClC structural comparison. Figure 9 (A) depicts the matrix for the complete ClC dimer. As each ClC structure taken from the PDB displayed different start and end residues, only residues 18 to 458 of each ClC subunit were compared in this study. Keeping this in mind, residues 1-441 and 442-882 along either axis of the matrix correspond to each subunit of the ClC dimer. These subunits can be discerned along the diagonal of the matrix and are delineated in Figure 9 (A) as 1 and 2. The domain motions within each ClC subunit are equivalent. Thus, the $C\alpha$-difference distance matrix for the single subunit was examined in more depth (see Figure 9 B). As with the previous protein families, domains were identified as spatially connected regions with interatomic distances varying by less than 0.5 Å. Overall sixteen subdomains were identified in the ClC monomer. Combined these subdomains form four rigid domains per ClC Cl^- channel subunit (see Figure 10). Domain 1 of the ClC structure is formed by residues 18-27; domain 2 by residues 33-37, 109-117, 121-141, 148-167, 172-188 and 193-232; domain 3 by residues 75-100 and 277-

307; and lastly, domain 4 by residues 244-273, 320-350, 357-376, 386-401, 405-416, 421-436 and 442-458.

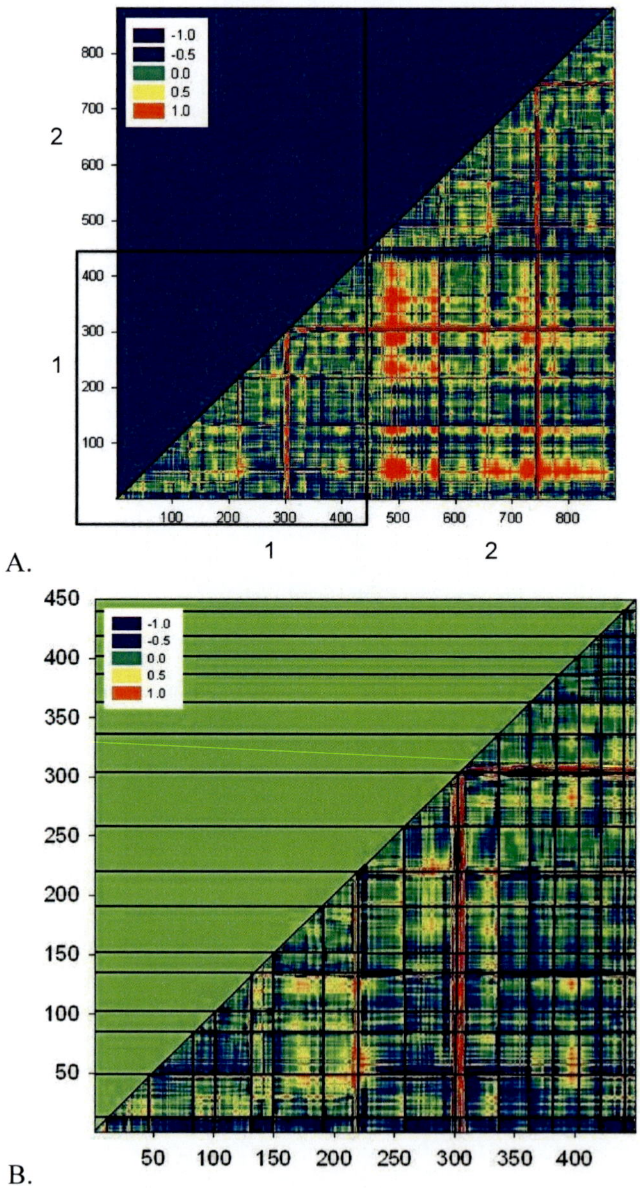

Figure 9. $C\alpha$-difference distance matrix between a ClC mutant bound to a monoclonal Fab fragment (1otu) and unbound ClC (1kpk). (A) Corresponds to the complete ClC dimer: residues 1-441 and 442-882 along either axis of the matrix correspond to each subunit of the dimer. The boxed area in the graph, representative of a single ClC subunit, is amplified in (B) to illustrate the domain and subdomain movements occurring within a single ClC subunit.

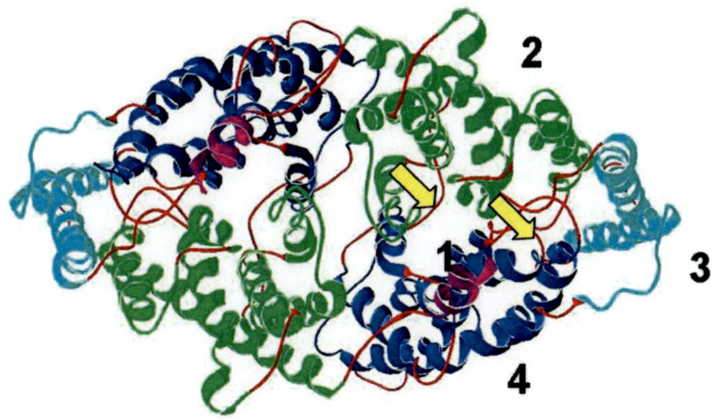

Figure 10. Domain motions of the ClC dimer. Each monomer consists of four domains made up of sixteen subdomains: domain 1 (18-27) shaded in purple; domain 2 (33-37; 109-117; 121-141; 148-167; 172-188; 193-232) shaded in green; domain 3 (75-100; 277-307) shaded in teal; and domain 4 (244-273; 320-350; 357-376; 386-401; 405-416; 421-436; 442-458) shaded in blue. Residues comprising the interdomain interfaces are shaded in red. The most flexible hinges in the ClC monomer are indicated with yellow arrows (Figure generated using Swiss-PdbViewer [7]).

Significant movements between the mutant ClC-Fab complex and the wild-type ClC are represented on the matrixes in Figure 9 by distance differences greater than 0.5 Å. In Figure 9 (A), the red regions in the square quadrant of the matrix correspond to distance differences greater than 0.5 Å. These regions are indicative of movements between subdomain 1 of one subunit with respect to all subdomains in the adjacent subunit. Subdomain 1 which corresponds to the N-terminus α-helix located near the cytoplasmic surface of the ClC channel is thus more mobile relative to the rest of the transmembrane α-helices. This can be attributed to the fact that the N-terminus from one of the ClC subunits in the reference structure (1kpk) is disordered while the other subunit N- terminus is visible as an α-helix near the cytoplasmic surface [42]. This particular movement was observed for all the bound versus unbound structural comparisons (see Table 4). From Figure 9 (B) it is evident that small movements within as well as between all subdomains of the ClC subunit occur. The most prominent movements are however attributed to two flexible hinges within the ClC Cl^- channel. These flexible hinges, indicated by the yellow arrows in Figure 10, correspond to residues 233-243 and 308-319 in each ClC subunit. Both hinges, but particularly the former, are in the vicinity of the ClC pore. The $C\alpha$-difference distance matrix is thus capable of distinguishing the movement of these hinges when comparing the ClC channel with bound Cl^- ions in the selective-filter relative to the ClC channel in the apo-form. Overall, similar patterns of domain movements in the ClC Cl^- channel were observed for all the structural comparisons listed in Table 4. Generally, however, the wild-type structures displayed more rigid domains within each subunit relative to the mutant structures.

Conformational Analyses of Other Protein Families

Calmodulin and Associated Domain Motions

Calmodulin (CaM) is an intracellular calcium binding protein found ubiquitously in eukaryotes [50-52]. It is responsible for the regulation of a multitude of different protein targets and as such affects a wide range of biological processes (e.g. smooth muscle contraction, neurotransmitter release and cytoskeletal organization) [50-51]. Many of these target proteins which require Ca^{2+} to function are actually unable to bind these ions. As a result, they utilize CaM as a sensor and mediator of the initial Ca^{2+}-signal. The binding and regulation of CaM/Ca^{2+} to such protein targets is thus achieved in a Ca^{2+}-dependent manner [52]. CaM also shows Ca^{2+}-independent binding to a number of proteins. In this type of interaction, a repeated IQ motif in the target protein plays an important role [50].

CaM is a small soluble protein made up of 148 amino acid residues [51-52]. It is characterised by two globular N- and C- terminal domains connected by a long α-helix linker [51-52]. Located on each globular domain are two characteristic EF-hand (helix-loop-helix) Ca^{2+}-binding motifs [51-52]. Each EF-hand motif pair consists of four helices arranged in a twisted four helix bundle which lie in close proximity to one another [52]. Four Ca^{2+} binding sites are observed in CaM, one in each EF-hand motif. Upon Ca^{2+} binding at these sites, the α-helices of the EF-hand motifs are observed to change their position relative to each other, forming an almost perpendicular open conformation [51-52]. This rearrangement leads to the exposure of several hydrophobic residues on the surface of each terminal domain, thus forming a large hydrophobic channel on each domain [52]. The exposure of these surfaces allows for binding of CaM to non-polar regions on the target proteins. As a consequence of this, CaM is able to behave as a versatile regulatory protein with its target proteins not requiring specific amino acid sequences or structural binding motifs. While the exact conformation and function of the central linker between the two terminal domains of CaM is unclear [52 and reference therein], it does provide sufficient flexibility allowing CaM/Ca^{2+} to find the best fit around each particular protein target [50, 52].

Several crystal and NMR structures of CaM in complex with fragments derived from CaM-regulated proteins are available from the PDB. Here, we focussed on only two CaM complex types: (1) CaM in complex with the anthrax exotoxin (edema factor, EF) from *Bacillus anthracis* and (2) CaM in complex with a peptide corresponding to the IQ domain of the cardiac $Ca_V1.2$ Ca^{2+} channel. The anthrax EF is a calmodulin-dependent adenylate cyclase which when inside the host cell, acquires CaM and becomes activated. Following activation, the toxin is then able to convert large amounts of ATP to cAMP [52]. In the heart, muscle contraction is triggered by Ca^{2+} entering through the $Ca_V1.2$ Ca^{2+} channel in cardiac myocytes and subsequently releasing the Ca^{2+} ions into the sarcoplasmic reticulum [53]. The binding of CaM/Ca^{2+} to the calcium channel alpha subunit at the IQ domain serves to regulate the Ca^{2+}-dependent inactivation and facilitation of the $Ca_V1.2$ Ca^{2+} channel [53].

The $C\alpha$-difference distance matrix approach was employed to monitor the domain motions occurring within each group of CaM complexes. Overall, 10 CaM-EF complexes and 2 CaM-Ca^{2+} channel alpha subunit complexes were taken from the PDB to be assessed (Table 5). Superimposition of the test CaM structures onto the reference CaM structures resulted in

RMSDs for the $C\alpha$-atoms ranging from 0.27 Å to 1.31 Å (see Table 5). These deviations indicate minor conformational differences between the studied CaM structures.

Table 5. CaM structures used for the computation of the difference distance matrixes. The PDB identification number of the structure, name of the ligand bound to the structure and RMS deviations from the reference structure.

	PDB codes[a]	Ligand	RMSD(Å)[b]
CaM-EF complexes	1k90_1k93	Ca^{2+}, 3AT vs. Ca^{2+}	0.81
CaM-EF complexes	1lvc_1k93	Ca^{2+}, YB^{3+}, DOT vs. Ca^{2+}	0.54
CaM-EF complexes	1pk0_1k93	Ca^{2+}, YB^{3+}, EMA vs. Ca^{2+}	0.46
CaM-EF complexes	1s26_1k93	Ca^{2+}, YB^{3+}, APC vs. Ca^{2+}	0.27
CaM-EF complexes	1sk6_1k93	Ca^{2+}, YB^{3+}, CMP vs. Ca^{2+}	1.31
CaM-EF complexes	1xfu_1xfz	Ca^{2+}, Mg^{2+} vs. Ca^{2+}, Mg^{2+}	0.36
CaM-EF complexes	1xfv_1xfz	Ca^{2+}, Mg^{2+}, 3AT vs. Ca^{2+}, Mg^{2+}	0.37
CaM-EF complexes	1xfw_1xfz	Ca^{2+}, Mg^{2+}, CMP vs. Ca^{2+}, Mg^{2+}	0.33
	2f3y_2f3z[c]	Ca^{2+}, Mg^{2+} vs. Ca^{2+}	0.52

[a] The CaM structural pairs, each in different conformations, compared using the $C\alpha$-difference distance matrix approach. For all structural comparisons, the first PDB code corresponds to the test CaM structure and the second PDB code corresponds to the reference CaM structure.

[b] The RMSD-values were calculated by superimposition of the $C\alpha$-atoms of the test PDB structure onto the reference structure and calculating the root-mean-square-deviation of the $C\alpha$-atoms from the reference structure. This was done using Swiss-PdbViewer [7].

[c] In this comparison each CaM structure has the voltage-gated calcium channel alpha subunit bound to it.

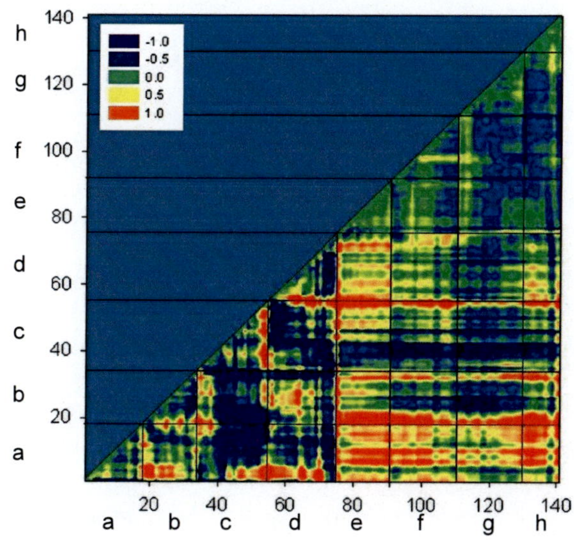

Figure 11. $C\alpha$-difference distance matrix for CaM bound to the adenylyl cyclase domain of anthrax edema factor and 3' deoxy-ATP (1k90) relative to CaM bound to the adenylyl cyclase domain of anthrax edema factor only (1k93). Residues 1-142 along either axis of the matrix correspond to the complete monomer.

Figure 11 shows the representative *Cα*-difference distance matrix for the $Ca^{2+}$and 3' deoxy-ATP bound CaM-EF complex versus the Ca^{2+} bound CaM-EF complex. As the CaM structures taken from the PDB displayed different start and end residues, only residues 5 to 146 of the CaM polypeptide were compared in this study. Taking this into account, residues 1-142 along either axis correspond to the CaM polypeptide chain. Domains in the CaM structure were identified as spatially connected regions with interatomic distances varying by less than 0.5 Å. Eight subdomains, delineated as **a-h** in the matrix, form two rigid domains in CaM (see Figure 12). Domain 1 is formed by subdomains **a-d** and domain 2 by subdomains **e-h**. Each domain essentially comprises an EF-hand pair and a portion of the flexible linker.

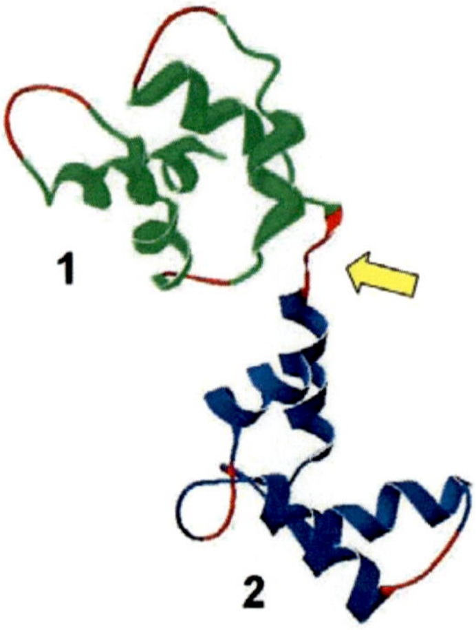

Figure 12. Domain motions of CaM in the open conformation. CaM consists of two domains made up of eight subdomains: domain 1 (5-20; 27-38; 42-57; 63-77) shaded in green; and domain 2 (82-95; 98-114; 118-133; 137-146) shaded in blue. Residues comprising the interdomain interfaces are shaded in red. The most flexible hinge in CaM is indicated with a yellow arrow (Figure generated using Swiss-PdbViewer [7]).

Evaluation of the *Cα*-difference distance matrix indicates several conformational changes in the two CaM-EF complexes under investigation, corresponding to distance differences greater than 0.5 Å in the matrix. Significant movements between equivalent residues in domain 1 were observed in the structural pair under investigation. Also, movements between subdomains of domain 1 with respect to subdomains of domain 2 were detected. The first EF-hand motif of the N-terminal domain displays some mobility with respect to the rest of the CaM structure. This can be observed by the red regions in the matrix corresponding to subdomain **a** on the y-axis with respect to subdomains **b-h** on the x-axis, and again subdomain **b** on the y-axis with respect to subdomains **e-h** on the x-axis. Significant movement is also detected in the matrix between subdomain **e** on the x-axis with subdomains **a-d** on the y-axis. The movements here occur between the C-terminal EF-motif closest to the flexible linker with respect to the entire N-terminal domain. When bound to the anthrax EF, CaM is reported to bind only two Ca^{2+} ions at the C-terminal domain. The Ca^{2+}-loaded C-terminal adopts a similar conformation to other Ca^{2+}-loaded CaM complexes while the Ca^{2+}-free N-terminal domain adopts a conformation typical for apo-CaM. [52]. The *Cα*-difference distance matrix approach is thus capable of detecting the conformational differences in the N- and C- terminals of CaM when bound to the anthrax toxin.

Hinge motions, associated with the intradomain loops connecting each EF-hand motif in the globular domains, were also detected from the matrix. The most prominent hinge movement however can be attributed to residues 78-81 of the flexible linker within the CaM structure (see Figure 12). When bound to the EF toxin, CaM is observed to adopt an extended conformation with the central section (residues 79-81) being partially unwound [52]. The residues here are reported to be uninvolved in the Ca^{2+} binding process [52]. The exact conformation and function of the central linker region following Ca^{2+}-binding to the globular domains is undefined. In solution, this region has been reported to be very flexible and unstructured. Consequently, the two globular domains were found to adopt various orientations relative to each other in solution [52].

For the CaM-Ca^{2+} channel alpha subunit complexes, movement between domain 1 with respect to domain 2 was predominantly observed. The flexible linker was also identified as the most flexible hinge. Overall, the structural conformational changes detected in both CaM complexes using the $C\alpha$-difference distance matrix approach can be attributed to the domain motions between domains 1 and 2 (i.e. N- and C- terminal domains) within the CaM structure, with the central flexible linker contributing to these movements. As is evident, the approach used here is more informative and accurate relative to looking at the overall RMSDs.

CDK-2 and Associated Domain Motions

Cyclin-dependent kinases (CDK) belong to a group of protein kinases responsible for the orderly progression of cells through various phases of the cell-cycle [54-56]. The ability of CDKs to trigger cell-cycle events is completely dependent on associated cyclin partners, whose fluctuating concentrations underlie the stage specific timing of CDK activity [54]. It is the deregulation of CDKs and their associated cyclins that underlies the molecular pathology of cancer [55-57].

CDK2 is probably the best characterised CDK. It associates with cyclin E at the beginning of S phase to promote the initiation of DNA synthesis, and subsequently relocates into complexes with cyclin A that are essential for entry into S phase [54-55]. Crystal structures of the catalytically active and inactive CDK2 enzyme have allowed for the elucidation of the complex mechanism of CDK2 activation by either cyclin E or cyclin A. The catalytically inactive CDK2 monomer consists of a smaller N-terminal domain and a larger α-helical C-terminal domain. Located on the N-terminal domain is an α-helix (C-helix) whose correct orientation is important for catalysis [54, 56]. This helix includes the large well-conserved PSTAIRE motif (residues 45-51) which is important for cyclin binding [58]. Found in the cleft formed between the N- and C- domains is the ATP binding site [54, 56].

Activation of the CDK2 monomer requires two components: (1) the association of the regulatory cyclin subunit (cyclin E at the G1/S transition, cyclin A during the S phase) at the catalytic cleft and (2) the phosphorylation of the kinase on a threonine residue (Thr160) [54-56]. The first step results in the formation of a CDK2/cyclin complex with basal catalytic activity [56]. Few structural changes occur in the cyclin partner but significant structural modifications are observed in CDK2. The most significant change involves the movement of the C-helix towards the active site cleft thus allowing for the correct positioning of catalytically important residues (Lys33, Glu51, Asp145, Phe146, Gly147) [56]. This serves to configure the CDK2 active site for ATP binding. The optimal alignment of ATP at the active site allows for phosphorylation of the kinase Thr160, consequently yielding the fully active CDK2/cyclin complex [54-56]. The formation of this heterodimeric complex allows for cell proliferation to proceed.

The molecular pathology of cancer involves unrestrained cellular proliferation. This can be attributed to aberrant CDK control and consequently loss of cell cycle checkpoint function [55-57]. It is believed that by selectively interrupting cell cycle regulation in cancer cells, the cell will die. It is along this line of thinking that numerous therapeutic agents targeting cancer have been developed. Several structural classes of compounds that are selective CDK inhibitors have been identified (e.g. staurosporin, purvalanol, olomoucine, roscovitine and indirubin) [55-56, 59]. Many of these inhibitors, which function by competing with ATP for binding to the catalytic active site, have been co-crystallised with CDK2 [59 and references therein].

Table 6. CDK2 structures used for the computation of the difference distance matrixes. The PDB identification number of the structure, name of the ligand bound to the structure and RMS deviations from the reference structure 1pw2.

	PDB code	Ligand	RMSD(Å)[a]
ATP	1b38	ATP	0.41
	1b39	ATP	0.42
	1hck	ATP	0.43
	1fin	cyclin A-ATP	0.81
	1jst	cyclin A-ATP	0.91
Inhibitors	1aq1	Staurosporine	0.68
	1ckp	Purvalanol b	0.48
	1w0x	Olomoucine	0.72
	2a0c	Olomoucine II	0.53
	2a4l	Roscovitine	0.46
	2bhe	BRY[b]	0.57
	2bhh	RYU[b]	0.39

[a] The RMSD-values were calculated by superimposition of the *Cα*-atoms of the PDB structure onto 1pw2 and calculating the root-mean-square-deviation of the *Cα*-atoms from 1pw2. This was done using Swiss-PdbViewer [7].

[b] Indirubin analogues.

The conformational changes associated with CDK2 activation or inhibition was assessed using the *Cα*-difference distance matrix approach. Thirteen CDK2 structures in different conformations were chosen from the PDB (Table 6). These included 1 CDK2 structure in the apo-form, 3 CDK2-ATP complexes, 2 CDK2-cyclin A-ATP complexes and 7 CDK2-inhibitor complexes. The bound vs. unbound structural conformations were compared for this group of proteins. Residues 36 to 43 which are reported to be disordered in several CDK2 structures [60] were not considered in the analyses.

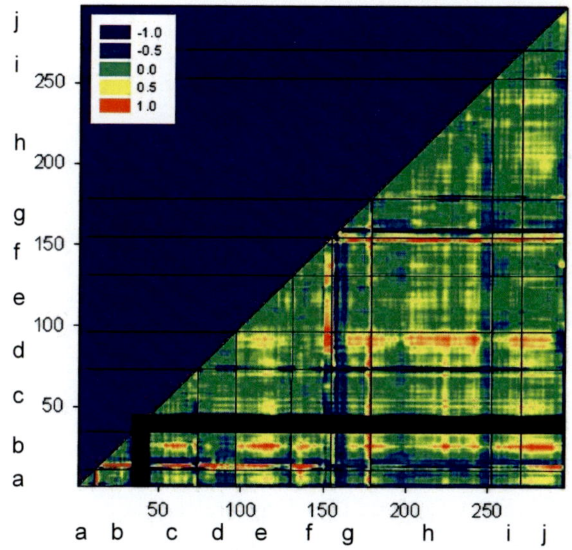

Figure 13. $C\alpha$-difference distance matrix between the CDK2-ATP bound (1b39) and CDK2 unbound (1pw2) structures. Residues 1-298 along either axis of the matrix correspond to the complete monomer. Residues 36 to 43 corresponding to the black areas in the matrix were not considered in the analysis as they are reported to be disordered in CDK2 [60].

Superimposition of the bound CDK2 structures onto the unbound structure resulted in RMSDs for the $C\alpha$-atoms ranging from 0.41 Å to 0.91 Å (see Table 6). These values indicate minor conformational differences between the bound and unbound CDK2 structures. The $C\alpha$-difference distance matrix approach however indicated subtle differences between the CDK structures compared. The representative $C\alpha$-difference distance matrix illustrated in Figure 13 is for the CDK2-ATP bound and CDK2 unbound structural comparison. Residues 1-298 along either axis correspond to the complete CDK2 polypeptide chain. Residues 36 to 43 correspond to the black areas in the Figure. Domains in the CDK2 structure were identified as spatially connected regions with interatomic distances varying by less than 0.5 Å. Ten subdomains were identified in CDK2. These are delineated in the matrix as **a-j**. Combined, these subdomains form three rigid domains in the CDK2 structure (see Figure 14). Subdomains **a-c** form domain 1. Located on subdomain **b** (residues 16-71) is the PSTAIRE motif (residues 45-51) which plays an important role in cyclin binding [58]. The two catalytically important residues, Lys33 and Glu51, which coordinate the α-phosphate of ATP at the CDK2 active site, are also located here [56]. Subdomains **d-f** encompass domain 2. Residues Asp145, Phe146 and Gly147 which form a "DFG" motif on β9, responsible for inducing the correct conformation in the triphosphate moiety of ATP for catalytic transfer, is found on subdomain **f** (132-151) [56]. Domain 3 constituting subdomains **g-j** corresponds to the large C-terminal domain of CDK2.

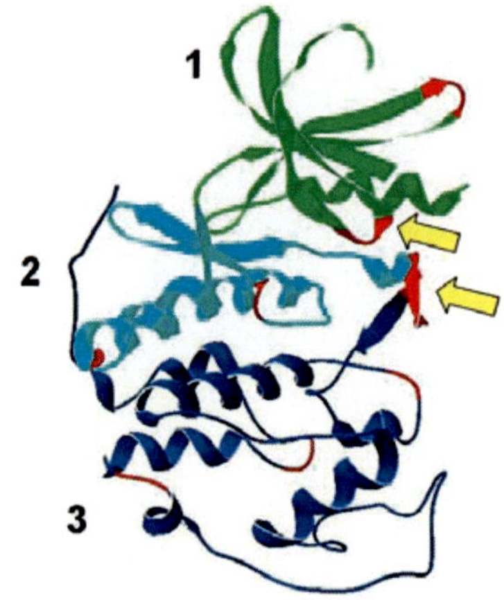

Figure 14. Domain motions of CDK2. The three rigid body domains identified: domain 1 (1-12; 16-71; 75-84) shaded in green; domain 2 (85-96; 101-129; 132-151) shaded in teal; and domain 3 (162-177; 180-251; 254-268; 272-298) shaded in blue. Residues comprising the interdomain interfaces are shaded in red. The most flexible hinges in CDK2 are indicated with yellow arrows (Figure generated using Swiss-PdbViewer [7]).

Subdomains displaying distance differences less than 0.5 Å are considered to move concertedly with respect to one another. Generally, concerted movements between all the equivalent subdomains of the CDK2 monomer are observed (i.e. **a** with **a**, **b** with **b**, etc.). Evaluation of the matrix in Figure 13 does however indicate several conformational differences between the two CDK2 structures under comparison. These differences are illustrated in the matrix in red and correspond to distance differences greater than 0.5 Å. The principal movements between the CDK2 structures in the bound form relative to CDK2 in the apo-form were observed between domain 1 and domain 2 (e.g. **a** with **d/c/f**; **b** with **d/c/f**). Also, two flexible hinges within the CDK2 structure were identified. These are indicated in Figure 14 and correspond to residues 13-15 and 152-161 in the CDK2 monomer. The movements described here were more prominent for the CDK2-cyclin A-ATP, CDK2-staurosporine and CDK2-olomoucine complexes. As described earlier, it is the binding of cyclin A to CDK2 which results in substantial conformational changes in the kinase active site [54-55]. In the case of the staurosporine inhibitor, binding to the CDK2 active site causes a slight relative opening of the N- and C- terminal domains [55]. Overall, the structural conformational changes detected using the $C\alpha$-difference distance matrix approach are attributed to the domain motions between domains 1 and 2 within the CDK2 structure, as well as the two flexible hinges.

Cyclooxygenases and Associated Domain Motions

Cyclooxygenases (COXs) are bifunctional enzymes that catalyse the first committed step in prostanoid synthesis [61-62]. Two COX isoforms exist, COX-1 and COX-2. The expression pattern, mode of regulation and biological function of each one is different [63 and references therein]. COX-1 is constitutively expressed in a broad range of tissues and relays basic housekeeping messages throughout the body [63-64]. COX-2, which displays a radically different tissue distribution, is sensitive to induction in many tissues by specific physiological and pathological stimuli [63]. An increasing amount of evidence exists indicating the involvement of both isoforms in a range of pathologies. These include thrombosis (COX-1), and inflammation, pain and fever, various cancers, and neurological disorders such as Alzheimer's and Parkinson's diseases (COX-2) [61-62].

COX-1 and COX-2 are large homodimeric integral membrane proteins [61-62, 64]. Each COX monomer consists of three structural domains: an epidermal growth factor (EGF) domain of approximately 50 residues at the N-terminus, a membrane binding domain (MBD) of about 50 residues and a large C-terminal globular catalytic domain of about 460 residues [61-62]. The EGF domains form a portion of the dimer interface. It has been suggested that both EGF domains may play a role in the insertion of COX into the lipid bilayer [61]. The MBDs consist of four short, consecutive, amphipathic α-helices. The first three helices lie roughly in the same plane while the last D-helix bends up into the catalytic domain. Hydrophobic and aromatic residues protrude from all four helices creating a hydrophobic patch that interacts with only one face of the underlying lipid bilayer [61-62]. These helices also surround an opening which leads into a hydrophobic channel that extends into the core of the almost entire α-helical catalytic domain [61-64]. It is here that fatty acid substrates such as arachidonic acid (AA) and nonsteroidal anti-inflammatory drugs (NSAIDs) enter the enzyme [61-62].

Despite a high level of structural homology, the COX isoforms display major differences in primary structure. Both isoforms have distinctly different signal peptides which differ in length and amino acid composition. Substantial sequence differences are also found in the MBDs between the two isoforms. Lastly, an 18 amino acid insert that is six residues in from the C-terminus is observed in COX-2. The function of this insert is unknown, but sequence alterations in this region have been shown to markedly affect the expression of active protein [61-62].

Two distinct but structurally and functionally related active sites are found in the COX structure. The cyclooxygenase active site is located at the end of the hydrophobic channel that extends into the core of the protein. It is here that the cyclooxygenase reaction in which AA is converted to the prostaglandin G2 (PGG2) intermediate occurs [61-64]. The subsequent peroxidase reaction which involves the reduction of PGG2 to the hydroxy endoperoxide, prostaglandin H2 (PGH2), occurs in the peroxidase active site. This site is spatially distinct from the cyclooxygenase site in that it lies on the exterior of the protein, on the side of the molecule opposite from the MBD [61-64].

The AA metabolites derived from the cyclooxygenase and subsequently peroxidase reactions are powerful mediators of the inflammatory response. NSAIDs thus achieve their anti-inflammatory and antipyretic effects by blocking PGG2 formation via COX inhibition [61-62, 64]. Adverse effects such as peptic ulceration and dyspepsia have been associated with the use of classic NSAIDs (e.g. aspirin and ibuprofen). This is a result of the non-specificity of the inhibitors towards COX-1 and COX-2. The newer COX-2 selective NSAIDs (e.g. Celebrex® and Vioxx®) exhibit the necessary anti-inflammatory and antipyretic activities but with little or no gastrointestinal side-effects [61-62, 64]. The effectiveness of these inhibitors can be attributed to the exploitation of structural differences between the two COX isoforms. The substitution of two residues in COX-1, namely Ile434 and Ile523, with valine in COX-2 results in an approximate 20% increase in the COX-2 NSAID binding site

[62]. This effectively results in a reduction in steric and ionic crowding by Arg120 in COX-2 thereby enhancing the binding of nonacidic NSAIDs to COX-2 [62].

Table 7. Cyclooxygenase structures used for the computation of the difference distance matrixes. The PDB identification number of the structure, name of the ligand bound to the structure and RMS deviations from the reference structure.

	PDB code[a]	Ligand	Mutations	RMSD(Å)[b]
COX1	1diy	Arachidonic acid	-	0.60
	1eqg	Ibuprofen	-	0.40
	1pth	Salicylic acid	-	0.37
	2ayl	Flurbiprofen	-	0.40
COX2	1cvu	Arachidonic acid	H207A	0.57
	3pgh	Flurbiprofen	-	0.43

[a] For COX1 structural comparisons, the $C\alpha$-difference distance matrix approach was used to compare the test, ligand bound structures indicated with the reference unbound structure, 1prh. For COX2 structural comparisons, the $C\alpha$-difference distance matrix approach was used to compare the test ligand bound structures indicated, with the reference unbound structure, 5cox.

[b] The RMSD-values were calculated by superimposition of the $C\alpha$-atoms of the test PDB structure onto the reference structure and calculating the root-mean-square-deviation of the $C\alpha$-atoms from the reference structure. This was done using Swiss-PdbViewer [7].

The $C\alpha$-difference distance matrix approach was employed to investigate the domain motions occurring within the two COX isoforms. Five COX-1 and 3 COX-2 structures in various conformations were taken from the PDB (Table 7). Conformational differences between inhibitor bound and unbound structures were assessed. The RMSDs obtained following superimposition of the test structures onto the reference structures ranged from 0.37 Å to 0.60 Å for COX-1 and 0.43 Å to 0.57 Å for COX-2 (see Table 7). Minor conformational differences between the test and reference structures are indicated by these values. The $C\alpha$-difference distance matrix approach was thus employed to acquire a more detailed description as to these conformational differences.

Figure 15 shows the representative $C\alpha$-difference distance matrix for the COX-1 natural substrate bound and COX-1 unbound structural comparison. The matrix for the complete COX-1 homodimer is depicted in Figure 15 (A). The PDB structures reported for both COX isoforms displayed different start and end residues. Thus, only residues 33 to 583 of each COX subunit were compared in this study. Keeping this in mind, residues 1-551 and 552-1102 along either axis of the matrix correspond to each subunit of the dimer. These subunits can be discerned along the diagonal of the matrix and are delineated in Figure 15 (A) as 1 and 2. The domain motions within each COX-1 subunit are equivalent. Thus, the $C\alpha$-difference distance matrix for the single subunit was examined in more depth (see Figure 15 B). The domains here were identified as spatially connected regions with interatomic distances varying by less than 0.5 Å. Three domains comprising twenty subdomains were identified in the COX-1 monomer (see Figure 16). These are delineated in the matrix. Domain 1 of the COX-1 monomer is formed by residues 33-50, 54-82, 123-143, 174-180, 446-458, 464-495 and 503-509; domain 2 by residues 86-93, 97-103, 108-122 188-214, 220-234, 378-396, 430-434 and 521-535; and domain 3 by residues 238-319, 325-352, 409-427, 548-571and 576-583. Equivalent domains were identified in the structurally homologous COX-2 structures.

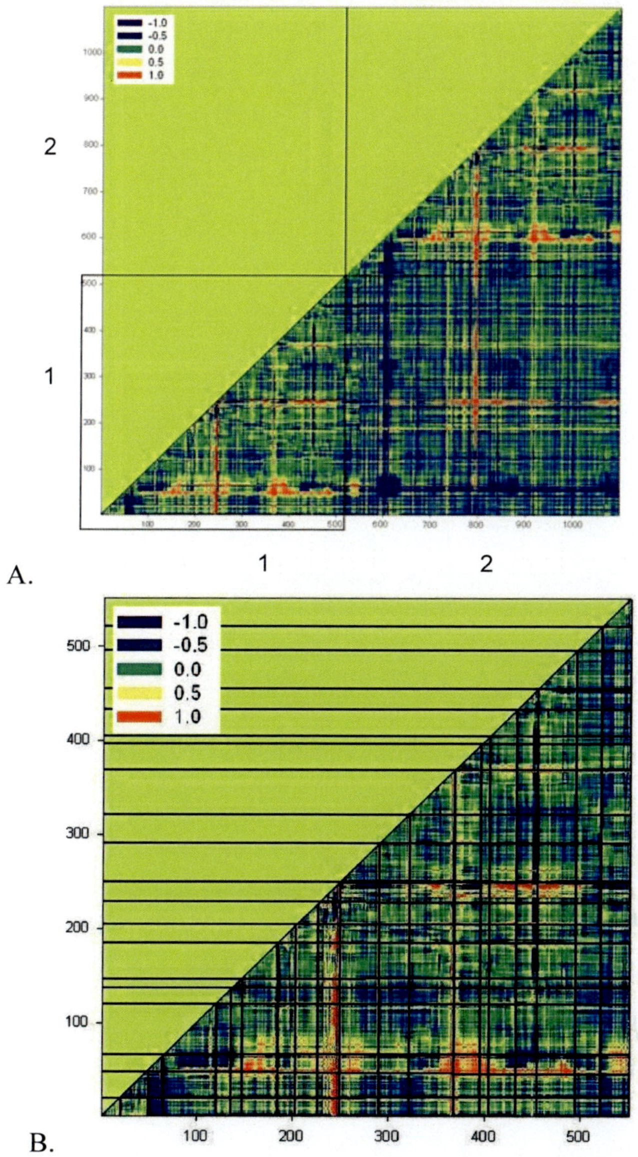

Figure 15. *Cα*-difference distance matrix between the arachidonic acid bound (1diy) and unbound (1prh) COX-1 structures. (A) Corresponds to the complete COX-1 dimer: residues 1-551 and 552-1102 along either axis of the matrix correspond to each subunit of the dimer. The boxed area in the graph, representative of a COX-1 monomer, is amplified in (B) to illustrate the domain and subdomain movements occurring within each COX-1 monomer.

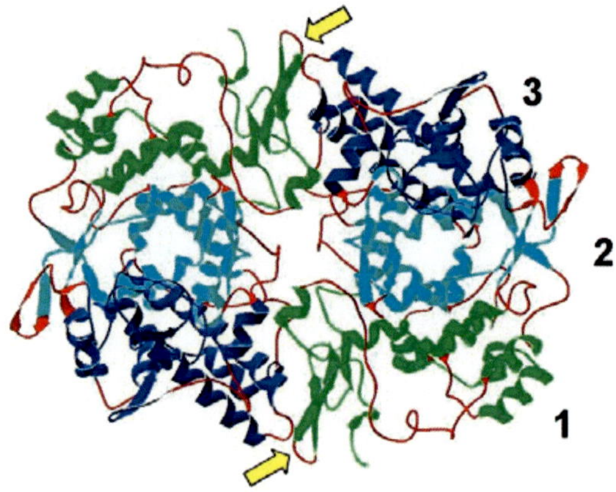

Figure 16. Domain motions of the COX homodimer. Each monomer consists of three domains made up of twenty subdomains: domain 1 (33-50; 54-82; 123-143; 174-180; 446-458; 464-495; 503-509) shaded in green; domain 2 (86-93; 97-103; 108-122; 188-214; 220-234; 378-396; 430-434; 521-535) shaded in teal; and domain 3 (238-319; 325-352; 409-427; 548-571; 576-583) shaded in blue. Residues comprising the interdomain interfaces are shaded in red. The most flexible hinges in the cyclooxygenase monomer are indicated with yellow arrows (Figure generated using Swiss-PdbViewer [7]).

The $C\alpha$-difference distance matrixes in Figure 15 indicate small but significant domain motions between the bound and unbound COX-1 structures. As everywhere in this work, these motions are represented in the matrixes by distance differences greater than 0.5 Å. In Figure 15 (A), the red regions in the square quadrant of the matrix correspond to movements predominantly within domain 3 of the equivalent monomers in the structural pair under investigation. An α-helix (residues 235-249 in COX-1) located on domain 3 is observed to be particularly mobile with respect to the rest of the structure. This particular movement was most noticeable in the matrix presented here. Analysis of Figure 15 (B) indicated small movements between several subdomains within the COX-1 monomer. Two predominant motions were recorded here: (1) between subdomain 3 (residues 86-93 in COX-1) and subdomains 7, 11, 13-18 and (2) between subdomain 10 (residues 238-319 in COX-1). This in effect corresponds to movements within/between all subdomains in each monomer. A flexible hinge was also identified in the COX-1 monomer. The hinges, indicated by yellow arrows in Figure 16, are located at the dimer interface of COX-1. Again the movements described here were more predominant for this structural comparison.

Overall, the moving fragments detected using the $C\alpha$-difference distance matrix approach identified three domains in the COX isoforms. These domains do not coincide with the structural domains reported in the literature. Furthermore, the mobile α-helix in domain 3 identified here has no reported function.

Conclusion

The difference distance matrix based method employed in this study allowed identification of domain and subdomain movements upon substrate and inhibitor binding. In the case of the HIV-1 protease, some drug resistant mutations are easily explainable by the proximity of the mutated residues to the inhibitors. Majority of the mutations outside of the HIV-1 protease binding site are situated in the interdomain interfaces, probably influencing domain flexibility and domain movements in the protease mutants. For the other families investigated, a detailed description of the conformational differences occurring between members of the same family was obtained thus providing insight into their modes of action.

In general, we found that the difference distance matrix based comparison of protein structures allows for observation of subtle atom shifts and is far superior to RMSD-based comparison. This method can aid in the understanding of the mechanism of action of proteins and their targets. In the case of the HIV-1 protease, the analysis of the domain motions associated with substrate or inhibitor binding may provide a useful strategy for designing new mutation resistant HIV-1 protease inhibitors.

Acknowledgements

We thank Fundação para a Ciência e Tecnologia (Portugal) for Grant SFRH/BD/24547/2005 attributed to C.S.A. and SFRH/BPD/20716/2004 attributed to V.K.

References

[1] Holm, L. & Sander, C. (1993) Protein structure comparison by alignment of distance matrices. *J Mol Biol.* **5**, 123-138.

[2] Rackovsky, S. & Goldstein, D. A. (1988) Protein comparison and classification: a differential geometric approach. *Proc. Natl. Acad. Sci. USA* **85**, 777-781.

[3] Richards, F. M. & Kundrot, C. E. (1988) Identification of structural motifs from protein coordinate data: secondary structure and first-level supersecondary structure. *Proteins* **3**, 71-84.

[4] Schneider, T. S. (2000) Objective comparison of protein structures: error-scaled difference distance matrices. *Acta Cryst. D* **56**, 714-721.

[5] Rose, R. B., Craik, C. S. & Stroud, R. M. (1998) Domain flexibility in retroviral proteases: structural implications for drug resistant mutations. *Biochemistry* **37** 2607-2621.

[6] Vondrasek, J., van Buskirk, C. P. & Wlodawer, A. (1997) Database of three dimensional structures of HIV proteinases. *Nature Struct. Biol.* **4**, 8.

[7] Guex, N. & Peitsch, M. C. (1997) SWISS-MODEL and the Swiss-PdbViewer: an environment for comparative protein modelling. *Electrophoresis.* **18**, 2714-2723.

[8] Laskowski, R. A., Chistyakov, V. V. & Thornton, J. M. (2005). PDBsum more: new summaries and analyses of the known 3D structures of proteins and nucleic acids. *Nucleic Acids Res.* **33**, D266-D268.

[9] Spinelli, S., Liu, Q. Z., Alzari, P. M., Hirel, P. H. & Poljak, R. J. (1991) The three-dimensional structure of the aspartyl protease from the HIV-1 isolate BRU. *Biochimie* **73**, 1391-1396.

[10] Wlodawer, A. & Erickson, J. W. (1993) Structure-based inhibitors of HIV-1 protease. *Annu. Rev. Biochem.* **62**, 543-585.

[11]Prabu-Jeyabalan, M., Nalivaika, E. & Schiffer, C. A. (2000) How does a symmetric dimer recognize an asymmetric substrate? A substrate complex of HIV-1 protease. *J. Mol. Biol.* **301**, 1207-1220.

[12]Wlodawer, A. & Vondrasek, J. (1998) Inhibitors of HIV-1 protease: a major success of structure-assisted drug design. *Annu. Rev. Biophys. Biomol. Struct.* **27**, 249-284.

[13]Surleraux, D. L., de Kock, H. A., Verschueren, W. G., Pille, G. M., Maes, L. J., Peeters, A., Vendeville, S., De Meyer, S., Azijn, H., Pauwels, R., de Bethune, M. P., King, N. M., Prabu-Jeyabalan, M., Schiffer, C. A. & Wigerinck, P. B. (2005) Design of HIV-1 protease inhibitors active on multidrug-resistant virus. *J Med Chem.* **48**, 1965-1973.

[14]Kurt, N., Haliloglu, T. & Schiffer, C. A. (2003) Structure-based prediction of potential binding and nonbinding peptides to HIV-1 protease. *Biophys J.* **85**, 853-863.

[15]Nauchitel, V., Villaverde, M. C. & Sussman, F. (1995) Solvent accessibility as a predictive tool for the free energy of inhibitor binding to the HIV-1 protease. *Protein Sci.* **4**, 1356-1364.

[16]King, N. M., Prabu-Jeyabalan, M., Nalivaika, E. & Schiffer, C. A. (2004) Combating susceptibility to drug resistance: lessons from HIV-1 protease. *Chem Biol.* **11** 1333-1338.

[17]Temesgen, Z. (2001) Current status of antiretroviral therapies. *Expert Opin. Pharmacother.* **2**, 1239-1246.

[18]Rodríguez-Barrios, F. & Gago, F. (2004) HIV protease inhibition: limited recent progress and advances in understanding current pitfalls. *Curr Top Med Chem.* **4**, 991-1007.

[19]Johnson, V. A., Brun-Vézinet, F., Clotet, B., Kuritzkes, D. R., Pillay, D., Schapiro, J. M. & Richman, D. D. (2006) Update of the drug resistance mutations in HIV-1: Fall 2006. *Special Contribution - Resistance Mutations* **14**, 125-130.

[20]Roberts, N. A. (1995) Drug-resistance patterns of saquinavir and other HIV proteinase inhibitors. *AIDS* **9**, S27-S32.

[21]Markowitz, M. M., Mo, H., Kempf, D. J., Norbeck, D. W., Bhat, T. N., Erickson, J. W. & Ho, D. D. (1995) Selection and analysis of HIV-1 variants with increased resistance to ABT-538, a novel protease inhibitor. *J. Virol.* **69**, 701-706.

[22]Nar, H., Bauer, M., Schmid, A., Stassen, J.- M., Wienen, W., Priepke, H. W. M., Kauffmann, I. K., Ries, U. J. & Hauel, N. H. (2001) Structural basis for inhibition promiscuity of dual specific thrombin and factor Xa blood coagulation inhibitors. *Structure* **9**, 29-37.

[23]Matter, H., Defossa, E., Heinelt, U., Blohm, P.- M., Schneider, D., Müller, A., Herok, S., Schreuder, H., Liesum, A., Brachvogel, V., Lönze, P., Walser, A., Al-Obeidi, F. & Wildgoose, P. (2002) Design and quantitative structure-activity relationship of 3-amidinobenzyl-1H-indole-2-carboxamides as potent, nonchiral, and selective inhibitors of blood coagulation factor Xa. *J. Med. Chem.* **45**, 2749-2769.

[24]Venkateswarlu, D., Perera, L., Darden, T. & Pedersen, L. G. (2002) Structure and dynamics of zymogen human blood coagulation factor X. *Biophys J.* **82**, 1190-1206.

[25]Maignan, S., Guilloteau, J.- P., Choi-Sledeski, Y. M., Becker, M. R., Ewing, W. R., Pauls, H. W., Spada, A. P. & Mikol, V. (2003) Molecular structures of human factor Xa complexed with ketopiperazine inhibitors: preference for a neutral group in the S1 pocket. *J. Med. Chem.* **46**, 685-690.

[26]Thiec, F., Cherel, G. & Christophe, O. D. (2003) Role of the Gla and first EGF-like domains of factor X in the prothrombinase and tissue factor-factor VIIa complexes. *J Biol Chem.* **278**, 10393-10399.

[27]Yavin, Y. Y., Wolozinsky, M. & Cohen, A. T. (2005) New antithrombotics in the prevention of thromboembolic disease. *Eur. J. Intern. Med.* **16**, 257-266.

[28]Kamata, K., Kawamoto, H., Honma, T., Iwama, T., & Kim, S.- H. (1998) Structural basis for chemical inhibition of human blood coagulation factor Xa. *Proc. Natl. Acad. Sci. USA* **95**, 6630-6635.

[29]Padmanabhan, K., Padmanabhan, K. P., Tulinsky, A., Park, C. H., Bode, W., Huber, R., Blankenship, D. T., Cardin, A. D. & Kisiel, W. (1993) Structure of human des(1-45) factor Xa at 2.2 Å resolution. *J. Mol. Biol.* **232**, 947-966.

[30]Brandstetter, H., Kuhne, A., Bode, W., Huber, R., von der Saal, W., Wirthensohn, K. & Engh, R. A. (1996) X-ray structure of active site-inhibited clotting factor Xa. Implications for drug design and substrate recognition. *J. Biol. Chem.* **271** 29988-29992.

[31]Agre, P., Brown, D. & Nielsen, S. (1995) Aquaporin water channels: unanswered questions and unresolved controversies. *Curr. Opin. Cell Biol.* **7**, 472-483.

[32]Verkman, A. S., van Hoek, A. N., Ma, T., Frigeri, A., Skach, W. R., Mitra, A., Tamarappoo, B. K. & Farinas, J. Water transport across mammalian cell membranes. (1996) *Am. J. Physiol.* **270**, C12-C30.

[33]Agre, P., King, L. S., Yasui, M., Guggino, W. B., Ottersen, O. P., Fujiyoshi, Y., Engel, A. & Nielsen, S. (2002) Aquaporin water channels - from atomic structure to clinical medicine. *J. Physiol.* **542**, 3-16.

[34]Borgnia, M., Nielsen, S., Engel, A. & Agre, P. (1999) Cellular and molecular biology of the aquaporin water channels. *Annu. Rev. Biochem.* **68**, 425-458.

[35]De Groot, B. L., Engel, A. & Grubmüller, H. (2001) A refined structure of human aquaporin-1. *FEBS Letters* **504**, 206-211.

[36]Sui, H., Han, B.- G., Lee, J. K. Walian, P. & Jap, B. K. (2001) Structural basis of water-specific transport through the AQP1 water channel. *Nature* **414**, 872-878.

[37]Murata, K., Mitsuoka, K., Hirai, T., Walz, T., Agrek, P., Heymann, J. B., Engel, A. & Fujiyoshi, Y. (2000) Structural determinants of water permeation through aquaporin-1. *Nature* **407**, 599-605.

[38]Ren, G., Reddy, V. S., Cheng, A., Melnyk, P. & Mitra, A. K. (2001) Visualization of a water-selective pore by electron crystallography in vitreous ice. *Proc. Natl. Acad. Sci. USA* **98**, 1398-1403.

[39]Savage, D. F., Egeal, P. F., Robles-Colmenares, Y., O'Connell III, J. D. & Stroud, R. M. (2003) Architecture and selectivity in aquaporins: 2.5 Å X-ray structure of aquaporin Z. *PLoS Biology* **1**, 334-340.

[40]Nielsen, S., Frøkiær, J., Marples, D., Kwon, T.- H., Agre, P. & Knepper, M. A. (2002) Aquaporins in the kidney: from molecules to medicine. *Physiol Rev* **82**, 205-244.

[41]Harries, W. E. C., Akhavan, D., Miercke, L. J. W., Khademi, S. & Stroud, R. M. (2004) The channel architecture of aquaporin 0 at a 2.2-Å resolution. *Proc. Natl. Acad. Sci. USA* **101**, 14045-14050.

[42]Dutzler, R., Campbell, E. B., Cadene, M., Chait, B. T. & MacKinnon, R. (2002) X-ray structure of a ClC chloride channel at 3.0 Å reveals the molecular basis of anion selectivity. *Nature* **415**, 287-294.

[43]Lobet, S. & Dutzler, R. (2006) Ion-binding properties of the ClC chloride selectivity filter. *EMBO J* **25**, 24-33.

[44]Dutzler, R., Campbell, E. B. & MacKinnon, R. (2003) Gating the selectivity filter in ClC chloride channels. *Science* **300**, 108-112.

[45]Steinmeyer, K., Lorenz, C., Pusch, M., Koch, M. C. & Jentsch, T. J. (1994) Multimeric structure of ClC-1 chloride channel revealed by mutations in dominant myotonia congenita. *EMBO J.* **15**, 737-743.

[46]Maduke, M., Miller, C. & Mindell, J. A. (2000) A decade of ClC chloride channels: structure, mechanism and many unsettled questions. *Annu Rev Biophys Biomol Struct* **29**, 411-438.

[47]Jentsch, T. J., Stein, V., Weinreich, F. & Zdebik, A. A. (2002) Molecular structure and physiological function of chloride channels. *Physiol. Rev.* **82**, 503-568.

[48]Mindell, J. A. & Maduke, M. (2001) ClC chloride channel. *Genome Biology* **2**, 3003.1-3003.6.

[49]Matulef, K. & Maduke, M. (2007) The ClC 'chloride channel' family: revelations from prokaryotes. (Review) *Molecular Membrane Biology* **24**, 342-350.

[50]Stirling, D. A., Welch, K. A. & Stark, M. J. R. (1994) Interaction with calmodulin is required for the function of Spc110p, an essential component of the yeast spindle pole body. *EMBO J.* **13**, 4329-4342.

[51]Wriggers, W., Mehler, E., Pitici, F., Weinstein, H. & Schulten, K. (1998) Structure and dynamics of calmodulin in solution. *Biophys J.* **74**, 1622-1639.

[52]Vetter, S. W. & Leclerc, E. (2003) Novel aspects of calmodulin target recognition and activation. *Eur. J. Biochem.* **270**, 404-414.

[53]Fallon, J. L., Halling, D. B., Hamilton, S. L. & Quicho, F. A. (2005) Structure of calmodulin bound to the hydrophobic IQ domain of the cardiac $Ca_v1.2$ calcium channel. *Structure* **13**, 1881-1886.

[54]Morgan, D. O. (1997) Cyclin-dependent kinases: engines, clocks and microprocessors. *Annu. Rev. Cell Dev. Biol.* **13**, 261-291.

[55]Noble, M. E. M. & Endicott, J. A. (1999) Chemical inhibitors of cyclin-dependent kinases: insights into design from X-ray crystallographic studies. *Pharamcol. Ther.* **82**, 269-278.

[56]Hardcastle, I. R., Golding, B. T. & Griffin, R. J. (2002) Designing inhibitors of cyclin-dependent kinases. *Annu. Rev. Pharmacol. Toxicol.* **42**, 325-348.

[57]Johnson, D. G. & Walker, C. L. (1999) Cyclins and cell cycle checkpoints. *Annu. Rev. Pharmacol. Toxicol.* **39**, 295-312.

[58]Báratová, I., Otyepka, M., Kříž, Z., Koča, J. (2004) Activation and inhibition of cyclin-dependent kinase-2 by phosphorylation; a molecular dynamics study reveals the functional importance of the glycine-rich loop. *Protein Sci.* **13**, 1449-1457.

[59]Knockaert, M., Gray, N., Damiens, E., Chang, Y.- T., Grellier, P., Grant, K., Fergusson, D., Mottram, J., Soete, M., Dubremetz, J.- F., Le Roch, K., Doerig, C., Schultz, P. G. & Meijer, L. (2000) Intracellular targets of cyclin-dependent kinase inhibitors: identification by affinity chromatography using immobilised inhibitors. *Chem. Biol.* **7**, 411-422.

[60]Wu, S. Y., McNae, I., Kontopidis, G., McClue, S. J., McInnes, C., Stewart, K. J., Wang, S., Zheleva, D. I., Marriage, H., Lane, D. P., Taylor, P., Fischer, P. M. & Walkinshaw, M. D. (2003) Discovery of a novel family of CDK inhibitors with the program

LIDAEUS: structural basis for ligand-induced disordering of the activation loop. *Structure*. **11**, 399-410.

[61]Smith, W. L., DeWitt, D. L. & Garavito, R. M. (2000) Cyclooxygenases: structural, cellular, and molecular biology. *Annu. Rev. Biochem*. **69**, 145-182.

[62]Garavito, R. M. & Mulichak, A. M. (2003) The structure of mammalian cyclooxygenases. *Annu. Rev. Biomol. Struct*. **32**, 183-206.

[63]Rowlinson, S. W., Kiefer, J. R., Prusakiewicz, J. J., Pawlitz, J. L., Kozak, K. R., Kalgutkar, A. S., Stallings, W. C. & Kurumbail, R. G. (2003) A novel mechanism of cyclooxygenase-2 inhibition involving interactions with Ser-530 and Tyr-385. *J. Biol. Chem*. **46**, 45763-45769.

[64]Selinsky, B. S., Gupta, K., Sharkey, C. T. & Loll, P. J. (2001) Structural analysis of NSAID binding by prostaglandin H2 synthase: time-dependent and time-independent inhibitors elicit identical enzyme conformations. *Biochemistry*. **40**, 5172-5180.

In: Protein Conformation: New Research
Editor: Linda B. Roswell, pp. 201-210
ISBN: 978-1-60456-219-4

Chapter 5

Hierarchical Order of Initiator Protein Assembly in Pre-replication Complex

Kenichi Yoshida*

Department of Life Sciences, Meiji University School of Agriculture, 1-1-1 Higashimita, Tama-ku, Kawasaki, Kanagawa 214-8571, Japan

Abstract

In eukaryotes, the initiation of DNA replication is a fundamental process that occurs prior to cell division, assuring the transfer of genomic information. For replication to occur, dozens of proteins must gather in a hierarchical and timely manner to establish the pre-replication complex (pre-RC) at DNA sites where DNA replication occurs. These proteins eventually form large replisome progression complexes. The six subunits of the origin recognition complex (ORC) serve as landmarks for the initiation of pre-RC assembly on the DNA. Together with CDC6 and CDT1, the ORC recruits six mini-chromosome maintenance proteins. These factors are essential for proper DNA replication initiation and licensing by the central unit to form replisomes. Regulators of pre-RC formation, such as kinases and a unique protein called geminin, are important topics of research. For the initiation of DNA replication to progress to elongation, several other factors including CDC45, GINS, and replication protein A are required. Once the protein assembly process of the DNA replication apparatus is understood in greater detail, its regulation could be tweaked to enable the molecular dissection of other cellular functional apparatuses, such as DNA-protein interaction-based transcriptional regulation and protein-protein interaction-based signaling pathways, eventually leading to a greater understanding of complicated cellular systems at a molecular level. In this review article, we summarize recent knowledge of the protein assembly process that occurs during the metazoan initiation of DNA replication and discuss future perspectives.

Introduction

For a protein to become functional, it must be localized at a site where the synthesized protein can work and it must be able to physically associate with its binding partners. In the case of

* E-mail address: yoshida@isc.meiji.ac.jp, Tel. & Fax.: +81-44-934-7107.

the initiation of eukaryotic DNA replication, the origin recognition complex (ORC) serves as a landmark for the initiation of DNA replication. Once DNA binds to the ORC, the ORC recruits helicases, polymerases, and accessory cofactors. This ordered process, which depends on the cell cycle, ensures that the genome is accurately duplicated once per cell cycle. DNA replication initiation has best been characterized in yeasts, and most of the involved processes are highly conserved among eukaryotes [1]. In yeast, autonomously replicating sequences (ARS), where the ORC binds and initiates DNA replication, have been well characterized [2]. ARS contain an 11-bp ARS consensus sequence (ACS) that is required for the binding of the ORC. Surprisingly, although many biochemical and genetic experiments have been performed in metazoan systems, the essential DNA sequences or sequences that are homologous to yeast ARS have never been defined in the metazoan genome. Although mammalian replication initiation sites have been reported, the initiation zones had a low sequence preference and were not specific DNA sequences [3]. The identified specific zones include the 5' region of the human c-myc gene containing a DNA unwinding element [4], the human lamin B2 gene [5], and the INK4/ARF locus [6] as putative origins of DNA replication. Metazoan ORC complexes may initiate DNA replication independent of a specific sequence [7-9], suggesting that protein complex formation and degradation processes that depend on the cell cycle should be noteworthy with regard to determining the initiation sites of DNA replication. In other words, tightly regulated assemblies of the DNA replication initiation apparatus recognize the DNA topology or chromatin context, resulting in the determination of initiation zones. Unlike the situation for replication, much is known about the initiation site of transcription. Recent studies have suggested the possible involvement of DNA replication initiation factors, especially the ORC, in the regulation of the heterochromatin topology, arguing a link between DNA replication initiation and transcriptional regulation.

In this review, we discuss the current architectural concept of the metazoan, especially mammalian, DNA replication initiation apparatus. When considering the mechanisms of multiple protein complexes involving processes that are centered on DNA, one may speculate that the formation of multiple protein complexes is not only regulated by protein-protein interactions, but is also affected by key molecules like nucleotides or their derivatives.

ORC as a Landmark for Initiator Assembly

The hetero-hexameric ORC is well known for its role in localizing the origin of DNA replication. In eukaryotes, homologues of all six ORC subunits (ORC1-6) have been identified in all eukaryotic species examined, including humans [10,11]. Uniquely, in mammalian cells, the amount of chromatin-bound ORC2-6 remains constant throughout the cell cycle, although the interactions between ORC1 and the other ORC subunits can vary [11]. This observation may explain why ORC1 can suffer from various post-translational modifications, including ubiquitination, degradation, and phosphorylation. These findings indicate that the status of ORC1 determines the timing when the pre-RC is established to initiate DNA replication. The interaction between mammalian ORC1 and the ORC core assembly, especially whether the assembly depends on the cell cycle, remains controversial [11].

Localization/interaction experiments have also implicated the ORC in ribosomal biogenesis and in centrosome and kinetochore function [12]. The ORC is known to interact with heterochromatin protein 1 (HP1) [13] and to participate in the establishment of transcriptionally silent chromatin domains. HP1 is an essential heterochromatin-associated protein known to play an important role in the organization of heterochromatin as well as in transcriptional regulation [14]. ORC1 recruited by HP1 associates with heterochromatin, and this requires a functional ATP binding site and a protein region partially overlapping the bromo-adjacent homology domain at the N-terminus of ORC1. ORC2 also localizes to the centrosomes throughout the entire cell cycle and is tightly bound to heterochromatin and HP1 during the G1 and early S phases [15]. Heterochromatin has been suggested to recruit the ORC to facilitate the replication of heterochromatic regions, where replication hardly occurs [16]. To support this notion, CDC6, a replication initiator, also interacts with HP1 [17]. One may speculate that transcription and DNA replication are functionally connected because both of these events occur on DNA.

Various experiments have suggested three possible intermediary states during the process of mammalian ORC assembly. In the first model, ORC2-5 forms the core complex and ORC1 and ORC6 become weakly associated with it [18-22]. In the second model, ORC1 is involved in the ORC2- 5 core complex, resulting in a less stable complex that can assemble the pre-RC and can also start DNA replication [19,23,24]. In the third model, which remains hypothetical, the complete assembly of ORC1-6 exists within the cells. The above evidence suggests that ORC6 exists in the mammalian genome, but that all of the ORC subunits, including ORC6, are not required for functional ORC complexes. Indeed, ORC6 neither interacts with the other ORC subunits nor is required for the initiation of DNA replication [18-21]. Accumulating evidence suggests that ORC6 is not an essential component of the initiator, but is rather involved in other cellular machinery. Indeed, human ORC6 has been reported to be an essential gene that coordinates chromosome replication and segregation during cytokinesis [25].

ORC recognizes specific DNA sequences in an ATP-dependent manner and serves as a scaffold for the assembly of other initiation factors [9,21]. Interestingly, ORC1, 4, and 5 are known to contain AAA+ ATPase domains, as does CDC6 [26]. In the case of human ORC, which can actually bind to DNA without the involvement of ATP, the presence of ATP increases the efficiency of DNA binding [9]. Human ORC1-5 possess ATPase activity, whereas a mutated Walker A motif in ORC1, but not in the ORC4 or ORC5 subunits, showed no ATPase activity. On the other hand, a DNA replication assay using Xenopus laevis revealed that the Walker A motifs of the ORC1, ORC4, and ORC5 subunits were all essential for replication activity [19]. The ATPase activity of ORC1 may be required for other cellular functions or to recruit unidentified factors into the pre-RC.

MCM Plays a Central Role in Initiation Licensing

As with the ORC, the MCM complex includes a family of six structurally related proteins, MCM2-7, which form a hetero-hexameric complex. MCM2-7 proteins are key licensing factors regulated by cell cycle-dependent protein kinases; this method of regulation assures DNA replication once per cell cycle [27]. MCM proteins are a distinct subgroup of the large AAA ATPase family and share a common sequence in their nucleotide-binding domains [28].

Equal numbers of each MCM molecule are known to be gathered into a ring-shaped structure that surrounds the DNA, with the DNA passing through its central channel [29-31]. The dimeric complex of MCM subunits 4, 6, and 7 has an intrinsic ATP-dependent DNA helicase activity [32] [33]. One can simply imagine that the improper assembly of DNA replication initiators results in a lack of genome stability and accelerated cell proliferation. Either over-expression or under-expression could cause such a situation. MCM4 hypomorphic mutations induce the destabilization of the MCM2-7 complex and impair DNA replication, suggesting that hypomorphic mutations in any of the MCM2-7 genes may increase the risk of cancer in humans [34]. MCM7 has been reported as a useful diagnostic marker in cerebral glioblastoma [35]. Moreover, MCM7 is upregulated in a variety of tumors, including prostate, cervical and hypopharyngeal carcinomas [36]. In addition to its role in carcinogenesis, MCM is also a marker of embryogenesis [37].

To become a pre-RC by incorporating MCM2-7, ORC should bind directly to the replication origin and recruit CDC6 and CDT1 [38]. The detailed scenario is as follows: ORC requires CDC6 to target replication initiation at specific DNA sequences in chromosomes; together with CDT1, it loads the ring-shaped MCM2-7 DNA helicase onto DNA [39-41]. In addition to MCM, ORC and CDC6 combine to form a ring-shaped complex that contains six AAA+ subunits. This activity is essential for MCM loading and results in a stable ORC, CDC6, and DNA complex that actually promotes MCM loading [42]. ATP hydrolysis by ORC is required for multiple rounds of MCM2-7 loading. CDC6 is preferentially bound to ORC1 and ORC2, indicating that CDC6 connects ORC1 and ORC2-5 via ORC2. CDC6 also binds to MCM7 proteins. CDC6 is a member of the AAA+ ATPase family and exhibits significant sequence homology to the eukaryotic clamp loader, replication factor C [43,44]. Taken together, these observations suggest that CDC6 may act as a scaffold clamp to sustain the pre-RC architecture, acting mainly as a clamp loader for the MCM. Another member of the MCM protein family, MCM8, is involved in the pre-RC by associating with CDC6 [45]. Further protein-protein interaction assays regarding whether an assembled protein complex is required for proper DNA unwinding are required. A recent study reported that caspase-3-mediated cleavage of CDC6 yielded dominant-negative inhibitors of replication initiation that can induce DNA damage [46], indicating that post-translational protein processing dramatically transforms the protein's properties.

Crucial Determinants of the Initiator Assembly

Other than ATP utilization, cyclin-dependent kinase levels, protein turnover, and the CDT1/geminin balance are known to regulate the activity of pre-RC [47,48]. Histone acetylation is also involved in the regulation of pre-RC. HBO1, a member of the MYST histone acetyltransferase family, inhibited the association of MCM2-7 with chromatin, even though ORC and CDC6 loading were normal [49]. This implies that the assembly of pre-RC is tightly regulated by the DNA topology as well as by trans-acting factors. Besides the stepwise establishment of pre-RC at the DNA replication origins in the G1 phase of the cell cycle, the activation of two S phase-promoting kinases, cyclin-dependent kinases (CDKs)/cyclins and CDC7/DBF4, during the G1/S transition is important [1,50,51]. The downstream targets of S phase-promoting kinases are thought to play a pivotal role in the initiation of DNA replication. Indeed, MCM proteins are downstream targets of CDC7/DBF4 phosphorylation and are known to be essential for the initiation of DNA replication in mammalian cells [52,53]. CDK phosphorylation of CDC6 stabilizes it by preventing its association with the anaphase promoting complex/cyclosome (APC/C) [54]. Genetic studies have shown that CDC7/DBF4 interacts with the subunits of ORC as well as with MCM proteins. The pre-RC is activated by CDKs and CDC7/DBF4 kinase. CDC7/DBF4 can phosphorylate the MCM4 protein, and this phosphorylation could be important in the loading of CDC45 onto the chromatin [55]. Besides CDKs/cyclins and CDC7/DBF4, ORC2 and the MCM subunits also interact with the polo-like protein kinase, PLK1 [56].

To establish the pre-RC, CDT1 is also required between the sequential assembly of CDC6 and MCM2-7 onto the origin regions [57]. The overexpression of CDT1 induces chromosomal instability in normal human cells. The accumulation of CDT1 causes multiple rounds of DNA replication without an intervening mitosis. Geminin is an inhibitor of CDT1 and is important for preventing inappropriate origin firing [58,59]. Thus, the level of CDT1 periodically fluctuates because of ubiquitination and proteolysis [60]. The binding of CDT1 to proliferating cell nuclear antigen (PCNA) facilitates CDT1 degradation by E3 ubiquitin ligase [61]. Xenopus laevis and mammalian cells have a unique protein, called geminin, that strictly controls licensing by binding to Cdt1 and inhibiting its function [62]. This inhibits the chromatin loading of the MCM complex. Geminin is destroyed at the end of mitosis, and this triggers a new pre-RC formation [63]. Similar to CDC6, geminin is cleaved by caspase-3, and the truncated forms of geminin can promote apoptosis [64].

When thinking about protein function, it is helpful to analyze the protein structure. Geminin has been analyzed at the molecular level; it comprises a tetramer formed by two dimers with monomers interacting via coiled-coil domains [65]. Oligomerization properties mediated by the coiled-coil domains of geminin have been implicated in the inhibition of DNA replication [66]. Other experiments suggest that geminin forms a homodimer, with each dimer binding one molecule of CDT1 [67]. A crystal structure analysis revealed that the N-terminal region of geminin may anchor CDT1, and the C-terminal region of geminin prevents the MCM complex's access to CDT1 [68]. Geminin also functions as a transcriptional repressor of the Hox gene [69]. This is reminiscent to the case of ORC, which participates in the regulation of heterochromatin. Different parts of the coiled-coil domain of geminin are required for its interaction with HoxA11 [70].

From the establishment of the pre-RC to the large replisome progression complexes, a four-subunit complex called GINS associates with MCM helicase at replication forks [71]. The integrity of helicase activity is regulated by a macromolecular structure, Cdc45-MCM-GINS complex, during S phase [72]. Searching for the roles of GINS in tumor biology may be promising [73]. CDC45 interacts with ORC1 and ORC6 and weakly with MCM3, 6, and 7, indicating that CDC45 integrates the DNA replication initiation machinery [22].

Conclusion

The assembly of the pre-RC complex is strictly regulated, beginning with ORC and followed by the recruitment of CDC6, CDT1, and MCM. For this to occur, cell cycle regulation upstream of pre-RC formation, such as S phase CDKs, are likely to play a crucial role. The phosphorylation status of each component of the pre-RC should be unveiled by comprehensive proteome analysis. Because the pre-RC is formed directly on the DNA, DNA modifications and histones probably affect the efficiency of protein assembly. Indeed, the timing of initiation of DNA replication at each origin is different. DNA replication initiation is a fundamental process for all organisms, but higher eukaryotes have evolved unique genes like geminin. Although the minimal common elements required for the initiation of DNA replication have been defined in yeast, they have not been accurately defined in mammalian genome sequences. Obviously, the genomes of mammals are much larger than the yeast genome. Therefore, more detailed regulatory systems are likely to be required for protein-protein interactions as well as DNA-protein interactions.

References

[1] Bell, S.P., and Dutta, A. (2002). DNA replication in eukaryotic cells. *Annu Rev Biochem* **71**, 333-374.

[2] Bell, S.P., and Stillman, B. (1992). ATP-dependent recognition of eukaryotic origins of DNA replication by a multiprotein complex. *Nature* **357**, 128-134.

[3] Todorovic, V., Falaschi, A., and Giacca, M. (1999). Replication origins of mammalian chromosomes: the happy few. *Front Biosci* **4**, D859-D868.

[4] Ghosh, M., Kemp, M., Liu, G., Ritzi, M., Schepers, A., and Leffak, M. (2006). Differential binding of replication proteins across the human c-myc replicator. *Mol Cell Biol* **26**, 5270-5283.

[5] Abdurashidova, G., Danailov, M.B., Ochem, A., Triolo, G., Djeliova, V., Radulescu, S., Vindigni, A., Riva, S., and Falaschi, A. (2003). Localization of proteins bound to a replication origin of human DNA along the cell cycle. *EMBO J* **22**, 4294-4303.

[6] Gonzalez, S., Klatt, P., Delgado, S., Conde, E., Lopez-Rios, F., Sanchez-Cespedes, M., Mendez, J., Antequera, F., and Serrano, M. (2006). Oncogenic activity of Cdc6 through repression of the INK4/ARF locus. *Nature* **440**, 702-706.

[7] Cvetic, C., and Walter, J.C. (2005). Eukaryotic origins of DNA replication: could you please be more specific? *Semin Cell Dev Biol* **16**, 343-353.

[8] Remus, D., Beall, E.L., and Botchan, M.R. (2004). DNA topology, not DNA sequence, is a critical determinant for Drosophila ORC-DNA binding. *EMBO J* **23**, 897-907.

[9] Vashee, S., Cvetic, C., Lu, W., Simancek, P., Kelly, T.J., and Walter, J.C. (2003). Sequence-independent DNA binding and replication initiation by the human origin recognition complex. *Genes Dev* **17**, 1894-1908.

[10]Bell, S.P. (2002). The origin recognition complex: from simple origins to complex functions. *Genes Dev* **16**, 659–672.

[11]DePamphilis, M.L. (2005). Cell cycle dependent regulation of the origin recognition complex. *Cell Cycle* **4**, 70-79.

[12]Chesnokov, I.N. (2007). Multiple functions of the origin recognition complex. *Int Rev Cytol* **256**, 69-109.

[13]Lidonnici, M.R., Rossi, R., Paixao, S., Mendoza-Maldonado, R., Paolinelli, R., Arcangeli, C., Giacca, M., Biamonti, G., and Montecucco, A. (2004). Subnuclear distribution of the largest subunit of the human origin recognition complex during the cell cycle. *J Cell Sci* **117**, 5221-5231.

[14]Li, Y., Kirschmann, D.A., and Wallrath, L.L. (2002). Does heterochromatin protein 1 always follow code? *Proc Natl Acad Sci USA* **99**, 16462-16469.

[15]Prasanth, S.G., Prasanth, K.V., Siddiqui, K., Spector, D.L., and Stillman, B. (2004). Human Orc2 localizes to centrosomes, centromeres and heterochromatin during chromosome inheritance. *EMBO J* **23**, 2651-2663.

[16]Leatherwood, J., and Vas, A. (2003). Connecting ORC and heterochromatin: why? *Cell Cycle* **2,** 573-575.

[17]Auth, T., Kunkel, E., and Grummt, F. (2006). Interaction between HP1alpha and replication proteins in mammalian cells. *Exp Cell Res* **312**, 3349-3359.

[18]Dhar, S.K., Delmolino, L., and Dutta, A. (2001). Architecture of the human origin recognition complex. *J Biol Chem* **276**, 29067-29071.

[19]Giordano-Coltart, J., Ying, C.Y., Gautier, J., and Hurwitz, J. (2005). Studies of the properties of human origin recognition complex and its Walker A motif mutants *Proc. Natl. Acad. Sci. USA* **102**, 69-74.

[20]Vashee, S., Simancek, P., Challberg, M.D., and Kelly, T.J. (2001). Assembly of the human origin recognition complex. *J Biol Chem* **276**, 26666-26673.

[21]Ranjan, A., and Gossen, M. (2006). A structural role for ATP in the formation and stability of the human origin recognition complex. *Proc Natl Acad Sci USA* **103**, 4864-4869.

[22]Kneissl, M., Putter, V., Szalay, A.A., and Grummt, F. (2003). Interaction and assembly of murine pre-replicative complex proteins in yeast and mouse cells. *J Mol Biol* **327**, 111-128.

[23]Chesnokov, I., Remus, D., and Botchan, M. (2001). Functional analysis of mutant and wild-type Drosophila origin recognition complex. *Proc Natl Acad Sci USA* **98**, 11997-12002.

[24]Ohta, S., Tatsumi, Y., Fujita, M., Tsurimoto, T., and Obuse, C. (2003). The ORC1 cycle in human cells: II. Dynamic changes in the human ORC complex during the cell cycle *J Biol Chem* **278**, 41535-41540.

[25]Prasanth, S.G., Prasanth, K.V., and Stillman, B. (2002). Orc6 involved in DNA replication, chromosome segregation, and cytokinesis. *Science* **297**, 1026-1031.

[26]Neuwald, A.F., Aravind, L., Spouge, J.L., and Koonin, E.V. (1999). AAA+: a class of chaperone-like ATPases associated with the assembly, operation, and disassembly of protein complexes. *Genome Res* **9**, 27-43.

[27]Tye, B.K. (1999). MCM proteins in DNA replication. *Annu Rev Biochem* **68**, 649-686.

[28]Lupas, A.N., and Martin, J. (2002). AAA proteins. *Curr Opin Struct Biol* **12**, 746-753.

[29]Davey, M.J., Indiani, C., and O'Donnell, M. (2003). Reconstitution of the Mcm2-7p heterohexamer, subunit arrangement, and ATP site architecture *J Biol Chem* **278** 4491-4499.

[30]Sato, M., Gotow, T., You, Z., Komamura-Kohno, Y., Uchiyama, Y., Yabuta, N., Nojima, H., and Ishimi, Y. (2000). Electron microscopic observation and single-stranded DNA binding activity of the Mcm4,6,7 complex. *J Mol Biol* **300**, 421-431.

[31]Ying, C.Y., and Gautier, J. (2005). The ATPase activity of MCM2–7 is dispensable for pre-RC assembly but is required for DNA unwinding. *EMBO J* **24**, 4334-4344.

[32]Ishimi, Y. (1997). A DNA helicase activity is associated with an MCM4, -6, and -7 protein complex *J Biol Chem* **272**, 24508-24513.

[33]You, Z., Ishimi, Y., Mizuno, T., Sugasawa, K., Hanaoka, F., and Masai, H. (2003). Thymine-rich single-stranded DNA activates Mcm4/6/7 helicase on Y-fork and bubble-like substrates. *EMBO J* **22**, 6148-6160.

[34]Shima, N., Buske, T.R., and Schimenti, J.C. (2007). Genetic screen for chromosome instability in mice: Mcm4 and breast cancer. *Cell Cycle* **6**, 1135-1140.

[35]Facoetti, A., Ranza, E., Benericetti, E., Ceroni, M., Tedeschi, F., and Nano, R. (2006). Minichromosome maintenance protein 7: a reliable tool for glioblastoma proliferation index. *Anticancer Res* **26**, 1071-1075.

[36]Honeycutt, K.A., Chen, Z., Koster, M.I., Miers, M., Nuchtern, J., Hicks, J., Roop, D.R., and Shohet, J.M. (2006). Deregulated minichromosomal maintenance protein MCM7 contributes to oncogene driven tumorigenesis. *Oncogene* **25**, 4027-4032.

[37]Ryu, S., and Driever, W. (2006). Minichromosome maintenance proteins as markers for proliferation zones during embryogenesis. *Cell Cycle* **5**, 1140-1142.

[38]DePamphilis, M.L., Blow, J.J., Ghosh, S., Saha, T., Noguchi, K., and Vassilev, A. (2006). Regulating the licensing of DNA replication origins in metazoa. *Curr Opin Cell Biol* **18**, 231-239.

[39]Labib, K., and Diffley, J.F. (2001). Is the MCM2–7 complex the eukaryotic DNA replication fork helicase? *Curr Opin Genet Dev* **11**, 64-70.

[40]Pacek, M., and Walter, J.C. (2004). A requirement for MCM7 and Cdc45 in chromosome unwinding during eukaryotic DNA replication. *EMBO Journal* **23**, 3667-3676.

[41]Shechter, D., Ying, C.Y., and Gautier, J. (2004). DNA unwinding is an MCM complex-dependent and ATP hydrolysis-dependent process. *J Biol Chem* **279**, 45586-45593.

[42]Speck, C., and Stillman, B. (2007). Cdc6 ATPase activity regulates ORC x Cdc6 stability and the selection of specific DNA sequences as origins of DNA replication. *J Biol Chem* **282**, 11705-11714.

[43]Perkins, G., and Diffley, J.F. (1998). Nucleotide-dependent prereplicative complex assembly by Cdc6p, a homolog of eukaryotic and prokaryotic clamp-loaders. *Mol Cell* **2**, 23-32.

[44]Shin, J.H., Grabowski, B., Kasiviswanathan, R., Bell, S.D., and Kelman, Z. (2003). Regulation of minichromosome maintenance helicase activity by Cdc6. *J Biol Chem* **278**, 38059-38067.

[45]Volkening, M., and Hoffmann, I. (2005). Involvement of human MCM8 in prereplication complex assembly by recruiting hcdc6 to chromatin. *Mol Cell Biol* **25**, 1560-1568.

[46]Yim, H., Hwang, I.S., Choi, J.S., Chun, K.H., Jin, Y.H., Ham, Y.M., Lee, K.Y., and Lee, S.K. (2006). Cleavage of Cdc6 by caspase-3 promotes ATM/ATR kinase-mediated apoptosis of HeLa cells. *J Cell Biol* **174**, 77-88.

[47]Sivaprasad, U., Machida, Y.J., and Dutta, A. (2007). APC/C - the master controller of origin licensing? *Cell Div* **2**, 8.

[48]Tada, S. (2007). Cdt1 and geminin: role during cell cycle progression and DNA damage in higher eukaryotes. *Front Biosci* **12**, 1629-1641.

[49]Iizuka, M., Matsui, T., Takisawa, H., and Smith, M.M. (2006). Regulation of replication licensing by acetyltransferase Hbo1. *Mol Cell Biol* **26**, 1098-1108.

[50]Mendez, J., and Stillman, B. (2003). Perpetuating the double helix: molecular machines at eukaryotic DNA replication origins. *Bioessays* **25**, 1158-1167.

[51]Forsburg, S.L. (2004). Eukaryotic MCM proteins: beyond replication initiation. *Microbiol Mol Biol Rev* **68**, 109-131.

[52]Tsuji, T., Ficarro, S.B., and Jiang, W. (2006). Essential role of phosphorylation of MCM2 by Cdc7/Dbf4 in the initiation of DNA replication in mammalian cells. *Mol Biol Cell* **17**, 4459-4472.

[53]Masai, H., and Arai, K. (2002). Cdc7 kinase complex: a key regulator in the initiation of DNA replication. *J Cell Physiol* **190**, 287-296.

[54]Mailand, N., and Diffley, J.F. (2005). CDKs promote DNA replication origin licensing in human cells by protecting Cdc6 from APC/C-dependent proteolysis. *Cell* **122**, 915-926.

[55]Masai, H., Taniyama, C., Ogino, K., Matsui, E., Kakusho, N., Matsumoto, S., Kim, J.M., Ishii, A., Tanaka, T., Kobayashi, T., et al. (2006). Phosphorylation of MCM4 by Cdc7 kinase facilitates its interaction with Cdc45 on the chromatin. *J Biol Chem* **281** 39249-39261.

[56]Stuermer, A., Hoehn, K., Faul, T., Auth, T., Brand, N., Kneissl, M., Pütter, V., and Grummt, F. (2007). Mouse pre-replicative complex proteins colocalise and interact with the centrosome. *Eur J Cell Biol* **86**, 37-50.

[57]Cook, J.G., Chasse, D.A., and Nevins, J.R. (2004). The regulated association of Cdt1 with minichromosome maintenance proteins and Cdc6 in mammalian cells. *J Biol Chem* **279**, 9625-9633.

[58]Tada, S., Li, A., Maiorano, D., Mechali, M., and Blow, J.J. (2001). Repression of origin assembly in metaphase depends on inhibition of RLF-B/Cdt1 by geminin. *Nat Cell Biol* **3**, 107-113.

[59]Wohlschlegel, J.A., Dwyer, B.T., Dhar, S.K., Cvetic, C., Walter, J.C., and Dutta, A. (2000). Inhibition of eukaryotic DNA replication by geminin binding to Cdt1. *Science* **290**, 2309-2312.

[60]Nishitani, H., Sugimoto, N., Roukos, V., Nakanishi, Y., Saijo, M., Obuse, C., Tsurimoto, T., Nakayama, K.I., Nakayama, K., Fujita, M., et al. (2006). Two E3 ubiquitin ligases, SCF-Skp2 and DDB1-Cul4, target human Cdt1 for proteolysis. *EMBO J* **25**, 1126-1136.

[61]Senga, T., Sivaprasad, U., Zhu, W., Park, J.H., Arias, E.E., Walter, J.C., and Dutta, A. (2006). PCNA is a cofactor for Cdt1 degradation by CUL4/DDB1-mediated N-terminal ubiquitination. *J Biol Chem* **281**, 6246-6252.

[62]Nishitani, H., and Lygerou, Z. (2002). Control of DNA replication licensing in a cell cycle. *Genes Cells* **7**, 523-534.

[63]Yoshida, K. (2007). Geminin organizes the molecular platform to balance cellular proliferation and differentiation. *Front Biosci* **12**, 2984-2992.

[64]Roukos, V., Iliou, M.S., Nishitani, H., Gentzel, M., Wilm, M., Taraviras, S., and Lygerou, Z. (2007). Geminin cleavage during apoptosis by caspase-3 alters its binding ability to the SWI/SNF subunit Brahma. *J Biol Chem* **282**, 9346-9357.

[65]Okorokov, A.L., Orlova, E.V., Kingsbury, S.R., Bagneris, C., Gohlke, U., Williams, G.H., and Stoeber, K. (2004). Molecular structure of human geminin. *Nat Struct Mol Biol* **11**, 1021-1022.

[66]Thépaut, M., Maiorano, D., Guichou, J.F., Augé, M.T., Dumas, C., Méchali, M., and Padilla, A. (2004). Crystal structure of the coiled-coil dimerization motif of geminin: structural and functional insights on DNA replication regulation. *J Mol Biol* **342**, 275-287.

[67]Ferenbach, A., Li, A., Brito-Martins, M., and Blow, J.J. (2005). Functional domains of the Xenopus replication licensing factor Cdt1. *Nucleic Acids Res* **33**, 316-324.

[68]Lee, C., Hong, B., Choi, J.M., Kim, Y., Watanabe, S., Ishimi, Y., Enomoto, T., Tada, S., Kim, Y., and Cho, Y. (2004). Structural basis for inhibition of the replication licensing factor Cdt1 by geminin. *Nature* **430**, 913-917.

[69]Kroll, K.L. (2007). Geminin in embryonic development: coordinating transcription and the cell cycle during differentiation. *Front Biosci* **12**, 1395-1409.

[70]Saxena, S., Yuan, P., Dhar, S.K., Senga, T., Takeda, D., Robinson, H., Kornbluth, S., Swaminathan, K., and Dutta, A. (2004). A dimerized coiled-coil domain and an adjoining part of geminin interact with two sites on Cdt1 for replication inhibition. *Mol Cell* **15**, 245-258.

[71]Gambus, A., Jones, R.C., Sanchez-Diaz, A., Kanemaki, M., van Deursen, F., Edmondson, R.D., and Labib, K. (2006). GINS maintains association of Cdc45 with MCM in replisome progression complexes at eukaryotic DNA replication forks. *Nat Cell Biol* **8**, 358-366.

[72]Aparicio, T., Ibarra, A., and Méndez, J. (2006). Cdc45-MCM-GINS, a new power player for DNA replication. *Cell Div* **1**, 18.

[73]Labib, K., and Gambus, A. (2007). A key role for the GINS complex at DNA replication forks. *Trends Cell Biol* **17**, 271-278.

In: Protein Conformation: New Research
Editor: Linda B. Roswell, pp. 211-224

ISBN: 978-1-60456-219-4

Chapter 6

ANALYTICAL APPROACHES TO COMPARATIVE MODELING OF PROTEIN STRUCTURES*

Aleksandar Poleksic
Department of Computer Science, University of Northern Iowa, Cedar Falls
IA 50614

Abstract

Comparative modeling builds a three-dimensional structure of a target protein based on one or several related proteins of known structure. The underlying assumption is that the sequence homology between two proteins implies structural similarity, which in turn implies a similarity in their function. Most algorithms for aligning and scoring pairs of proteins are based on the dynamic programming techniques in conjunction with the residue-residue mutation scores. However, the pair-wise alignment techniques that use amino-acid substitution matrices and residue-type information are often unable to recognize the relationships between proteins that are known to be homologous. Profile-profile methods utilize sequence information contained in multiple sequence alignments corresponding to both target and the template family of proteins. By taking advantage of the distribution of amino-acid counts in the columns of the alignment profiles, these methods can often recognize the similarity between remotely related sequences, as evidenced by large scale benchmarking experiments such as LiveBench, CAFASP, and CASP. We give an overview of the existing profile-profile techniques, concentrating on the key aspects of the alignment algorithms. A novel approach to scoring pairs of profiles is presented that focuses on the alignment scoring function and the estimation of the statistical parameters for the alignment score statistics.

Introduction

The function of a protein can be derived from its particular shape and structure. However, determining protein structure by X-ray crystallography or Nuclear Magnetic Resonance is lengthy and resources consuming process. With the increasing gap between known protein

* Reviewed by Natarajan Kannan, Ph.D., Howard Hughes Medical Institute, Department of Chemistry and Biochemistry, University of California San Diego, La Jolla, CA 92093.

sequences and experimentally solved structures, the only practical solution to gaining structural insight for majority of proteins is in computational modeling of their structures.

Comparative modeling (or homology modeling) is a class of methods for predicting the structure of a protein from its primary sequence. A comparative modeling method approximates the 3D structure of a target protein by identifying a protein of known structure (template) that is likely to resemble the structure of the target. Every comparative modeling method comprises the following three steps:

1) Identifying the template structure and optimally aligning the target and the template.
2) Generating the backbone structure by copying the backbone atoms from the template.
3) Adding loops and side chains.

In order to identify the best template for a given target, the target sequence must be optimally aligned and scored against representative sequences from the database of all known structures (templates). An optimal alignment for a pair of amino acid sequences represents the most likely evolutionary relationship between the sequences for a given scoring scheme. Since the number of all possible alignments between two protein sequences is exponential in the lengths of the sequences, finding an optimal alignment by enumerating all possible alignments is computationally prohibitive. To circumvent this combinatorial explosion in the number of alignments, most homology modeling methods incorporate a dynamic programming technique, such as Needleman-Wunsch [1] or Smith-Waterman [2]. By breaking the alignment problem into subproblems and combining the solutions of the subproblems, the dynamic programming is able to compute an optimal alignment between two amino acid sequences in real time.

An essential part of any alignment method is the alignment scoring scheme. S*equence-sequence* methods use residue mutation matrices and fixed gap penalties. The algorithms in this category, such as BLAST [3] or FASTA [4], are generally very fast but they often fail to recognize proteins that are known to be homologous.

P*rofile-sequence* methods utilize the information contained in the protein families to compare sequence profiles to single sequences. A profile is derived from a multiple alignment of sequences from a protein family. More specifically, a profile can be viewed as a probability distribution matrix containing, at each sequence position i, the distribution of twenty amino acid letters occurring in the i-th column of the corresponding multiple alignment.

Profile-profile alignment algorithms employ sequence profiles in place of both the query and the template sequence. By incorporating information about aligned residues in the multiple alignments corresponding to both sequences, profile-profile methods are often able to recognize similarity between sequences sharing less than 20% of identical residues.

A significant difference in the performance of the algorithms in these three categories was evident in the CAFASP4 benchmark of fully automated structure prediction methods [5]. Figure 1 summarizes the performance of representative methods from each group: BLAST (*sequence-sequence*), PSI-BLAST [6] (*profile-sequence*) and STRUCTFAST [7] (*profile-profile*). The servers were ranked by the total MaxSub score [8]. More details about the CAFASP experiment can be found at http://www.cs.bgu.ac.il/~dfischer/CAFASP4.

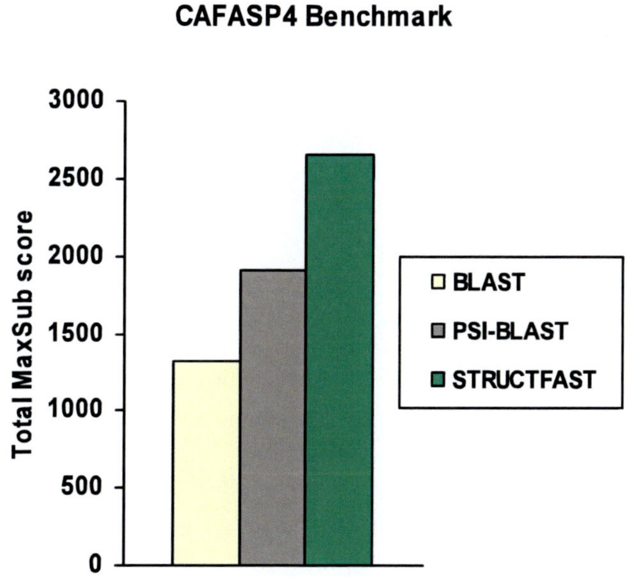

Figure 1. The results of the CAFASP4 benchmark on 70 target proteins. The Eidogen's STRUCTFAST program (*sfst*) was one of the best performing methods in CAFASP4 overall (data not shown).

In the rest of this paper we give an overview of the existing profile-based methods for protein sequence comparison. A new approach to scoring pairs of profiles, called STRUCT_PROF, is presented along with a statistical procedure for estimating the significance of profile-profile alignment scores (PPCIS).

Preliminaries

Scoring Function

The alignment scoring scheme is of critical importance for the algorithm's sensitivity and the alignment accuracy. The scoring scheme specifies the scores for matching an arbitrary pair of positions from two sequences and the penalties for residue insertions and deletions (gaps).

In the *sequence-sequence* alignment algorithms, such as BLAST or FASTA, the pairwise scores for matching amino acid letters are given in a 20x20 lookup table (*substitution matrix*). The substitution matrix (such as BLOSUM62 [9]) is typically derived from observations of the frequencies of mutations of amino-acids in the consensus alignment blocks. Frequently observed mutations are assigned positive scores and rare substitutions are given negative scores.

It is well known that BLAST and other algorithms that use only amino acid mutation scores and fixed gap penalties are often unable to recognize pairs of homologous proteins. The problem is that the alignment becomes unreliable as the protein similarity enters the "twilight zone", i.e. the subspace of sequences sharing not more than 20% sequence identity.

A number of successful efforts have been made to detect weak similarity between proteins in the twilight zone by fundamentally changing the definition of sequence homology.

Instead of comparing pairs of amino-acid strings, profile-based methods utilize the information contained in protein families to compare sequence profiles to single sequences or other profiles. The NCBI's PSI-BLAST algorithm computes the "position specific scoring matrix" (PSSM) by collecting database sequences related to the target. In the next iteration PSSM is used in place of the target sequence to search for new homologues. The procedure is repeated until no new sequences are found. The score for aligning profile position q with the residue r is computed by comparing the probability of the occurrence of r at the position q to the overall (background) frequency of the residue r.[1] By utilizing family dependent substitution scores, PSI-BLAST is often able to capture remote similarities missed by BLAST and other sequence-sequence procedures.

Hidden Markov model (HMM) based algorithms, such as HMMER (Eddy 1997) and SAMT02 [10], use a probabilistic model instead of the position specific scoring matrix. This explicit probabilistic treatment of protein models allows the HMM based methods to accurately compute the likelihood that an arbitrary sequence is generated from each model.

In the profile-profile methods, the score for aligning two positions q^1 and q^2 reflects the similarity between the distributions of amino acid letters at those positions. Some methods in this category, including FFAS [11] and ORFeus [12], use the dot product as the measure of similarity of two distributions.[2] Other algorithms, such as PROF_SIM [13], COMPASS [14], and STRUCTFAST consider background amino acid frequencies [15] as the reference point. PROF_SIM uses an information theory measure called Jensen–Shannon entropy. The score for matching two profile positions q^1 and q^2 in PROF_SIM is given by:

$$\mathrm{Score}(q^1,q^2) = (1 - D^{JS}(q^1,q^2))(1 + D^{JS}(\mathrm{Avg}(q^1,q^2),b)) \tag{1}$$

where

$$D^{JS}(x,y) = (1/2)(D^{KL}(x,\mathrm{Avg}(x,y)) + D^{KL}(y,\mathrm{Avg}(x,y))) \tag{2}$$

$$D^{KL}(x,y) = \sum_{k=1}^{20} x_k \log_2 \frac{x_k}{y_k} \tag{3}$$

and

$$\mathrm{Avg}(x,y) = \left\{ \frac{x_k + y_k}{2} \right\}_{k=1}^{20} \tag{4}$$

In the equations above b denotes the background (overall) distribution of amino-acid letters and x_k is the probability of the residue k at the profile position x.

[1] Such as, for example, Robinson & Robinson background frequency [15]

[2] The straightforward application of the dot product results in the alignments over the entire lengths of the two sequences. However, in order for an algorithm to generate more accurate, local alignments, the expected score for aligning two profile positions must be negative. One way to "adjust" the expected score is to subtract a constant positive value ("zero shift") from each score in the scoring matrix [11]

The scoring functions used in STRUCTFAST and COMPASS both extend the scoring function developed for the PSI-BLAST method. COMPASS computes the relative entropy score:

$$\mathrm{Score}(q^1,q^2)=c_1\sum_{k=1}^{20}n_k^1\ln\frac{q_k^1}{b_k}+c_2\sum_{k=1}^{20}n_k^2\ln\frac{q_k^2}{b_k} \tag{5}$$

where n_k^1 and n_k^2 are the "effective counts" for the amino acid k at the positions q^1 and q^2, respectively. The scoring function in STRUCTFAST is similar, but it does not contain the weighting terms c_1 and c_2. In addition, the "effective counts" in STRUCTFAST are computed using the PSI-BLAST method. [3]

Statistical Significance of Alignment Scores

The raw alignment score is of little value when ranking the database hits. Part of the problem is that the long sequences tend to have high alignment scores even if they are not related. In order to accurately assess the statistical significance of an alignment score, the score has to be compared against the (background) distribution of scores for random sequences of the same lengths and compositions.

For local sequence-sequence alignments that are not allowed to contain gaps it is well established that the expected number of locally optimal sub-alignments with score of at least x is approximately Poisson distributed with mean value

$$E = Kmne^{-\lambda x} \tag{6}$$

In the equation above, m and n are the lengths of the sequences, K is the natural scale for the size of the search space, and λ is the scale parameter for the scoring system [16,17]. Using the equation 6, it can be easily shown that the probability of finding at least one alignment with score greater than or equal x is

$$P(S \geq x) = 1 - e^{-E} \tag{7}$$

The last value is called the p-value of the alignment score x and is a proper measure of the statistical significance of x.[4]

There is strong evidence that the theory presented above applies to local alignments with gaps. Many experiments have shown that the distribution of random alignment scores generated by gapped sequence-sequence algorithms is approximately Gumbel, specified by two parameters λ and K. However, for gapped alignment methods (such as gapped BLAST), λ and K must be estimated (rather than computed analytically) from a comparison of large number of random sequences.

[3] STRUCTFAST uses a modified version of PSI-BLAST to output the effective counts in the profile columns.

Recently, Olsen *et al.* proposed the Island Method [18,19], a procedure for fitting the Gumbel parameters using a relatively small number of sequence shuffles. The Island Method takes advantage of the scores of the "alignment islands" which can be extracted with a slight modification of the dynamic programming engine. Although the method proposed by Olsen *et al.* is much faster than the brute force approach of shuffling and aligning the sequences, it alone still does not allow for quick estimates of λ and K in the context of a large database search.

Computing the background score distribution for profile-based algorithms is much more challenging problem. In order to quickly estimate the significance of database matches, HMMER method pre-computes the Gubmel parameters for each Hidden Markov model in its profile library. The distribution parameters for a given HMM are computed by aligning and scoring HMM against the random sequences.

The score significance in PSI-BLAST is computed "on the fly" by constructing the residue scores within each profile column to the same "scale" as those in BLOSUM62. The underlying assumption is that, after rescaling, the background distribution of PSI-BLAST alignment scores is the same as the distribution of scores in the gapped BLAST program. Indeed, many experiments suggest that this assumption is valid and that the rescaling of the profiles yields accurate p-values.

Estimating the significance of profile-profile alignment scores is a very difficult problem. In absence of a rigorous analytical theory, some profile-profile algorithms use Z-statistics to measure the score significance. The problem with Z-score measure is that it makes an incorrect assumption about the Gaussian form of the background score distribution. An efficient procedure similar to PSI-BLAST statistics is implemented in the COMPASS algorithm. However, the method presented in Sadreyev *et al.* works only in the context of the scoring scheme used in COMPASS.

Methods

Profile Construction

In our STRUCT_PROF method, the sequence profiles are generated by running six iterations of PSI-BLAST against the non-redundant protein sequence database *nr*. By default, the input to PSI-BLAST program is a single string of amino-acid letters. However, PSI-BLAST can be started from an arbitrary PSSM or an arbitrary multiple sequence alignment. To enhance the quality of the profiles for proteins of known structures (templates), we "jump-start" PSI-BLAST from the multiple structural alignment of the template's structural neighbors. The structural alignments given to PSI-BLAST are generated with the Dali program [20]. In case where the number of structural neighbors of a template is small, the Dali structural alignments are replaced with the alignments from the PFAM database.

[4] In database search algorithms, another measure, called *E-value*, is often used to report the score significance. The *E-value* is defined as the p-value multiplied by the size of the database.

Scoring Function

Aside from comparing the distributions of amino-acid letters, STRUCT_PROF takes account of the structural features of the target and the template proteins. More specifically, the score for aligning positions q^1 and q^2 in the target and the template sequence, respectively, is given by:

$$\mathrm{Score}(q^1, q^2) = ss_{q^1} \cdot ss_{q^2} \cdot \left(\log\frac{P_{q^1}(q^2)}{P_b(q^2)} + \log\frac{P_{q^2}(q^1)}{P_b(q^1)}\right) \tag{8}$$

where

$$P_{q^i}(q^j) = N_j! \prod_{k=1}^{20} \frac{(q_k^i)^{n_k^j}}{n_k^j!} \tag{9}$$

and

$$P_b(q^i) = N_i! \prod_{k=1}^{20} \frac{b_k^{n_k^i}}{n_k^i!}. \tag{10}$$

The values $\{q_k^i\}_{k=1}^{20}$ are the amino acid frequencies at the position q^i, $\{b_k\}_{k=1}^{20}$ are the background amino acid frequencies, $\{n_k^i\}_{k=1}^{20}$ are the "effective" residue counts observed at the position q^i, and $N_i = \sum_{k=1}^{20} n_k^i$, $i = 1,2$. Note that, if the amino acid distributions are viewed as multinomial models then $P_{q^j}(q^i)$ and $P_b(q^i)$ represent the probability that the effective residue counts at the profile position q^i are generated from the models q^j and b, respectively.

The terms ss_{q^1} and ss_{q^2} are used to weight the similarity score $\log\frac{P_{q^1}(q^2)}{P_b(q^2)} + \log\frac{P_{q^2}(q^1)}{P_b(q^1)}$ according to secondary structure states of residues at profile positions q^1 and q^2. To account for a greater variability of amino acid distributions in the loop regions of the proteins, we down-weight the scores between the residues in loop regions and up-weight the scores between the residues in the regular secondary structure regions. More specifically,

$$ss_{q^i} = \frac{1 + p_{q^i}}{2}, \quad i = 1,2 \tag{11}$$

where p_{q^i} is the probability that the residue at the position q^i is located in an *α-helix* or a *β-sheet*. The probabilities p_{q^1} are computed using the PSIPRED algorithm [21]. For the proteins of known structures (templates), we use the observed secondary structure states to compute p_{q^2}. In other words, if a template residue is part of a loop region then p_{q^2} is set to 0; otherwise, $p_{q^2} = 1$. The secondary structure for template residues is assigned using the DSSP algorithm [22].

Gap Penalties

The gap opening and extension penalties at each profile position are weighted according to the gap content in the multiple alignments used to build the profiles. The idea is to propagate the gap information contained in the PSI-BLAST alignments. In other words, if the alignment positions *i* (for the query) and *j* (for the template) are highly conserved, then the gap penalty at *(i, j)* will be high. A mild penalization is used for positions having higher frequency of gap characters.

More precisely, the penalties for opening and extending the gap at *(i, j)* are computed as:

$$\mathrm{Open(i, j)} = \mathrm{O} \cdot \mathrm{w}_\mathrm{i}^1 \cdot \mathrm{w}_\mathrm{j}^2 \tag{12}$$

$$\mathrm{Extend(i, j)} = \mathrm{E} \cdot \mathrm{w}_\mathrm{i}^1 \cdot \mathrm{w}_\mathrm{j}^2 \tag{13}$$

O and *E* are the "initial" opening and extension penalties (set to 10 and 1 respectively) and

$$\mathrm{w}_\mathrm{l}^\mathrm{k} = 1 - \frac{\mathrm{f}_\mathrm{l}^\mathrm{k}}{2} \tag{14}$$

In the formula 14, f_l^k is the observed frequency of the gap character ("-") at the position *l* in the profile *k, k = 1, 2.*

The Significance of Profile-Profile Scores

The statistical significance of the alignment scores is computed using the Profile-Profile Convergent Island Statistics (PPCIS) [23]. The parameters of the Gumbel-type extreme value distribution are estimated by random simulations anew for all pairs of sequences being compared. To take account of the length and composition of the profiles, both profiles are repeatedly shuffled (by shuffling the profile positions) and aligned. The extreme value distribution is fit to the island scores computed from the alignments of profiles shuffles. An overview of the Convergent Island Statistics procedure [24] is given below. For more details on PPCIS method, the reader is referred to Poleksic and Danzer [23].

The CIS Method

The Convergent Island Statistics (CIS) builds upon the Island Method [18,19], but is capable of recognizing the lack of sequence similarity early in the shuffling process and thus save on the CPU time. Full blown shuffling is needed only when there is enough evidence that the sequences are related.

To determine whether the p-value of a score x is below the threshold value p_0, CIS performs a series of statistical tests t_1, ..., t_k. In each test t_j, the sequences are repeatedly shuffled and aligned until the cumulative number of alignment islands reaches the threshold value N_j. The conservative estimates of the Gumbel parameters λ_j and K_j are then computed from the set of N_j scores such that

$$p(\lambda < \lambda_j) = p(K > K_j) = s_i \quad (15)$$

[5]

where s_i is the one tailed confidence level for i-th standard deviations above the mean in the Gaussian distribution. Finally, the estimates λ_j and K_j are used to determine whether the score x is likely to have the p-value lower than p_0.

According to the equation 15, the probability of CIS making an error (i.e. discarding a significant hit) in each test is $1 - s_i^2$. For instance, if $i = 3$, then the probability of an error is only about 0.003.

Figure 2 gives the theoretical accuracy of the parameter estimates for different values of i and k.

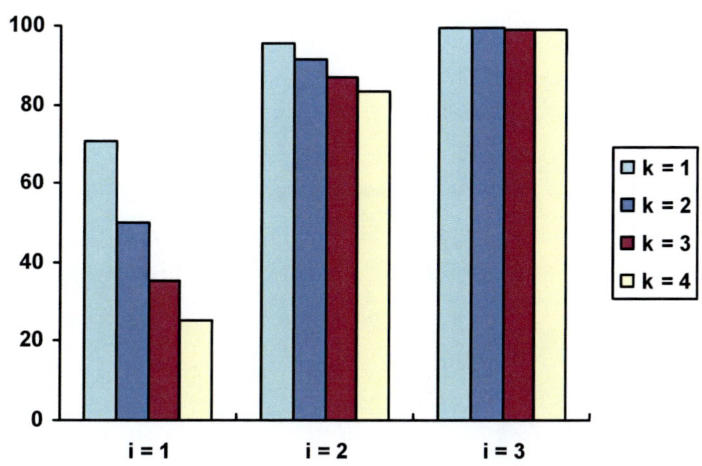

Figure 2. Algorithm's accuracy (% accuracy) as a function of i (the number of standard deviations in the Normal distribution) and k (the number of tests).

[5] p(A) denotes the probability of the event A.

In the CIS method, the number of island N_j increases from one step to another. Typically, the insignificant alignments are quickly discarded (in early tests), after computing only few sequence shuffles. But, if the alignment score x is not discarded (i.e. if it passes all tests), the final p-value for x is computed using the maximum likelihood estimates of the distribution parameters computed from the final set of N_k island scores.

Results

We tested the scoring system in STRUCT_PROF against the scoring scheme used in our older STRUCTFAST algorithm. For reference, we also include the results of the popular PSI-BLAST method. We used two benchmarks to test the performance of STRUCT_PROF: Lindahl benchmark (Figure 3) and CAFASP4 set of targets (Figure 4).

The Lindahl benchmark [25] is used to measure the sensitivity of a fold recognition method in detecting weak relationships between protein sequences. The Lindahl test set is composed of 1310 pairs of proteins classified in three groups according to SCOP [26] hierarchy. We tested ability of PSI-BLAST, STRUCTFAST, and STRUCT_PROF to place a correct member of the SCOP group (family, superfamily, and fold) on the top of its ranked list. As seen in Figure 3, STRUCT_PROF is able to recognize 26%, 28%, and 13% relationships missed by PSI-BLAST in SCOP fold, family and superfamily categories, respectively.

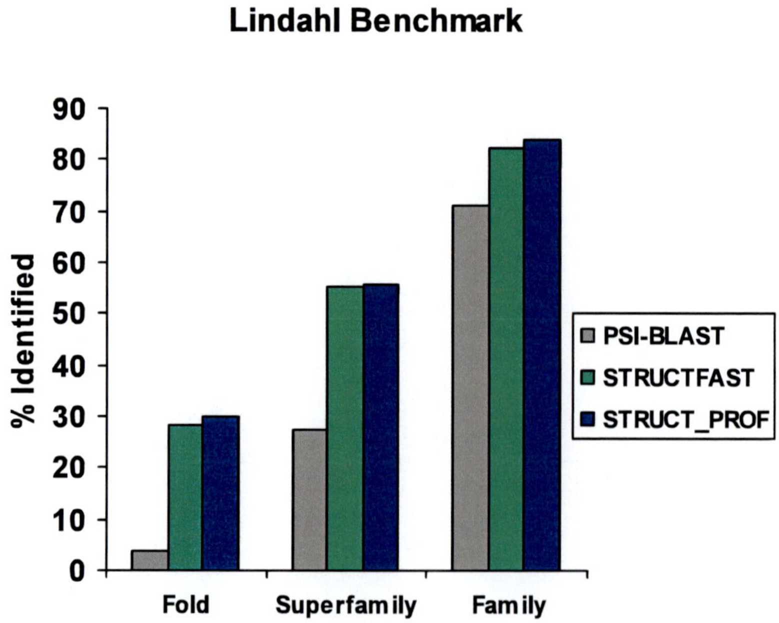

Figure 3. Lindahl Benchmark of fold recognition servers. The results for PSI-BLAST are taken from Shi *et al.* [27].

Figure 4 summarizes the performance of STRUCT_PROF on the set of 70 CAFASP4 targets. In order to perform a head-to-head comparison with STRUCTFAST, the same templates (i.e. those identified by the STRUCTFAST algorithm in CAFASP4) are given to both programs as input. We emphasize that the results of this benchmark should be taken with caution, because the protein sequence and structure information available to STRUCT_PROF at the time of this testing exceeds the information used by the STRUCTFAST program at CAFASP4.

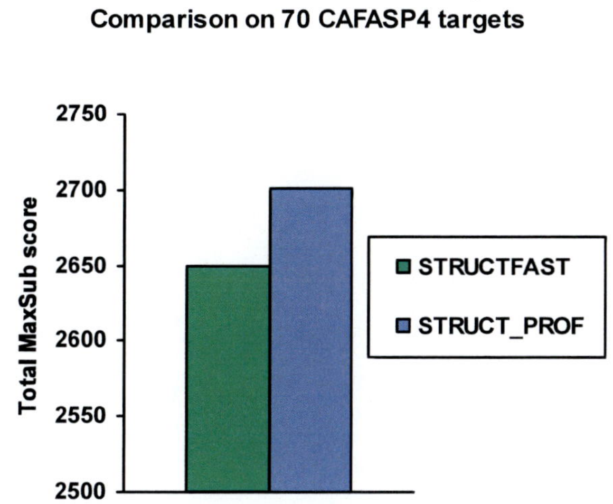

Figure 4. The comparison of STRUCT_PROF and STRUCTFAST on the set of 70 CAFASP4 targets. An in-house developed algorithm for computing MaxSub scores is used to measure the quality of the models produced by two methods. The MaxSub scores are multiplied by 100.

We also emphasize that the sensitivity of STRUCT_PROF is partially due to the rigorous procedure for computing the significance of the alignment scores. On the negative side, this procedure is very slow, in particular when tested against methods that use pre-computed background distribution parameters. However, STRUCT_PROF is still much faster than the regular Island Method. Figure 5 shows the speed advantage of STRUCT_PROF over the Island Method on a representative test set consisting of 2000 SCOP pairs using different cutoffs p_0 for discarding the alignment scores. More details about this analysis can be found in [23].

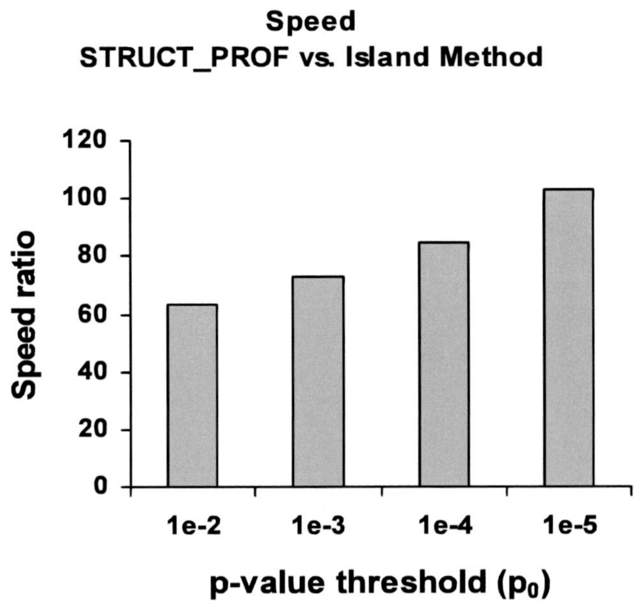

Figure 5. The speed advantage of STRUCT_PROF over the Island Method as a function of the p-value threshold p_0.

Conclusion

There are many different methods for comparing and aligning protein sequences. The widely used BLAST and FASTA algorithms utilize dynamic programming technique in conjunction with residue-residue substitution scores specified in the mutation matrices. The simplicity of the scoring system used in BLAST and FASTA allows these methods to quickly and accurately estimate the statistical significance of the alignment scores. However, the pairwise alignment methods that use only amino-acid mutation matrices and fixed gap parameters are often unable to recognize the relationships between proteins that are known to be homologous.

Profile-based algorithms, such as PSI-BLAST, compare sequence profiles to single sequences of amino acids or other profiles. By utilizing information about aligned residues at the profile positions, these methods are able to improve the inference of remote homologues from protein sequences in the twilight zone. Profile-profile methods enjoy additional enhancement in sensitivity by employing sequence profiles in place of both target and the template sequence. The results of recent benchmarking experiments, such as CASP, CAFASP, and Live Bench, suggest that the profile-profile methods can often recognize the relationships between sequences that share less than 20% identity.

We gave an overview of several well established profile-based methods for protein sequence alignment. We also presented a new scoring system consisting of a new scoring function and a statistical procedure for estimating the significance of profile-profile alignment scores.

Some techniques described in this paper have been incorporated in the protein structure prediction method called UNI-EID. UNI-EID has shown very good performance in the CASP7 blind prediction benchmark. To keep this note concise, we omit the details of the PPCIS algorithm, which will be published elsewhere. In the meantime we will be happy to provide this information per request.

References

[1] Needleman S. B. and Wunsch C. D. (1970) A general method applicable to the search for similarities in the amino acid sequence of two proteins. *J. Mol. Biol.* **48**, 443-53.

[2] Smith TF, Waterman MS (1981). Identification of Common Molecular Subsequences. *J. Mol. Biol.* **147**, 195-197.

[3] Altschul,S.F., Gish,W., Miller,W., Myers,E.W. and Lipman,D.J. (1990) Basic local alignment search tool. *J. Mol. Biol.*, **215**, 403-410.

[4] Pearson,W.R. and Lipman,D.J. (1988) Improved tools for biological sequence comparison. *Proc. Natl. Acad. Sci. USA*, **85**, 2444-2448.

[5] Fischer D, Rychlewski L, Dunbrack RL Jr, Ortiz AR, Elofsson A. (2003) CAFASP3: the third critical assessment of fully automated structure prediction methods.*Proteins*. **53**(S6), 503-516.

[6] Altschul,S.F., Madden,T.L., Schäffer,A.A., Zhang,J.., Zhang,Z., Miller,W. and Lipman,D.J. (1997) Gapped BLAST and PSI-BLAST: a new generation of protein database search programs. *Nucleic Acids Res.*, **25**, 3389-3.

[7] Debe DA, Danzer JF, Goddard WA, Poleksic A. (2006). STRUCTFAST: protein sequence remote homology detection and alignment using novel dynamic programming and profile-profile scoring. *Proteins*, **64**, 960-7.

[8] Siew N, Elofsson A, Rychlewski L, Fischer D. (2000) MaxSub: an automated measure for the assessment of protein structure prediction quality. *Bioinformatics*, **16**, 776-785.

[9] Henikoff S, Henikoff JG. (1992) Amino acid substitution matrices from protein blocks. *Proc Natl Acad Sci U S A*. **89**, 10915-9.

[10]Karplus,K., Karchin,R., Barrett,C., Tu,S., Cline,M., Diekhans,M., Grate,L., Casper,J. and Hughey,R. (2001) What is the value added by human intervention in protein structure prediction? *Proteins* (Suppl. 5), 86–91.

[11]Jaroszewski, L., Rychlewski, L., Li, Z., Li, W. & Godzik, A. (2005) FFAS03: a server for profile-profile sequence alignments. *Nucl. Acids Res.* **33**, W284-W288.

[12]Ginalski K, Pas J, Wyrwicz LS, von Grotthuss M, Bujnicki JM, Rychlewski L. (2003) ORFeus: Detection of distant homology using sequence profiles and predicted secondary structure. *Nucleic Acids Res*. **31**, 3804-7.

[13]Yona,G. and Levitt,M. (2002) Within the twilight zone: a sensitive profile–profile comparison tool based on information theory. *J. Mol. Biol.*, **315**, 1257–1275.

[14]Sadreyev, R.I. and Grishin, N.V. (2003) COMPASS: A tool for comparison of multiple protein alignments with assessment of statistical significance. *J. Mol. Biol.* **326,** 317–336.

[15]Robinson, A.B. & Robinson, L.R. (1991) Distribution of glutamine and asparagine residues and their near neighbors in peptides and proteins. *Proc. Natl. Acad. Sci. USA* **88**, 8880-8884.

[16]Karlin,S. and Altschul,S.F. (1990) Methods for assessing the statistical significance of molecular sequence features by using general scoring schemes. *Proc. Natl Acad. Sci. USA*, **87**, 2264–2268.

[17]Dembo,A.., Karlin,S. And Zeitouni,O. (1994) Critical phenomena for sequence matching with scoring. *Ann. Prob.*, **22**, 1993-2021.

[18]Olsen,R., Bundschuh,R. And Hwa,T. (1999) Rapid assessment of extremal statistics for gapped local alignment. In Lengauer,T., Schneider,R., Bork,P., Brutlag,D., Glasgow,J., Mewes,H.-W. And Zimmer,R (eds), *Proceedings of the Seventh International Conference on Intelligent Systems for Molecular Biology,* AAAI Press, Menlo Park, CA, pp. 211-222.

[19]Altschul,S.F., Bundschuh,R., Olsen,R., Hwa,T. (2001) The estimation of statistical parameters for local alignment score distributions. *Nucleic Acids Res.* **29**, 351-61.

[20]Holm L, Sander C. (1993) Protein structure comparison by alignment of distance matrices. *J Mol Biol.* **233**, 123-38.

[21]Jones DT. (1999) Protein secondary structure prediction based on position-specific scoring matrices. *J. Mol. Biol.* **292**, 195-202.

[22]Kabsch W, Sander C (1983). Dictionary of protein secondary structure: pattern recognition of hydrogen-bonded and geometrical features. *Biopolymers* **22**, 2577-637.

[23]Poleksic A. and Danzer JF. A fast algorithm for estimating the statistical significance of protein sequence alignments generated by profile-based methods, *in review.*

[24]Poleksic A, Danzer JF, Hambly K, Debe DA. (2005) Convergent Island Statistics: a fast method for determining local alignment score significance. *Bioinformatics*. **21**, 2827-31.

[25]Lindahl E., Elofsson A. (2000) Identification of related proteins on family, superfamily and fold level. *J Mo. Biol.* **295**, 613-625.

[26]Murzin A. G., Brenner S. E., Hubbard T., Chothia C. (1995). SCOP: a structural classification of proteins database for the investigation of sequences and structures. *J. Mol. Biol.* **247**, 536-540.

[27]Shi J, Blundell TL, Mizuguchi K. (2001). FUGUE: sequence-structure homology recognition using environment-specific substitution tables and structure-dependent gap penalties. *J Mol Biol*. **310**, 243–257.

In: Protein Conformation: New Research
Editor: Linda B. Roswell, pp. 225-238
ISBN: 978-1-60456-219-4

Chapter 7

Modification of Ubiquitin Structure by the Cytotoxic Drug Pentamidine

Miguel A. Fuertes**, Carlos Alonso and José M. Pérez
Centro de Biología Molecular "Severo Ochoa" (CSIC-UAM)
Facultad de Ciencias, Universidad Autónoma de Madrid
28049-Madrid, Spain.

Abstract

Pentamidine is a bisbenzamidine endowed with antiparasitic and antitumor properties, which induces cell death by apoptotic mechanisms. Its cellular targets are both DNA and proteins. In this communication, we analyze the interaction of pentamidine with Ubiquitin (Ub), which is a protein involved through the Ub-proteasome pathway in the turn-over of cellular proteins. Hence, we report herein that the interaction of pentamidine with Ub produces a 6% increase in the β-sheet content of the protein as detected by Far-UV circular dichroism spectroscopy. Moreover, fluorescence emission studies show that pentamidine induces in Ub a significant decrease in fluorescence of both Phe(4) and Phe(45) residues located on the β-cluster of the molecule but not of Tyr(59) residue located on the α-cluster. Because the β-cluster of Ub is composed by a large number of positively charged or basic aminoacids our data indicate that pentamidine specifically modifies the β-cluster or "basic face" of Ub. Thus, since pentamidine has two positive charges at physiological pH it is not surprising that electric repulsion may be involved in the modification of the β-cluster of Ub by pentamidine. It has been reported that the "basic" face of Ub is involved in recognition and poly-ubiquitinylation of different protein substrates. So, due to the fact that pentamidine specifically modifies the β-cluster or "basic" face of Ub, it is likely that pentamidine may inhibit the biochemical functions of Ub.

Keywords: Pentamidine, Ubiquitin, Anti-tumor, Anti-parasitic, Apoptosis, Necrosis.

* Centro de Biología Molecular "Severo Ochoa" (CSIC-UAM), Facultad de Ciencias, Universidad Autónoma de Madrid, 28049-Madrid, Spain.

Introduction

Pentamidine whose chemical name is [1,5-di(4-amidinophenoxy)pentane] belongs to the pharmacological group of molecules known as bisbenzamidines. The drug pentamidine, clinically formulated as pentamidine isothionate (Fig. 1), has been broadly used for the treatment of several diseases caused by protozoan parasites including pneumocystosis, babesiosis, trypanosomiasis and leishmaniasis [1]. Moreover, pentamidine derivatives such as piperazine-linked bisbenzamidines and bisguanylhydrazones are presently being explored as drugs against pneumocystosis and trypanosomiasis, respectively [1, 2].

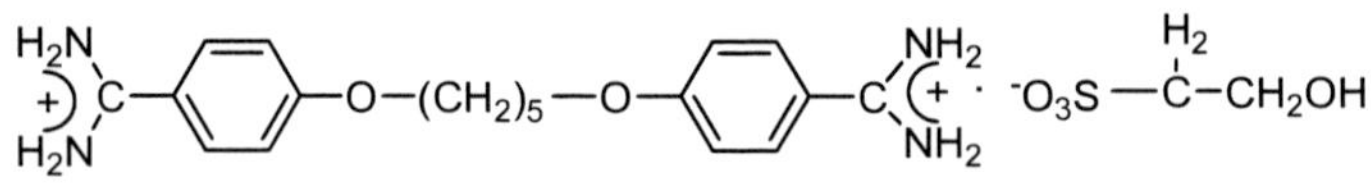

Pentamidine isothionate
MW=593

Figure 1.Structure of pentamidine isothionate. MW= molecular weight.

Pentamidine destroys non-replicating *Pneumocystis carinii* as well as hampers DNA biosynthesis in Trypanosomatids, which are known to actively transport pentamidine intracellularly [3]. Furthermore, diverse lines of evidence indicate that the action of pentamidine against leishmaniasis, an illness provoked by a number of species of the genus *Leishmania,* whose amastigote forms are able to reside in host macrophages, might be mediated through host cellular targets and the host immune system [4]. Pentavalent antimonials [Sb(V)] such as sodium antimony gluconate (SAG) are the basis of treating all forms of leishmaniasis. Nevertheless, pentamidine isothionate has been used as second-line chemotherapy to circumvent the problem of clinical resistance which show certain forms of the disease to pentavalent antimonials [5]. Interestingly, in the treatment of cutaneous leishmaniasis, the high cure rate associated with a low dose of pentamidine given for a short period makes it an attractive alternative to [Sb(V)] compounds [6].

In the last years the mechanism of programmed cell death (PCD) or apoptosis-like cell death has been discovered in diverse single-celled parasites including *Leishmania* species [7]. So, PCD could be exploited as a pharmacological target to fight against leishmaniasis. In fact, it has been recently reported that in promastigotes of *Leishmania infantum,* pentamidine cytotoxicity leads to programmed cell death (PCD), which is associated with cellular cycle alterations in G1, S and G2/M phases [8]. Pentamidine has also been administered to HIV patients coinfected with *Leishmania* [6]. Severe presentation of visceral leishmaniasis is feasible in patients with AIDS, and relapses are usual. Indeed, *Leishmania* amastigotes are commonly found in Kaposi's sarcoma and herpes zoster cutaneous lesions concomitant with visceral leishmaniasis. So, the search for pentamidine analogs more active than the parental drug against these opportunistic infections is also a current field of research [5].On the other hand, it has been recently found that pentamidine is an inhibitor of proteins of regenerating liver (PRLs). Besides, pentamidine possesses anti-cancer potential inhibiting the growth of

WM9 human melanoma tumors in nude mice through induction of tumor cell death by necrosis [9].

The biochemical mechanisms of cytotoxic activity of pentamidine still remain elusive. In Trypanosomatids, some researchers have proposed that pentamidine acts on the kinetoplast (mitochondrion) DNA and inhibits its functions [10]. In fact, pentamidine is a DNA minor groove binder, with characteristic binding sites at AT base pair stretches. In the pentamidine-DNA binding motif an optimization of van der Waals forces and hydrogen bonding between the drug and DNA bases is induced [11].

a

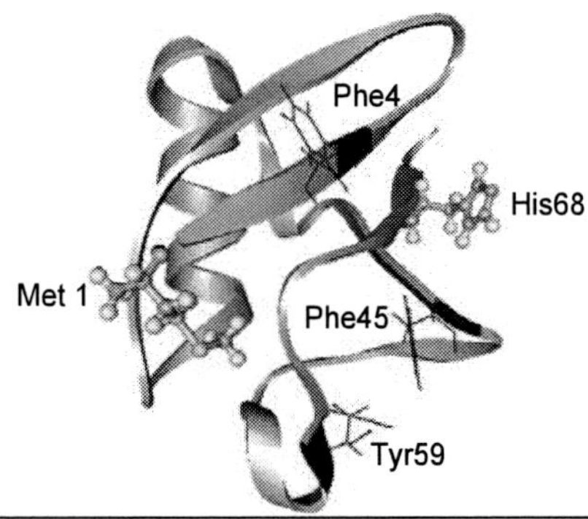

b

↓

MQIFVKTLTG KTITLEVEPS DTIENVKAKI
QDKEGIPPDQ QRLIFAGKQL EDGRTLSDYN
IQKESTLHLV LRLRGG

↑

Figure 2. (a) Ribbon diagram of the Ub molecule with the location of Phe(4), Phe(45), Tyr(59) (lines), and amino (Met(1)) and carboxy (His(68)) ends (sticks). (b) Aminoacid sequence of Ub. The arrows indicate the location of Met(1) and His(68) and the Phe(4), Phe(45), Tyr(59) are underlined.

Ubiquitin is a small protein that is probably present in all eukaryotic cells [12]. It consists of a single 8565 M polypeptide chain of 76 amino acid residues (Fig. 2). Ub has been isolated and sequenced from a variety of sources. The primary structures through Arg(74) are identical in insect [13], trout [14], bovine [15], and human [16] Ub. All of these sequencing studies indicated that Ub was composed of only 74 amino acid residues. However, subsequent studies demonstrated that, Ub has 76 residues and terminates with Gly-Gly [17]. Ub that lacks the COOH-terminal Gly-Gly is almost certainly a proteolytic artifact *in vitro*. The amino acid sequences of yeast [18] and oat [19] Ub differ in only three of 76 residues from that of Ub in higher eukaryotes. It appears to be one of the most conserved of all eukaryotic proteins. In the nucleus, Ub is bound largely to histone H2A via the carboxyl terminus of Ub and an internal lysine on histone H2A [20, 21]. The Ub-H2A complex is found in 10% to 20% of nucleosomes, and these ubiquitinylated nucleosomes are believed to be preferentially within transcribed regions of the chromosomes [22]. In the cytoplasm Ub forms conjugates with a large variety of proteins, and these conjugates are thought to be substrates for ATP-dependent protein degradation [23, 24]. Recently, a mouse cell line temperature-sensitive for Ub-protein conjugation was identified [25]. In these mutant cells the degradation of normally short-lived proteins is also temperature-sensitive, and it is suggested that Ub-dependent pathways are involved in the degradation of more than 90% of short-lived proteins in eukaryotic cells [26]. Ub acts through its post-translational attachment (ubiquitinylation) to other proteins, where these modifications alter the function, location or trafficking of the protein, or targets it for destruction by the 26S proteasome [27]. The terminal glycine in the C-terminal 4-residue tail of Ub can form an isopeptide bond with a lysine residue in the target protein, or with a lysine in another Ub molecule to form a Ub chain that attaches itself to a target protein. Ub has seven lysine residues, any one of which can be used to link Ub molecules together, resulting in different structures that alter the target protein in different ways. It appears that Lys(11)-, Lys(29) and Lys(48)-linked poly-Ub chains target the protein to the proteasome for degradation, while mono-ubiquitinylated and Lys(6)- or Lys(63)-linked poly-Ub chains signal reversible modifications in protein activity, location or trafficking [28]. For example, Lys(63)-linked poly-Ubiquitinylation is known to be involved in DNA damage tolerance, inflammatory response, protein trafficking and signal transduction through kinase activation [29]. In addition, the length of the Ub chain alters the fate of the target protein. Regulatory proteins such as transcription factors and histones are frequent targets of ubiquitinylation [30].

This has made Ub an ideal model system for both establishing the factors important in determining its stability [31-35] and in the development of computational approaches to designing hydrophobic cores [36-40]. In addition, the folding of Ub has been studied extensively using a number of different experimental approaches [41-45]. Recently, the kinetics of folding and unfolding have been measured by conventional methods and compared with those derived from the most slowly exchanging amide residues in native Ub. Rate constants, of folding and unfolding, from these two very different types of experiments showed good agreement [46]. Ub has also been used as a model system to test novel phage-display-based methodologies to select stable folded domains of proteins from libraries [47, 48].

To shed further light on the molecular mechanism of action of pentamidine, we report in this communication biophysical data that show that pentamidine not only modifies DNA but that also modifies the structure of a protein such as Ub. Because Ub is a crucial protein for cellular physiology which is involved through the Ub-proteasome pathway in the turn-over of cellular proteins (ubiquitinylation process), it is likely that the binding of pentamidine to Ub may contribute to its biochemical mechanisms of cytotoxicity by inhibiting Ub function.

Materials and Methods

Chemicals

Reagents and solvents were purchased and used as obtained from the suppliers. Pentamidine isothionate and Ub were purchased from Sigma-Aldrich, Alcobendas (Madrid), Spain. Stock solutions of Ub (MW= 8565 amu) were prepared at concentration of 5 mg/ml in 0.1 x PBS (phosphate-buffered saline) buffer and were diluted to a final concentration of 50 μM for the assays. Stock solutions of pentamidine isothionate were freshly prepared before use at a final concentration of 1 mg/ml in double distilled water.

Far-UV CD Spectroscopy of Ub:pentamidine Complexes.

Far-UV circular dichroism spectra of Ub:pentamidine complexes were recorded in a J-600 spectropolarimeter (JASCO) at a scan rate of 50 $nm.min^{-1}$. The response time was 2 s and the spectra were averaged from 4 scans. The path length of the CD cells used was 0.1 cm. The observed ellipticities were converted to mean residue ellipticities, $[\Theta_{222}]$, in $deg.cm^2.dmol^{-1}$ using an average residue weight for Ub of 108.43. All scans were recorded at 37°C. The samples for CD were prepared as follows: aliquots of 50 μM Ub in 0.1xPBS buffer (8 mM Na_2HPO_4, 2 mM NaH_2PO_4, 10 mM NaCl) were incubated 24 h at 37°C in the in the absence or presence of pentamidine at a molar ratio of pentamidine to Ub r_i of 1/50.

Secondary structure analysis of CD data.

Most of the secondary structural analyses were realized using DICHROWEB (http://www.cryst.bbk.ac.uk/cdweb), an interactive Web server [49] that permits analyses via the following software package CDPro [50-54]. The package permits analyses via the methods CONTINLL, SELCON3, and CDSSTR available at the website <lamar.colostate.edu/~sreeram> with a wide range of protein spectral databases all derived from soluble proteins [54]. The CDPro package was used for analyses comparing the effect of pentamidine on the secondary structure of Ub. As a means of comparison of the goodness of fit of the various methods, the normalized root-mean-square deviation (NRMSD) parameter [55] was calculated for all of the analyses. *NRMSD* is defined as $NRMSD = \sum \sqrt{\frac{(\theta_{\exp} - \theta_{cal})^2}{\theta^2{}_{\exp}}}$ summed over all wavelengths and where θ_{exp} and θ_{cal} are, respectively, the experimental ellipticities and the ellipticities of the back-calculated spectra for the derived structure. *NRMSD* values of <0.1 mean that the back-calculated and experimental spectra are in close agreement [56]. A low *NRSMD* is not sufficient to indicate a correct analysis, but a poor *NRMSD* (high values) generally indicates that the analysis is wrong.

Fluorescence Spectroscopy.

Fluorescence experiments were performed in an AMINCO-BOWMAN (Runcorn, Chesire, UK) series-2 luminescence spectrometer equipped with a Selecta Ultraterm temperature control unit. The emission wavelength was comprised in the range from 285 to 335 nm using an excitation wavelength of 275 nm for Tyr and 260 nm for Phe. The maximum emission wavelengths were 303 nm for Tyr and 300 nm for Phe. The path length cells used was 1.0 cm. The emission spectrum of pentamidine and the blank without protein were subtracted from the spectra. The samples for fluorescence spectroscopy were prepared as in the case of CD spectroscopy.

Results

Interaction of Pentamidine with Ub.

Our investigations were aimed at studying changes in both secondary and tertiary structures of the Ub interacting with pentamidine. It has been previously hypothesized that DNA-binding drugs such as pentamidine might exert their pharmacological effects by interfering with DNA-protein interactions [57]. We choose Ub for our structural studies because it is the most highly conserved protein known and is present in kinetoplastids parasites including all *Leishmania* species [58]. Furthermore, Ub is a crucial protein for cell functioning, which is involved through the Ub-proteasome pathway in the degradation of proteins involved in (i) cell cycle regulation, (ii) modulation of cell surface receptors and ion channels, and (iii) antigen processing and presentation [59]. Apart from its important physiological role, Ub is a structural model to study conformational changes induced in proteins by interacting drugs because it is a small and tightly folded molecule of 76 aminoacids arranged in 2 α-helices and 5 β-sheets (MW = 8565 molecular mass units, amu). Moreover, Ub contains three fluorophore aminoacid sites, which are Phe(4), Phe(45) and Tyr(59) (Fig. 2).

Changes in Ub Secondary Structure Upon Pentamidine Binding

Fig. 3 shows the Far-UV CD spectra (at pH=7) of native Ub and of Ub:pentamidine complexes formed at a molar ratio of Pentamidine per Ub molecule of r_i=1/50. It may observe that both CD spectra are relatively different because the CD spectrum of Ub: pentamidine complexes have a reduced ellipticity (at both 208 and at 222 nm) relative to that observed for native Ub spectrum. We also run the CD spectra of Ub and of Ub:pentamidine samples at pH 3.0 obtaining similar results (data not shown). The FAR UV CD spectra of Ub and of Ub:pentamidine complexes were analyzed with the CDPro software. The secondary structures were calculated from circular dichroism data for Ub using different algorithms and reference databases. As a general rule, the best fit, as assessed by the NRMSD parameter (Table 1), the correctness of the results, as assessed by the *r*-parameter (correlation coefficient) and a visual inspection of the back-calculated and measured spectra were good for all methods and reference databases of Ub. As indicated in Table 1, the method that best fit the CD curves is CONTINLL with NRMSD values for Ub and Ub modified by pentamidine of 0.041 and 0.059, respectively. The calculations by the CONTINLL method showed that the interaction of pentamidine with Ub increases in a 6% the beta structure (22.5% to 27.5%) and decreases in a ~6% the coil structure (36.3% to 30.8%) relative to the spectrum of native unmodified Ub. In contrast, the percentage of alpha structure did not vary in Ub:pentamidine complexes relative to native Ub. Therefore, these data indicate that pentamidine significantly modifies the secondary structure of Ub.

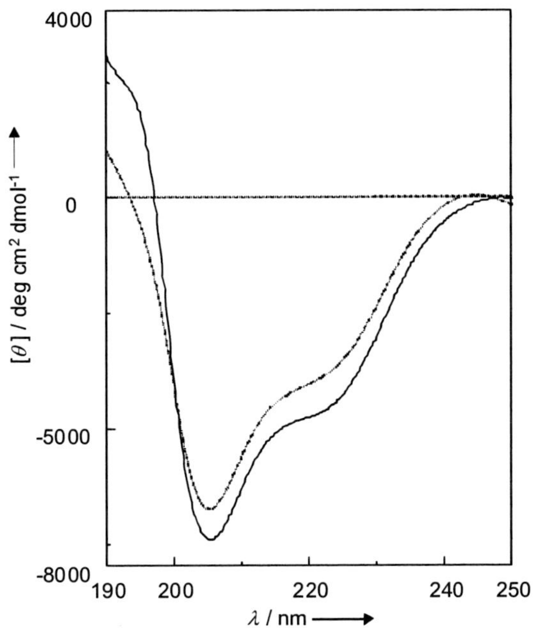

Figure 3. Effect of pentamidine at neutral pH and 37 °C on the FAR UV CD spectra of native Ub (r_i = 1/50).

Table 1. Calculated secondary structures and NRMSD* parameter for CD data obtained from Ub and Ub:pentamidine

(%)	Ub	Ub:pentamidine
Alpha	41.2	41.7
Beta	22.5	27.5
Coil	36.3	30.8
NRMSD	0.041	0.059

* *NRMSD*, Normalized Root-Mean-Square Deviation (see the Methods section)

Changes in Ub Tertiary Structure in Ub:Pentamidine Complexes

To detect changes in the tertiary structure of Ub after pentamidine binding, fluorescence emission spectra of Ub in the absence or in the presence of pentamidine (r_i = 1/50 and 1/100) were also run (Fig. 4). The results showed that the intensity of light emission for excited Phe residues (Phe(4) and Phe(45)) decreases strongly in the presence of pentamidine. Indeed, the fluorescence intensity of the maximum decreases about 25% in Ub:pentamidine complexes relative to native Ub value (Fig. 4a). This result is consistent with changes in the environment of the Phe residues upon interaction of pentamidine with the Ub molecule. In contrast, the intensity of the emitted light for excited Tyr(59) is equal in Ub:pentamidine complexes than in native Ub (Fig. 4b). Quenching experiments with Ub in the presence of the denaturing agent guanidine hydrochloride (GdnHCl) indicated that denatured Ub presents a maximum of Phe- or Tyr-fluorescence higher than that of native Ub (data not shown). Hence, a decrease of intensity in the intrinsic Phe-fluorescence would indicate that Phe(4) and Phe(45) residues are less exposed to the solvent. These results agree with those previously obtained by circular dichroism spectroscopy supporting that pentamidine changes substantially the tertiary structure of Ub.

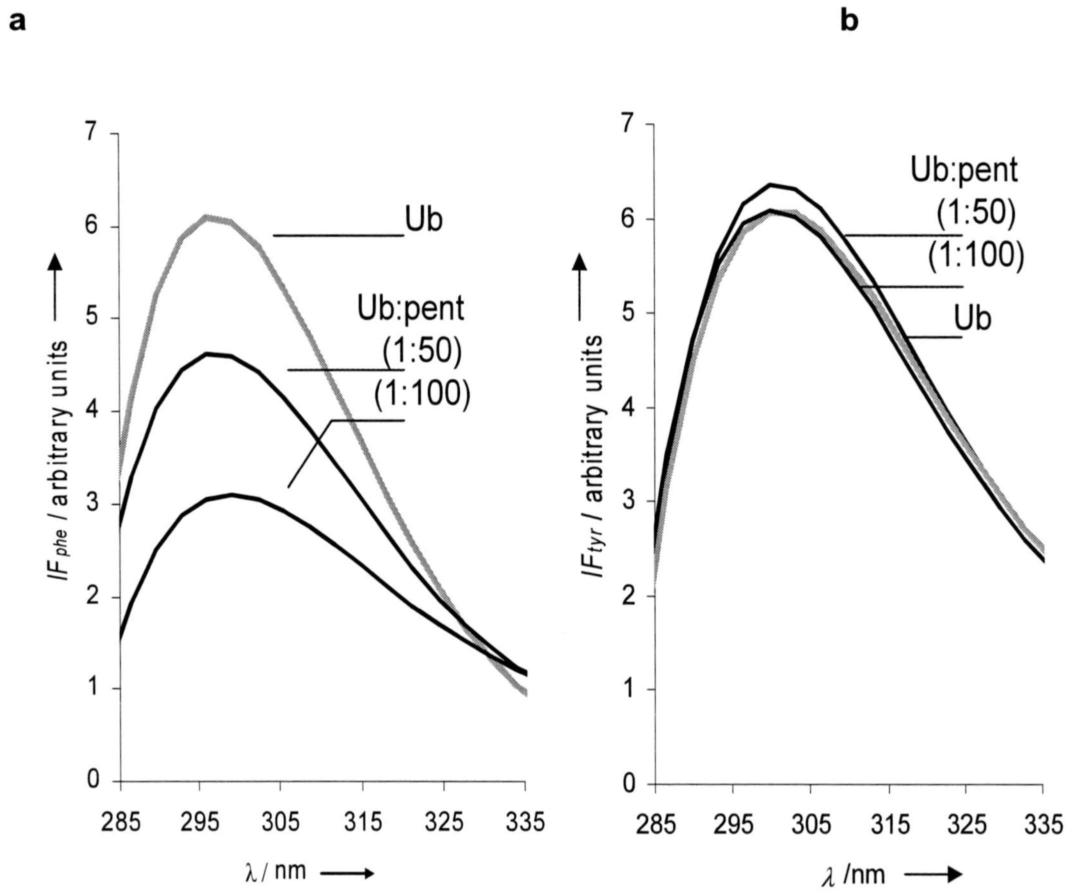

Figure 4. Effect on native Ub of pentamidine (ri = 1/50 and 1/100) at neutral pH and 37 °C as detected by intrinsic Tyr- (panel a) and Phe-fluorescence measurements (panel b).

Conclusion

It is known that DNA may be a feasible pharmacological target for pentamidine [10]. As DNA, proteins may be also cellular targets for cytotoxic drugs such as pentamidine [8, 59]. To gain new insights into the biochemical mechanism of cytotoxic activity of pentamidine, we have analyzed the conformational changes induced by pentamidine in the protein model Ub. As already, mentioned Ub plays a central role in turnover of cellular proteins through the Ub-proteasome pathway. As a matter of fact, Ub is involved in degradation of short-lived and regulatory proteins important in diverse basic cellular processes including cell cycle regulation and modulation of cell surface receptors and ion channels [60]. The Ub-proteasome route functions as an enzymatic cascade called "ubiquitinylation" whereby multiple Ub molecules are covalently bound to the protein that is going to be degraded. The PolyUb modification marks the protein for destruction and leads it to the 20S proteasome complex for destruction [61].

The biophysical data reported in this paper show that pentamidine alters the secondary and tertiary structures of Ub at low molar ratios of pentamidine per Ub molecule (r_i= 1/50 and 1/100). We think that these results are of considerable interest in view of the fact that Ub is a tightly folded protein, whose tertiary structure is highly resistant to denaturation by both cations and anions [62].

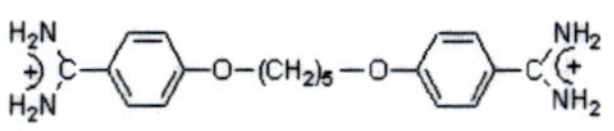

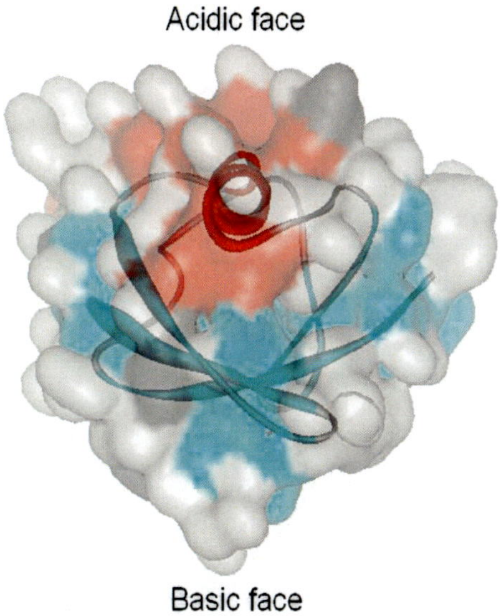

Figure 5. Ribbon diagram of Ub showing the basic (β-cluster) and acidic (α-cluster) faces of the molecule.

As shown in Fig. 2, Phe(4) and Phe(45) residues of Ub are located in a β cluster whereas the Tyr(59) residue is located in an α-helix cluster. The data obtained by intrinsic Phe-fluorescence suggest that the environment of the side chains of Phe-residues changes notably in the presence of increasing amounts of pentamidine but not the environment of the Tyr side chain. In addition, the circular dichroism data indicate that the α-helix content of Ub does not change in Ub:pentamidine complexes while the β-sheet content increases a 6%. Therefore, a conclusion deduced putting together the circular dichroism and fluorescence data is that the interaction of pentamidine with Ub is specific and confined to the β-sheet cluster of the protein (see ribbon diagram of Fig. 5). A maximum at around 300 nm is present in the intrinsic Phe-fluorescence emission spectra of native Ub. On the contrary, upon pentamidine binding, Phe(4) and Phe(45) fluorescence emission at 300 nm strongly decays. Bearing in mind that the denaturant agent GdnHCl increases fluorescence emission of Ub at 300 nm, it is reasonable to assume that pentamidine binding to Ub induces the burial in a hydrophobic environment of the β-cluster containing Phe(4) and Phe(45) residues [63]. In contrast, the fact that fluorescence emission of Tyr(59) located in the α-cluster does not change upon pentamidine binding to Ub, let us to conclude that the structure of the α-cluster is not changed by pentamidine. These data support the circular dichroism data analyzed by the CONTINLL method, which indicate that the amount of β- structure of Ub increases a 6% after pentamidine binding while the amount of random coil decreases a ~6% and the amount of α-structure remains the same.

Six basic lysine amino acid residues are located in the β-cluster of Ub and because of that this Lys(6) motif is called the "basic" face of Ub. In contrast, the α-cluster possesses acidic amino acid residues and is called the "acidic" face of Ub [8]. The α and β clusters of Ub are depicted in the ribbon diagram of Fig. 5. Taking into account that pentamidine is endowed with two positive charges at physiological pH which are supplied by is two benzamidine residues it is likely that electric repulsion may be involved in the modification of the β-cluster of Ub by pentamidine. The "basic" face of Ub is involved in recognition and polyubiquitinylation of different protein substrates [61]. Thus, because our results indicate that pentamidine specifically modifies the β-cluster or "basic" face of Ub, it may be concluded that pentamidine has the potential to inhibit the biochemical activity of Ub.

Summing up, the data reported in this work suggest that some of the cytototoxic effects of pentamidine could be related to its ability to induce conformational changes in the Ub molecule.

Acknowlegdments.

This work was supported by Spanish Comisión Interministerial de Ciencia y Tecnología (SAF 2004-03111). We also thank sponsorship by a CDTI grant to laboratorios CBF-Leti. An institutional grant from Fundación Ramón Areces is also acknowledged.

References

[1] Cushion, MT; Walzer, PD; Collins, MS; Rebholz, S; Vanden Eynde, JJ; Mayence, A; Huang, TL. Highly Active Anti-Pneumocystis carinii Compounds in a Library of Novel Piperazine-Linked Bisbenzamidines and Related Compounds. *Antimicrob Agents Chemother*. 2004, 48, 4209-4216.

[2] Borges, MN; Messeder, JC; Figueroa-Villar, JD. Synthesis, anti-Trypanosoma cruzi activity and micelle interaction studies of bisguanylhydrazones analogous to pentamidine. *Eur J Med Chem.* 2004, 39, 925-929.

[3] Sands, M; Kron, MA; Brown, RB. Pentamidine: a review. *Rev Infect Dis* 1985, *7* 625-634.

[4] Murray, HW; Hariprashad, J; Fichtl, RE. Treatment of experimental visceral leishmaniasis in a T-cell-deficient host: response to amphotericin B and pentamidine. *Antimicrob Agents Chemother.* 1993, 37, 1504-1505.

[5] Croft, SL; Coombs, GH. Leishmaniasis, current chemotherapy and recent advances in the search for novel drugs. *Trends Parasitol.* 2003, 19, 502-508.

[6] Singh, S; Sivakumar, R. Challenges and new discoveries in the treatment of leishmaniasis. *J Infect Chemother.* 2004, 10, 307-315.

[7] Nguewa, PA; Fuertes, MA; Valladares, B; Alonso, C; Perez, JM. Programmed cell death in trypanosomatids: a way to maximize their biological fitness? *Trends Parasitol.* 2004, 20, 375-380.

[8] Nguewa, PA; Fuertes, MA; Cepeda, V; Iborra, S; Carrión, J; Valladares, B; Alonso, C; Pérez, JM. Pentamidine is an antiparasitic and apoptotic drug that selectively modifies ubiquitin. *Chem. & Biodiver.* 2005, 2, 1387-1400.

[9] Pathak, MK; Dhawan, D; Lindner, DJ; Borden, EC; Farver, C; Yi, T. Pentamidine Is an Inhibitor of PRL Phosphatases with Anticancer Activity. *Mol Cancer Ther.* 2002, 1, 1255-1264.

[10]Peter, W. *Int. J. Med. Res.* 1981, 73 *(supl)*, 1.

[11]Reddy, BS; Sharma, SK; Lown, JW. Recent Developments In Sequence Selective Minor Groove DNA Effectors. *Curr Med Chem.* 2001, 8, 475-508.

[12]Goldstein, G; Scheid, M; Hammerling, U; Boyse, EA; Schlesinger, DH; Niall, HD. Isolation of a Polypeptide that has Lymphocyte-Differentiating Properties and is Probably Represented Universally in Living Cells *Proc. Nat. Acad. Sci., U.S.A.* 1975, 72, 11–15.

[13]Gavilanes, JG; de Buitrago, GG; Perez-Castells, R; Rodriguez, R. Isolation, characterization, and amino acid sequence of a ubiquitin- like protein from insect eggs. *J. Biol. Chem.* 1982, 257, 10267–10270.

[14]Watson, DC; Levy, WB; Dixon, GH. Free ubiquitin is a non-histone protein of trout testis chromatin . 1978. *Nature (London),* 276, 196–198.

[15]Schlesinger, DH; Goldstein, G; Niall, HD. The Complete Amino Acid Sequence of Ubiquitin, an Adenylate Cyclase Stimulating Polypeptide Probably Universal in Living Cells. *Biochemistry* 1975, 14, 2214–2218

[16]Schlesinger, DH; Goldstein, G. Molecular conservation of 74 amino acid sequence of ubiquitin between cattle and man.. *Nature (*London*)* 1975, 255, 423–424

[17]Wilkinson, KD; Audhya, TK. Stimulation of ATP-dependent proteolysis requires ubiquitin with the COOH-terminal sequence Arg-Gly-Gly. 1981. *J. Biol. Chem.* **256**, 9235–9241.

[18]Wilkinson, KD; Cox, MJ; O'Connor, LB; Shapira, R. Structure and Activities of a Variant Ubiquitin Sequence from Bakers' Yeast. 1986. *Biochemistry* **25**, 4999–5004.

[19]Vierstra, RD; Langan, SM; Schaller, GE. Complete Amino Acid Sequence of Ubiquitin from the Higher Plant Avena sativa. 1986. *Biochemistry* **25**, 3105–3108.

[20]Goldknopf, IL; Busch, H. Isopeptide linkage between nonhistone and histone 2A polypeptides of chromosomal conjugate-protein A24. *Proc. Nat. Acad. Sci. USA* 1977, 74, 864–868

[21]Hunt, LT; Dayhoff, MO. Amino-terminal sequence identity of ubiquitin and the nonhistone component of nuclear protein A24. *Biochem. Biophys. Res. Commun.* 1977, 74, 650–655

[22]Levinger, L; Varshavsky, A. Selective arrangement of ubiquitinated and D1 protein-containing nucleosomes within the Drosophila genome. *Cell* 1982, 28, 375–385.

[23] Hershko, A; Ciechanover, A; Heller, H; Haas, AL; Rose, IA. Proposed role of ATP in protein breakdown: conjugation of proteins with multiple chains of the polypeptide of ATP-dependent proteolysis. In: *Proc. Nat. Acad. Sci. USA* 1980, 77, 1783–1786.

[24]Wilkinson, KD; Urban, MK; Haas, AL. Ubiquitin is the ATP-dependent proteolysis factor I of rabbit reticulocytes. *J. Biol. Chem.* 1980, 255, 7529–7532

[25]Finley, D; Ciechanover, A; Varshavsky, A. Thermolability of ubiquitin-activating enzyme from the mammalian cell cycle mutant ts85. *Cell* 1984, 37, 43–55

[26]Ciechanover, A; Finley, D; Varshavsky, A. Ubiquitin dependence of selective protein degradation demonstrated in the mammalian cell cycle mutant ts85. *Cell* 1984, 37, 57–66

[27]Burger, AM; Seth, AK. The ubiquitin-mediated protein degradation pathway in cancer: therapeutic implications. *Eur. J. Cancer* 2004, 40, 2217-2229.

[28]Passmore, L.A. , Barford, D. Getting into position: the catalytic mechanisms of protein ubiquitylation. *Biochem J* 2004, 379, 513-525.

[29]Pickart, CM; Fushman, D. Polyubiquitin chains: polymeric protein signals. *Curr Opin Chem Biol* **8** 610-616 2004

[30]de Napoles, M; Mermoud, JE; Wakao, R; Tang, YA; Endoh, M; Appanah R; Nesterova, TB; Silva, J; Otte, AP; Vidal, M; Koseki, H; Brockdorff, N. Polycomb group proteins Ring1A/B link ubiquitylation of histone H2A to heritable gene silencing and X inactivation. *Dev. Cell* 2004, 7, 663-676

[31]Makhatadze, GI; Lopez, MM; Richardson, JM; Thomas, ST. Anion binding to the ubiquitin molecule, *Protein Sci.* 1998, 7, 689-697.

[32]Ibarra-Molero, B; Loladze, VV; Makhatadze, GI; Sanchez-Ruiz, JM. Thermal versus guanidine-induced unfolding of ubiquitin. An analysis in terms of the contributions from charge-charge interactions to protein stability, *Biochemistry* 1999, 38, 8138-8149.

[33]Loladze, VV; Ibarra-Molero, B; Sanchez-Ruiz, JM; Makhatadze, GI. Engineering a thermostable protein via optimization of charge-charge interactions on the protein surface, *Biochemistry* 1999, 38, 16419-16423.

[34]Thomas, ST; Makhatadze, GI. Contribution of the 30/36 hydrophobic contact at the C-terminus of the R-helix to the stability of the ubiquitin molecule, *Biochemistry* 2000, 39, 10275- 10283.

[35]Makhatadze, GI; Loladze, VV; Ermolenko, DN; Chen, X; Thomas, ST. Contribution of surface salt bridges to protein stability: guidelines for protein engineering, *J. Mol. Biol.* 2003, 327, 1135-1148.

[36]Lazar, GA; Desjarlais, JR; Handel, TM. De novo design of the hydrophobic core of ubiquitin. *Protein Sci.* 1997, 6, 1167- 1178.

[37]Lazar, GA; Handel, TM.- Hydrophobic core packing and protein design. *Curr. Opin. Chem. Biol.* 1998, 2, 675-679.

[38]Lazar, GA; Johnson, EC; Desjarlais, JR; Handel, TM. Rotamer strain as a determinant of protein structural specificity. *Protein Sci.* 1999, 8, 2598-2610.

[39]Johnson, EC; Handel, TM. Effect of hydrophobic core packing on side-chain dynamics, *J. Biomol. NMR* 1999, 15, 135- 143.

[40]Johnson, EC; Lazar, GA; Desjarlais, JR; Handel, TM. Solution structure and dynamics of a designed hydrophobic core variant of ubiquitin, *Struct. Fold Des.* 1999, 15, 967-976.

[41]Briggs, MS; Roder, H. Early hydrogen-bonding events in the folding reaction of ubiquitin, *Proc. Natl. Acad. Sci. USA*, 1992, 89, 2017-2021.

[42]Khorasanizadeh, S; Peters, ID; Butt, TR; Roder, H. Folding and stability of a tryptophan containing mutant of ubiquitin. *Biochemistry* 1993, 32, 7054-7063.

[43]Khorasanizadeh, S; Peters, ID;Roder, H. Evidence for a three-state model of protein folding from kinetic analysis of ubiquitin variants with altered core residues. *Nat. Struct. Biol.* 1996, 3, 193-205.

[44]Sabelko, J; Ervin, J; Gruebele, M. Observation of strange kinetics in protein folding. *Proc. Natl. Acad. Sci. USA*, 1999, 96, 6031-6036.

[45]Krantz, BA; Sosnick, TR. Distinguishing between two- and three-state models for ubiquitin folding. *Biochemistry* 2000, 39, 11696-11701.

[46]Sivaraman, T; Arrington, CB; Robertson, AD. Kinetics of unfolding and folding from amide hydrogen exchange in native ubiquitin, *Nat. Struct. Biol.* 2001, 8, 331-333.

[47]Finucane, MD; Tuna, M; Lees, JH; Woolfson, DN. Core-directed protein design. I. An experimental method for selecting stable proteins from combinatorial libraries, *Biochemistry* 1999, 38, 11604-11612.

[48]Finucane, MD; Woolfson, DN. Core-directed protein design. II. Rescue of a multiply mutated and destabilized variant of ubiquitin. *Biochemistry* 1999, 38, 11613-11623.

[49]Lobley, A; Wallace, BA. DICHROWEB: A Website for the Analysis of Protein Secondary Structure from Circular Dichroism Spectra. *Biophysical Journal.* 2001 80, 373A.

[50]Sreerama, N; Woody, RW. A self-consistent method for the analysis of protein secondary structure from circular dichroism. *Anal Biochem.* 1993, 209, 32-44.

[51]Sreerama, N; Venyaminov, SY; Woody, RW. Estimation of the number of a-helical and b-strand segments in proteins using circular dichroism spectroscopy. *Protein Sci.* 1999, 8, 370-380.

[52]Provencher, SW; Glockner, J. Estimation of Globular Protein Secondary Structure from Circular Dichroism. *Biochemistry.* 1981, 20, 33-37.

[53]van Stokkum, IH; Spoelder, HJ; Bloemendal, M; van Grondelle, R; Groen, FC. Estimation of protein secondary structure and error analysis from circular dichroism spectra. *Anal. Biochem.* 1990, 191, 110-118.

[54]Sreerama, N; Woody, RW. Estimation of Protein Secondary Structure from Circular Dichroism Spectra: Comparison of CONTIN, SELCON, and CDSSTR Methods with an Expanded Reference Set. *Anal. Biochem* 2000, 287, 252-260.

[55]Mao, D; Wachter, E; Wallace, BA. Folding of the Mitochondrial Proton Adenosinetriphosphatase Proteolipid Channel in Phospholipid Vesicles. *Biochemistry* 1982, 21, 4960-4968.

[56]Brahms, S; Brahms, J. Determination of protein secondary structure in solution by vacuum ultraviolet circular dichroism. *J. Mol. Biol.* 1980, 138, 149-178.

[57]Neidle, S. DNA minor-groove recognition by small molecules. *Nat. Prod. Rep.* 2001, 18, 291-309.

[58]Telles, S; Abate, T; Slezynger, T; Henriquez, DA. Trypanosoma cruzi ubiquitin as an antigen in the differential diagnosis of Chagas disease and leishmaniasis. *FEMS Immunol. Med. Microbiol.* 2003, 37, 23-28.

[59]Bach, I; Ostendorff, HP. Orchestrating nuclear functions: ubiquitin sets the rhythm. *Trends Biochem. Sci.* 2003, 28, 189-196.

[60]Larsen, CN; Price, JS; Wilkinson, KD. Substrate Binding and Catalysis by Ubiquitin C-Terminal Hydrolases: Identification of Two Active Site Residues. *Biochemistry.* 1996, 35, 6735-6744.

[61]Pickart, CM; Fushman, D. Polyubiquitin chains: polymeric protein signals. *Curr. Opin. Chem. Biol.* 2004, 8, 610-616.

[62]Makhatadze, GI; Lopez, MM; Richardson, JM: Thomas, ST. Anion binding to the ubiquitin molecule. *Protein Sci.* 1998, 7, 689-697.

[63]Nguewa, PA; Fuertes, MA; Iborra, S; Najajreh, Y; Gibson, D; Martinez, E; Alonso, C; Perez, JM. Water soluble cationic trans-platinum complexes which induce programmed cell death in the protozoan parasite Leishmania infantum. *J Inorg Biochem* 2005. 99, 727-736.

In: Protein Conformation: New Research
Editor: Linda B. Roswell, pp. 239-250
ISBN: 978-1-60456-219-4

Chapter 8

STRUCTURAL RELATEDNESS VIA SEQUENCE SPACE OF PROTEIN MODULES

Zakharia M. Frenkel [1,*] ***and Edward N. Trifonov*** [1,**]

[1] Genome Diversity Center, Institute of Evolution, University of Haifa
Haifa 31905, Israel

Abstract

A new approach to definition of relatedness of two protein sequences has been proposed recently. The main point of this approach is finding a chain of natural "intermediate" sequence fragments between two selected fragments, in which all closest neighbors are similar at (high enough) selected threshold. We have found, that by taking relatively short fragments (15-25 amino acids) one can build long (more than 100 steps) chains, "walks" in the sequence space of all fragments from 112 complete bacterial proteomes (about 300,000 sequences). Such chains can be combined into the trees and networks. Belonging of two given sequences to the same "walk" or network – is a new criterion of sequence relatedness. Large networks are observed which contain fragments from different protein families with very different sequences but with the similar 3D structure, as well as with different sequences but from the proteins with the same functions. One perspective future development of this criterion of the sequence relatedness is also a new definition of relatedness of 3D-structures via the networks. At present, a single commonly used method for quantitative estimation of difference between two protein 3D-structures remains RMSD (Root Mean Square Deviation) in spite of its well known substantial limitations. Our new approach can reveal that, sometimes, two RMSD-different but clearly similar structures correspond to sequences from the same network or sequence walk, separated by only several intermediate sequences. We believe, that this approach will provide a new point of view on protein structure evolution and classification as well as new potentials for protein structure prediction.

* E-mail address: zakharf@research.haifa.ac.il, Tel: (972-4)-8288097, (972-547)-959406 (Corresponding author)
** E-mail address: trifonov@research.haifa.ac.il, Tel: (972-4)-8288096; Fax: (972-4)-8246554

II. Introduction

1. Formatted Protein Sequence Space

The protein sequence space as multitude of all protein sequences, both extant and extinct, has been originally suggested by J. Meynard Smith [1] as an evolutionary concept. This notion later became popular in theoretical work [2-4]. It was also instrumental in many occasions for sequence comparisons [5] and, indeed, evolutionary considerations [6-9]. One difficulty in practical use of the space is a broad range of natural protein sizes often differing by several orders of magnitude, from about 30 residues for antimicrobial peptides [10], to over 25,000 residues for titin [11, 12]. When two sequences of different lengths are compared, the best match region usually guides the judgment about their possible relatedness. The match region is often rather small, 10 to 40 residues [13, 14]. The sequences beyond the match region are usually left unattended. It is assumed tacitly that the best match found and the function encoded in it, are sufficient to claim relatedness between the sequences, and to indicate where the main functional sites of the compared sequences are located. One function for one protein is typically assumed, while actually this would be only one *common* function for the sequences, and long flanking sequences may well harbor some other important sites, as is usual for many multifunctional proteins [15-17].

In order to give due attention to all parts of a sequence of interest one has to take its smaller fragments, and compare these fragments to other sequences in a database, to find, perhaps, more than one match, if more than one function is carried by the given sequence. This is best done when all the sequences of a database are presented in form of fragments of the same length, and a new database (formatted sequence space) is created - of the protein sequence fragments, each with the tag specifying the source. The standard fragment length can be taken equal to 20 residues, closer to the lower edge of the protein profile sizes [13, 14]. Such sequence space, constructed in our laboratory [18, 19] turned out to be a very powerful tool for sequence comparisons and protein functional, structural and evolutionary characterization [18-22]. In particular, the networks of sequence-wise related fragments formed in the space are shown to contain, primarily, fragments with the same structure characteristic for the given network. This opens new unexplored perspectives in structural characterization of sequenced proteins, for which no structural information is available.

Currently, proteins of over 112 fully sequenced prokaryotic species are used to form a space of about 10^8 fragments, points in the space [18-20]. All 20 residue long fragments present in each protein are taken. That is, a protein of length L contributes L-20+1 fragments to the space. Every fragment of this huge database is tagged, to specify from which species and from which protein it comes.

The formatted sequence space exhibits all its informational richness when the point (fragments) of the space are connected. The straightforward way to make connections between the fragments is to compare their sequences pair-wise, and connect them when the sequence identity is equal or above certain threshold, e. g. 60%. With this threshold the E-value (probability of having such match by chance [23]) is about 10^{-6}. If n+1 naturally occurring fragments are found that can be connected this way, one after another – the 'walk' in the sequence space [18], the realization of such walk in randomized sequences has a chance $E^n = 10^{-6n}$. The immediate conclusion is that the walks in the natural sequence space of lengths up to 300-400 connections are highly improbable to happen by chance and, therefore, the walks are likely to represent, indeed, actual relatedness of all the point along the walk.

The walks in the natural sequence space, indeed, may involve several hundred steps [18]. The striking observation is that the sequences at the ends of such walks accumulate so many changes at all intermediate steps, that they become very much different, such that none of modern sequence comparison techniques would suggest any degree of relatedness between them. The sequence space approach, thus, introduces principally new definition of the sequence relatedness, via chains of pair-wise close relatives.

The walks only rarely appear in the sequence space as isolated strings of connected dots. Since every fragment may have many close matches (at a given threshold), the corresponding point in the space has many immediate connections, up to several thousands [20]. As a result, almost every walk has many branches at almost every point. This makes together a network (cluster) of connected points, so that, actually, one can arrive from any point within the cluster to any other point of the same cluster, by many alternative ways. The new definition of sequence relatedness is, therefore: Two sequence fragments are related if they both belong to the same network (cluster) in the formatted sequence space.

This definition is principally different from currently used criteria of sequence relatedness. It does not use any arbitrarily introduced gaps and, respectively, penalties. The similarity of amino acids is not used either during the sequence match comparisons. Only direct match (identity) is accepted, when two sequence fragments are compared. The amino acid substitutions that do happen in natural sequences are taken care of by the network itself, where the substitutions appear already in the nearest connection fragments (at the thresholds lower than 100%), and only naturally occurring actual substitutions are integrated in the networks, without any need for combining amino acids in often artificial groups of similarity. The natural indels are reflected in the space by existence of separate networks for the fragments with indels, which often are fused with the respective original (no-indel) networks [19]. Any two sequences are related not because their amino acids are similar by whatever criteria, or because they match well when some gaps are introduced. They are related by the fact that they belong to the same network of pair-wise closely (and obviously) related fragments. Paradoxically as it sounds, two sequence fragments, in order to be related, do not have to have any appreciable sequence similarity.

One may ask what is good about such far from obvious sequence relatedness. The answer is that, as it turns out, the fragments belonging to the same network not only have immediate or very distant sequence relatedness. But also they, essentially, share the same functions and the same structure.

Here and below the terms network and cluster are used interchangeably. Strict meaning of the network is fully isolated group of connected points, while clusters are either isolated or connected to some network via one or few bonds, that can disappear when the fragment identity threshold is raised.

2. Conservation Properties of the Networks

Already in the first papers on the formatted sequence space it was observed that the long sequence walks keep the functions ascribed to the fragments unchanged. Such long walks are demonstrated in case of ABC transporters and histone-like proteins [18]. That implies that a whole network that is crossed by the walk may be associated with the same protein function. Later analyses of the networks amply demonstrated that [20-22]. Moreover, there are strong indications that the fragments of the same network share, essentially, the same structure, as is shown by identifying fragments from crystallized proteins within the walks and networks [19]. The following section provides further support to the notion of unique functional and structural identity of the sequence networks.

3. The Networks and Modular Structure of Proteins

The modularity of proteins already follows from their functions as often there is one-to-several correspondence between a protein and its functions [15-17]. The modules are also well detectable as separate different folds within one large protein. For example, beta-galactosidase consists of five distinct different folds [24]. Many proteins contain domains that are found also in other proteins [25]. In other words there are certain structural/functional units that are common for many different proteins, and the proteins may harbor several such units that would be natural to call modules. The problem is, however, that these units are often rather ill-defined. To begin with, the sizes of the modules discussed in literature may vary from about 30 amino acids, as, e. g., ATP/GTP binding P-loop that certainly qualifies as the module since it is found in a large variety of proteins [26], to 200-300 residues, as, e. g., heme-dependent peroxidases – one large fold with one annotated function (SCOP code a.93).

If one splits a protein sequence to 20 residue pieces, for almost each of the pieces a unique cluster in the sequence space is found [22], so that the clusters can be mapped on the protein sequence. Remarkably, several large clusters are identified in the space for a given protein, with different locations in the sequence, making a distinct occupancy profile with several peaks. The sequences under the peaks can be considered as modules, as many of them are shared by other proteins [21, 22]. Typical sizes of the modules defined this way are from 20 to 35 residues. Thus, the networks (clusters) in the formatted sequence space are assemblies of the naturally occurring sequence fragments, which share the same function and structure, with their sequences related via connections within the networks. The whole sequence space is a display of distinct objects - clusters - on which every protein may be projected, and *vice versa*. The clusters projected on the given protein sequence would provide full description of its modularity. The standard size of thus defined modules is a distinct advantage.

That size, 20 to 35 residues, is, essentially, the same as the size of the earlier suggested standard units of protein structure, closed loops, 25-30 residues [27-29]. Comparison of the maps of the closed loops [29] and the maps of the clusters of the sequence space [22], on the same protein sequence demonstrates that the closed loops and the clusters of the sequence space, are different manifestation of the same entity - protein modules of nearly standard size, each with distinct sequence, structure and function.

4. Protein Sequence Space and Evolution

The sequence space in its current form appears to be an ideal play-ground for evolutionary studies, as it was originally suggested by Maynard Smith. Indeed, should we know all protein sequences of extinct species, we could construct a *complete* sequence space, where every evolutionary change in the sequences would be reflected as connected pairs of the original fragments and their mutated versions. Some of the walks in such space would then represent complete chains of evolutionary events, from the earliest sequence forms to the extant sequences. The networks built only on the modern sequences would certainly miss those ancient fragments, though, perhaps, not all of them, as many would survive residing today in some protein variants, in some species. Whatever walks one may follow in the modern sequence space, these would be only hypothetical evolutionary paths, with many steps missing, and with many possible alternative ways. Besides, one has to realize that the points distant along the evolutionary path may well connect (short cut) in the sequence space due to close match (moderate divergence) between respective sequence fragments. In other words, the true evolutionary paths within the network of modern sequence space may have very convoluted trajectories, with points of return and interruptions.

Notwithstanding the above limitations, even the questionable possible scenarios suggested by the networks, are very appealing if not revealing. One example is the evolutionary tree built for so-called omnipresent sequence elements [21, 30]. These elements, from hexapeptides to nonapeptides, are present in every prokaryotic genome and, thus, presumably originate from last common ancestor of the prokaryotes. The analysis of the networks containing the omnipresent elements, at various levels of sequence identity, strongly indicates that all the elements have common origin. Indeed, at the level 60% of identity of respective 20 residue long sequence fragments containing the elements, they all belong to the same large network. At higher levels of identity the clusters containing the omnipresent elements separate, and a very clear tree is formed, suggestive of the evolution of the elements, from their ancestral form(s) to modern omnipresent sequences [21]. Remarkably, the tree describes the events of molecular evolution that presumably happened *before* the very first bifurcations in the tree of life, all at the stage of last common ancestor.

The examples shown in the Figures below give another illustration of possible evolutionary connections between proteins of various functions - connections between distinct almost isolated clusters in the networks.

5. Structural Relatedness via Sequence Space of Protein Modules

As has been indicated earlier (section 2), prolonged walks and networks in the protein sequence space can be built, with conservation of the 3D structure of the fragments [19, 20]. The walks and networks, containing structurally different fragments have not been analyzed before. On one hand, the presence of such different fragments in one network can be interpreted as "accidental touch" of two different networks. Several criteria to select nonrandom connections were proposed [20]. For example, if two clusters (i.e. group of points connected one to another rather than to other points of the network) are connected by several parallel bonds, or by several alternative paths through other clusters – such connection can be considered as 'nonrandom'. In other words, these clusters are expected to share the same properties, functional or structural, depending on the threshold of connections.

On the other hand, structurally different fragments within the same network can be considered as natural structural variety, that dependent, for example, on other sequences in the protein, on ligand associations etc. What is more interesting, it can reflect evolution of the fragments. In this case some overall structural properties of the fragments are expected to be conserved.

It is revealed in this work, that the fragments with the same structure are, indeed, clustered in the sequence space. Moreover, the fragments with different structure (i.e. RMSD is large), that are connected to lager network via only one or few walks, in many cases conserve a general 3D shape of the fragments, typical for the network. Thus, the networks in the formatted protein sequence space can serve as a new criterion for detection of the structural relatedness of the proteins.

III. Results

We selected all sequences of proteins with known 3D-structure (about 13,000; taken from Astral database [31, 32]), split them into 20 aa fragments (with overlapping) and mapped these fragment (i. e. found all similar ones at the threshold 60%) in the protein sequence space described above. That is, we detected in which networks (clusters) these fragments are located. After this, we selected several thousand networks with sizes from 200 to 2,000 points (fragments), connected with 60% identity threshold, that contained at least several fragments similar to PDB sequences of proteins with various protein folds. Several hundreds (randomly chosen) of these networks were visually analyzed with the help of Pajek [33] in Kamada-Kawai free energy layout. To highlight those points (vertices) in the networks, that correspond to the PDB sequences, they are presented in the networks as squares, at the background of small circles – other points of the networks. Those vertices, for which corresponding structure was similar (RMSD < 3Å) to the most common structure of the network, were presented as red squares, otherwise – as blue squares.

As it is shown in the Figures 1 and 2, the blue boxes (i.e., wrong structures) reside in almost isolated clusters connected to the networks by only a single chain of red circles (i.e. sequence fragments of unknown structure). Several hundreds of such cases are inspected, showing the same properties.

The clusters of Figure 1A form a ring (clusters 4, 9, 13, 7) with few or only a single bond connections (as between clusters 4 and 9) and correspond to the proteins with different sequences and functions (Fig. 1B). Nevertheless, they all correspond to the same 3D structure from the same fold e.3 (beta-lactamase/transpeptidase-like). Such result was expected, because every two clusters of the ring can be connected by at least two alternative ways. The clusters 11 and 12 do not belong to the ring, but as they are connected one to another and to the ring (13) by several bonds, it is not surprising, that they have the same 3D structure. In a similar way such prediction can be made for clusters 5 and 6, that also connected one to another and to cluster 4 of the ring by multiple bonds. Indeed, clusters 5 and 6 are annotated as "penicillin binding proteins" (see Fig. 1B), like cluster 4, and have conserved S and K (in red) present in all other clusters with the same structure.

Well isolated clusters 3 and 8 are connected to the network by single point-to-point walks, and, indeed, their structures are different (Fig. 1A). Moreover, the conserved pair of S and K is lost (cluster 3 contains only K). Similar prediction can be made for clusters 1, 2 and 10, that are connected by single bond and lost conserved S and K. Their respective structures (not available yet) would be expected to be different as well.

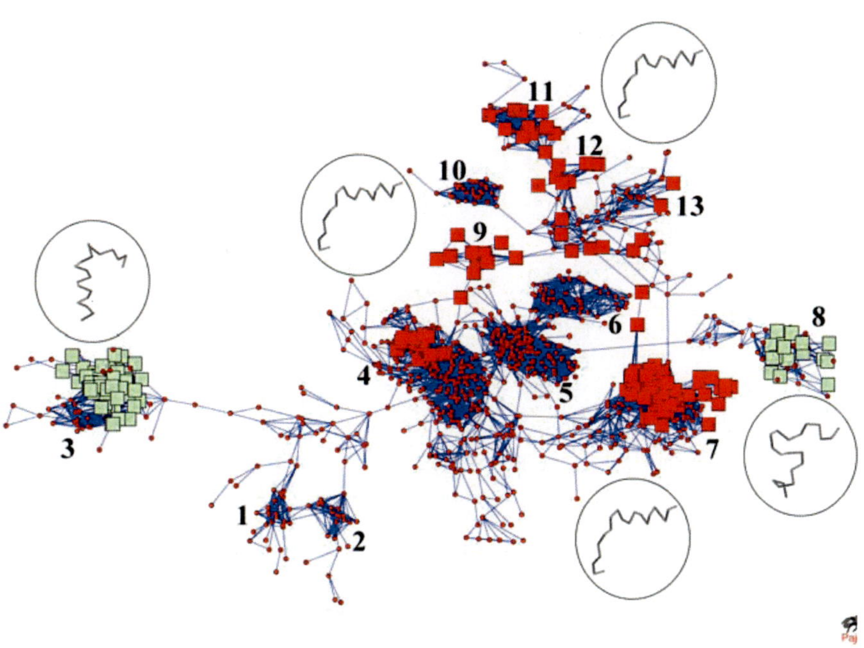

	sequence	Swiss-Prot/PDB code	Protein function
1	ATKATIAGAEFELQKADGTK	Q929J3	Putative peptidoglycan bound protein
2	ESGALLKGVTFEVQDEKGKV	Q81GX1	Collagen adhesion protein
3	TRAQLREIAEKKMPDLNAND	Q8DM28	50S ribosomal protein L11
	KRDKVREIAELKMPDLNAAS	1hc8a_	Ribosomal protein L11, C-terminal domain
4	HQGNFEPGSTMKVMTLAAAI	Q8K6C6	Putative penicillin binding protein 2X
	YQSNYEPGSTMKVMMLAAAI	1qmea4	Penicillin-binding protein 2x, transpeptidase domain
5	TQAWRQPGSSFKPFVYSAAL	Q81ZZ3	Penicillin-binding 1
6	TQIKRQPGSTIKPLAVYTPA	Q88X34	Penicillin binding protein 2A
7	AHQKLNPASLTKLMTSYVAG	Q9KMQ0	D-alanyl-D-alanine carboxypeptidase
	ADVRRDPASLTKMMTSYVIG	1nzoa2	Penicillin-binding protein 5, N-terminal domain
8	SPYEYLPFGGGNRRCIGLAF	Q8YN07	Cytochrome
	DPYIYTPFGSGPRNCIGMRF	1tqna_	Mammalian cytochrome P450 3a4
9	ADERFPVCSVFKTLAAAVL	Q82F11	Putative beta-lactamase
	ADELFPMCSVFKTLSSAAVL	1bsg__	Beta-Lactamase, class A
10	NALHFGAGNIGRGFIGPLLL	Q89A37	Mannitol-1-phosphate 5-dehydrogenase
11	ADQAFSIQSISKVFTLTLAL	Q98NB7	Glutaminase
	SDYRFALESISKVCTLALAL	1u60a_	Probable glutaminase YbaS
12	PETLFEIGSVSKTFTATLAG	P24735	Beta-lactamase precursor
	PQTLFELGSISKTFTGVLGG	1onha_	AMPC beta-Lactamase, class C
13	EDHLFRIYSMTKPLTSLACL	Q9RYU0	Hypothetical protein DRA0218
	EDTLFRLASVTKPIVALAVL	1ci9a_	Esterase EstB

Fig. 1B

Figure 1. Example of the network that contains several 3D structures (see text). A. Graphical presentation of the network. The points (vertices) that correspond to the PDB sequences, are presented as squares, at the background of small circles – other points of the network. Those vertices, for which corresponding structure is similar (RMSD < 3Å) to the most common structure of the network, are presented as red squares, otherwise – as blue squares. B. Description of the centers of clusters numbered in Fig. 1A.

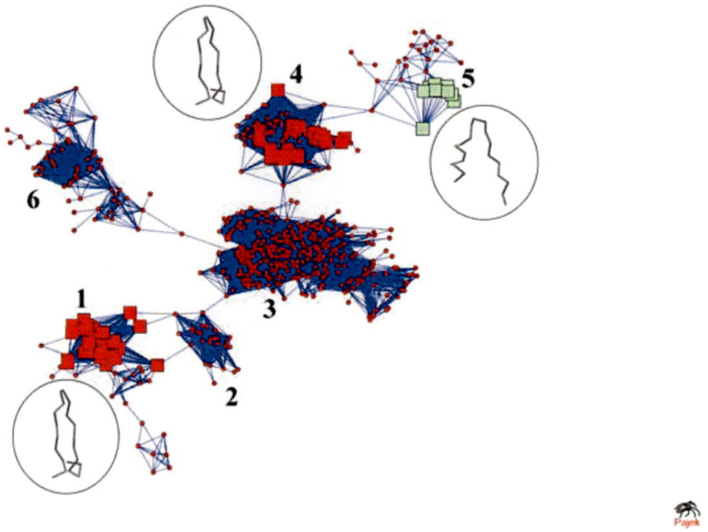

Fig. 2A

	sequence	Swiss-Prot/PDB code	Protein function
1	EAAFVDYGAERHGFLPLKEI	Q8X8J5	RNase E
	EAAFVDYGAERHGFLPLKEI	1smxa_	S1-domain of Ribonuclease E
2	QAAFVDIGLDKAAFLHASDI	P25537	Ribonuclease G
3	FGAFVEIFSGKDGLVHISEL	P50849	Polyribonucleotide nucleotidyltransferase
4	SFGAFVDLGGIDGLVHISEI	Q83HP3	30s ribosomal protein S1
	EFGSFVRLGPLDGLIHVSQI	1go3e1	C-terminal domain of RNA polymerase II subunit RBP7
5	SLDPFLTLLKRDGTLTLVGA	Q9PAV6	Alcohol dehydrogenase
	NLDDFTTLLKRDGTMTLVGA	1uufa2	Alcohol dehydrogenase-like prot. YahK
6	NFVVVNAGLKSESYIPIEEF	Q82TD6	Ribosomal protein S1

Fig. 2B

	sequence	Swiss-Prot/PDB code	Protein function
1	NLDDFTTLLKRDGTMTLVGA	1uufa2	Alcohol dehydrogenase-like prot. YahK
2	DLDAFLSLLKRDGTMTLVGA	Q882D1	Oxidoreductase
3	DFGAFLSLLRTDGALVNVGA	Q82I44	Putative NADP-dependent alcohol d
4	DFGAFLNLGGVDGLLHNHDA	Q9ZKF6	30S ribosomal protein S1
5	NFGAFVDLGGVDGLVHVSEL	Q8FPN4	Putative 30S ribosomal protein S1
6	EFGSFVRLGPLDGLIHVSQI	1go3e1	C-terminal domain of RNA polymerase II subunit RBP7

Fig. 2C

Figure 2. Second example of the network with several 3D structures (see text). A and B. As in Fig. 1. C. The chain of the points (walk) between two PDB sequences (1 and 6) that correspond to different structures from clusters 4 and 5 of Fig. 2A. The walk connects topmost of the cluster 4 with bottommost point of cluster 5.

Interestingly, although the clusters 3 and 8 have the structures different from the rest of the network, the general paths of the protein chains (Γ-shape) are the same. This similarity is apparently maintained due to the connection to the network, though by only single bond. Similar conservation of the overall shape of protein chain was also observed earlier [19, 28].

Another example, Fig. 2, demonstrates additionally that single bond connections between clusters also can "transmit" properties of protein fragments. All clusters (excluding 5, with a "wrong" structure; see Fig. 2A) consist of fragments from the proteins with RNA-binding activity (Fig. 2B). The structures of clusters 1 and 4 are very similar (β-hairpin). The prediction follows, that clusters 2 and 3 located between them have the same structure. They are from the same protein family b.40.4.5 (cold shock DNA-binding domain-like). In addition to functional relatedness they contain 'AFV'-motif (in cluster 4 it is shifted due to one-residue deletion, see Fig. 2B). But the strongest support for such prediction is that the probability of transition from correct structure to another correct structure via the "wrong" structure is negligible (as described previously [19]).

The cluster 5 is connected to the network by a single vertex, has no AFV motif, and its function is not connected with RNA binding. It is not surprising, therefore, that its structure is different. Nevertheless, like in case of Fig. 1, being connected to the network, it conserves general shape of the protein chain (hairpin-like). These two different structures from clusters 4 and 5 can be connected by a very short 'walk' of only four proteomic fragments, as described in Fig. 2C (sequences 2 – 5).

The examples presented in the Figs 1 and 2 demonstrate that nearly isolated clusters of the networks may possess different structure, and that similar structures are grouped together. Thus, the protein fragments of the networks (clusters) are characterized not only by sequence relatedness, but by structural (and functional) similarity as well.

The networks discussed above are only representatives examples of several thousands similar cases. To derive more general conclusions, the whole multitude of the cases should be analyzed in a massive computerized project. We are currently engaged in the first stages of the project. Preliminary data, including the examples above, make us confident that the major picture of joint clustering of sequences, functions and structures will remain as described.

IV. Perspectives

The partition of the sequence fragments in the formatted sequence space by similarity connections between the sequences, with formation of distinct networks, would not imply *a priori* also functional and/or structural similarities of respective protein fragments. On the contrary, considering wide divergence of the sequences within any given network, to nearly complete mismatch between many of them, one would expect, perhaps, only marginal correlation between sequences, functions and structures in the sequence space. Surely, given, say, two 20 aa fragments with sequences matching only in two-three positions (10-15%) one would not dare to claim neither sequence relatedness, nor functional or structural similarity between such protein fragments. It was a pleasant surprise to discover that, on the contrary, the clustering of the sequence fragments in the formatted sequence space, essentially, coincides with the clustering of functions and structures. The networks, thus, represent principally new objects for the protein research. Knowledge on the functional and structural characteristics of the networks would make the problem of protein characterization an easy task. Indeed, one would need only to distribute the fragments of the given protein in the networks of the sequence space, and the list of networks (with their functions and structures) would provide the best possible description of all specific functions and structural units of the protein. The better the networks are characterized, the more accurate would be the network based description. The major challenge for future studies becomes obvious: insufficient current description of the networks/modules. Often, several modules come from the same protein for which only single-function annotation is available. Thus, the same functional description is given to different types of the modules residing in the proteins, although their different sequences and structures would suggest also different specific functions. The modules, discovered originally as distinct structural units - closed loops [27], represented now by the networks in the formatted sequence space [22], should become focus of functional and structural studies, as distinct objects. That would help to unravel details of protein functions, to detect distant relatedness between various proteins, of which many remain unidentified and, last but not least, to predict the structure of the proteins for which the direct structural data are not yet available.

V. Conclusion

In this work distribution of the different structures into the networks of the formatted protein sequence space was investigated. It was found, that that vertex clusterization in the network (multiple independent connections between points or clusters) correspond to the fragments with the same 3D-structure. The points of different structure usually connected to another part of the network via only one walk, in many cases conserve a general 3D shape of other fragments of the network.

Acknowledgements

The work has been supported by the Center for Complexity Science grant GR2006-018.

References

[1] Maynard Smith, J. (1970). Natural selection and the concept of a protein space. *Nature* **225**, 563-564.

[2] Tiana, G., Broglia, R. A. & Shakhnovich, E. I. (2000). Hiking in the energy landscape in sequence space: A bumpy road to good folders. *Proteins-Structure Function and Genetics* **39**, 244-251.

[3] Heger, A. & Holm, L. (2000). Towards a covering set of protein family profiles. *Progress in Biophysics & Molecular Biology* **73**, 321-337.

[4] Xia, Y. & Levitt, M. (2004). Funnel-like organization in sequence space determines the distributions of protein stability and folding rate preferred by evolution. *Proteins-Structure Function and Bioinformatics* **55**, 107-114.

[5] Mrowka, R., Patzak, A., Herzel, H. & Holste, D. (2004). Sequence-related human proteins cluster by degree of evolutionary conservation. *Physical Review E* **70**.

[6] Ortlund, E. A., Bridgham, J. T., Redinbo, M. R. & Thornton, J. W. (2007). Crystal structure of an ancient protein: Evolution by conformational epistasis. *Science* **317**, 1544-1548.

[7] Bornberg-Bauer, E. & Chan, H. S. (1999). Modeling evolutionary landscapes: Mutational stability, topology, and superfunnels in sequence space. *Proceedings of the National Academy of Sciences of the United States of America* **96**, 10689-10694.

[8] Mirny, L. A., Abkevich, V. I. & Shakhnovich, E. I. (1998). How evolution makes proteins fold quickly. *Proceedings of the National Academy of Sciences of the United States of America* **95**, 4976-4981.

[9] Keefe, A. D. & Szostak, J. W. (2001). Functional proteins from a random-sequence library. *Nature* **410**, 715-718.

[10]Oren, Z. & Shai, Y. (1998). Mode of action of linear amphipathic alpha-helical antimicrobial peptides. *Biopolymers* **47**, 451-463.

[11]Barinaga, M. (1995). Molecular-Biology - Titanic Protein Gives Muscles Structure and Bounce. *Science* **270**, 236-236.

[12]Labeit, S. & Kolmerer, B. (1995). Titins - Giant Proteins in Charge of Muscle Ultrastructure and Elasticity. *Science* **270**, 293-296.

[13]Henikoff, S. & Henikoff, J. G. (1991). Automated Assembly of Protein Blocks for Database Searching. *Nucleic Acids Research* **19**, 6565-6572.

[14]Su, Q. J. J., Lu, L., Saxonov, S. & Brutlag, D. L. (2005). eBLOCKs: enumerating conserved protein blocks to achieve maximal sensitivity and specificity. *Nucleic Acids Research* **33**, D178-D182.

[15]Kirschner, K. & Bisswanger, H. (1976). Multifunctional Proteins. *Annual Review of Biochemistry* **45**, 143-166.

[16]Schwab, W. (2003). Metabolome diversity: too few genes, too many metabolites? *Phytochemistry* **62**, 837-849.

[17]Ramasarma, T. (1994). One Protein - Many Functions. Current *Science* **67**, 24-29.

[18]Frenkel, Z. M. & Trifonov, E. N. (2007). Walking through protein sequence space. *Journal of Theoretical Biology* **244**(1), 77-80.

[19]Frenkel, Z. M. & Trifonov, E. N. (2007). Walking through the protein sequence space: Towards new generation of the homology modeling. *PROTEINS*: *Structure*, *Function*, *and Bioinformatics* **67**, 271-284.

[20]Frenkel, Z. M. & Trifonov, E. N. (2007). Evolutionary networks in the formatted protein sequence space. *Journal of Computational Biology*, **14**, 1044-1057.

[21]Frenkel, Z. M. & Trifonov, E. N. (in press). Combinations of ancestral modules in proteins. *Journal of Molecular Evolution*, doi: 10.1007/s00239

[22]Frenkel, Z. M. & Trifonov, E. N. (in press). From protein sequence space to elementary protein modules. *Gene*, doi: 10.1016/j.gene.2007.10.024.

[23]Karlin, S. & Altschul, S. F. (1990). Methods for Assessing the Statistical Significance of Molecular Sequence Features by Using General Scoring Schemes. Proceedings of the National *Academy of Sciences of the United States of America* **87**, 2264-2268.

[24]Jacobson, R. H., Zhang, X. J., Dubose, R. F. & Matthews, B. W. (1994). 3-Dimensional Structure of Beta-Galactosidase from Escherichia-Coli. *Nature* **369**, 761-766.

[25]Doolittle, R. F. (1995). The Multiplicity of Domains in Proteins. *Annual Review of Biochemistry* **64**, 287-314.

[26]Leipe, D. D., Wolf, Y. I., Koonin, E. V. & Aravind, L. (2002). Classification and evolution of P-loop GTPases and related ATPases. *Journal of Molecular Biology* **317**, 41-72.

[27]Berezovsky, I. N., Grosberg, A. Y. & Trifonov, E. N. (2000). Closed loops of nearly standard size: common basic element of protein structure. *Febs Letters* **466**, 283-286.

[28]Berezovsky, I. N., Kirzhner, A., Kirzhner, V. M., Rosenfeld, V. R. & Trifonov, E. N. (2003). Protein sequences yield a proteomic code. *Journal of Biomolecular Structure & Dynamics* **21**, 317-325.

[29]Berezovsky, I. N., Kirzhner, A., Kirzhner, V. M. & Trifonov, E. N. (2003). Spelling protein structure. *Journal of Biomolecular Structure & Dynamics* **21**, 327-339.

[30]Sobolevsky, Y. & Trifonov, E. N. (2005). Conserved sequences of prokaryotic proteomes and their compositional age. *Journal of Molecular Evolution* **61**, 591-596.

[31]Chandonia, J. M., Hon, G., Walker, N. S., Lo Conte, L., Koehl, P., Levitt, M. & Brenner, S. E. (2004). The ASTRAL Compendium in 2004. *Nucleic Acids Research* **32** D189-D192.

[32]Brenner, S. E., Koehl, P. & Levitt, R. (2000). The ASTRAL compendium for protein structure and sequence analysis. *Nucleic Acids Research* **28**, 254-256.

[33]Batagelj, V. & Mrvar, A. (2002). Pajek - Analysis and visualization of large networks. *In Graph Drawing*, Vol. 2265, pp. 477-478.

INDEX

A

B

C

D

H

I

J

K

L

Q

R

S

T

U

V

W

X

Y